Sustainable Development Goals Series

World leaders adopted Sustainable Development Goals (SDGs) as part of the 2030 Agenda for Sustainable Development. Providing in-depth knowledge, this series fosters comprehensive research on these global targets to end poverty, fight inequality and injustice, and tackle climate change.

The sustainability of our planet is currently a major concern for the global community and has been a central theme for a number of major global initiatives in recent years. Perceiving a dire need for concrete benchmarks toward sustainable development, the United Nations and world leaders formulated the targets that make up the seventeen goals. The SDGs call for action by all countries to promote prosperity while protecting Earth and its life support systems. This series on the Sustainable Development Goals aims to provide a comprehensive platform for scientific, teaching and research communities working on various global issues in the field of geography, earth sciences, environmental science, social sciences, engineering, policy, planning, and human geosciences in order to contribute knowledge towards achieving the current 17 Sustainable Development Goals.

This Series is organized into eighteen subseries: one based around each of the seventeen Sustainable Development Goals, and an eighteenth subseries, "Connecting the Goals," which serves as a home for volumes addressing multiple goals or studying the SDGs as a whole. Each subseries is guided by an expert Subseries Advisor.

Contributions are welcome from scientists, policy makers and researchers working in fields related to any of the SDGs. If you are interested in contributing to the series, please contact the Publisher: Zachary Romano [Zachary.Romano@springer.com].

More information about this series at http://www.springer.com/series/15486

Somnath Hazra • Anindya Bhukta
Editors

Sustainable Development Goals

An Indian Perspective

Springer

Editors
Somnath Hazra
School of Oceanographic Studies
Jadavpur University
Kolkata, India

Anindya Bhukta
Netaji Mahavidyalaya
Arambagh
Hooghly, West Bengal, India

ISSN 2523-3084 ISSN 2523-3092 (electronic)
Sustainable Development Goals Series
ISBN 978-3-030-42487-9 ISBN 978-3-030-42488-6 (eBook)
https://doi.org/10.1007/978-3-030-42488-6

This Springer imprint is published by the registered company Springer Nature Switzerland AG
The registered company address is: Gewerbestrasse 11, 6330 Cham, Switzerland

Our homage to

Prof. Rabindra N. Bhattacharya
One of the pioneer environmental economists of India

Prof. Bhattacharya is an outstanding scholar and renowned teacher. He is one of the pioneers of studies in environmental economics in India. During his colourful teaching life, he has served several research institutes and universities of India and abroad. He has also chaired and delivered lectures in different national and international forums on different occasions. He will be remembered by his students for his scholarship, sagacity and kindness.

An Introductory Note

The beginning of a new millennium was a milestone in the history of civilization. To commemorate this occasion, the policymakers in the UN took an ambitious goal for the sustainable development of this planet. The set of goals is known as 'The Millennium Development Goals'. The objective was to attain these goals by 2015. The main objective of the MDG was to lead this world in a trajectory of development which would be sustainable.

The word 'sustainable' came to be pre-affixed with the word 'development' from the recognition of the policymakers of this world that development cannot be sustainable on its own, neither by automatic trickle down of the fruits of development on the poor nor by the 'invisible' Midas touch of the market forces. The early idea of development believed in trickle-down theory. However, the world rejected this idea in the last quarter of the twentieth century. After this rejection, an IMF-led growth policy, in the name of 'Structural Adjustment Programme' appeared in the early 1990s, which prescribed the withdrawal of all kinds of government support and free playing of market forces. The policymakers started speaking in terms of sustainable development only after the failure of this programme.

The MDG included eight different goals. They were: eradicate extreme hunger and poverty; achieve universal primary education; promote gender equality and empower women; reduce child mortality; improve maternal health; combat HIV/AIDS, malaria and other diseases; ensure environmental sustainability and develop a global partnership for development.

Although a considerable progress was reached towards the achievement of these goals, much remained to be achieved. Firstly, different countries made their progress in different aspects. For example, some of the Asian countries achieved significant success in income goals, but failed to achieve non-income goals. Failure to synchronize different goals led these countries to a distorted, unbalanced growth trajectory. Secondly, apart from wide intra-country disparities in achieving different goals, there are also discernible inter-country disparities. For example, success of countries in Sub-Saharan Africa is far below than that of Asian and Latin American countries. Thirdly, during this 15-year time period, the demand for development has changed its perspective a lot. For example, the environment of the world has become so fragile that it requires immediate attention. If energy and water crisis cannot be resolved on an urgent basis, the entire process of development will be compelled to go in vain.

At this juncture, failing to achieve the goals of MDG during the scheduled time frame in full scale, they were revisited, reviewed and restructured. The new set of goals has now been termed Sustainable Development Goals. This new set includes 17 goals altogether to make the development programme sustainable throughout the world. They are: end poverty in all its forms everywhere; end hunger, achieve food security and improved nutrition, and promote sustainable agriculture; ensure healthy lives and promote well-being for all at all ages; ensure inclusive and equitable quality education and promote life-long learning opportunities for all; achieve gender equality and empower all women and girls; ensure availability and sustainable management of water and sanitation for all; ensure access to affordable, reliable, sustainable, and modern energy for all; promote sustained, inclusive and sustainable economic growth, full and productive employment and decent work for all; build resilient infrastructure, promote inclusive and sustainable industrialization and foster innovation; reduce inequality within and among countries; make cities and human settlements inclusive, safe, resilient and sustainable; ensure sustainable consumption and production patterns; take urgent action to combat climate change and its impacts; conserve and sustainably use the oceans, seas and marine resources for sustainable development; protect, restore and promote sustainable use of terrestrial ecosystems, sustainably manage forests, combat desertification, halt and reverse land degradation and halt biodiversity loss; promote peaceful and inclusive societies for sustainable development, provide access to justice for all and build effective, accountable and inclusive institutions at all levels; strengthen the means of implementation and revitalize the global partnership for sustainable development.

Each of these goals has several targets, the total of which came to be 169 in number. As a signatory to the programme member of UN Body, India is also committed to achieve all the targets by the scheduled time frame. In the present volume, the authors will try to examine how far it is practicable for India to achieve the declared targets within the declared time frame.

The Sustainable Development Goal is essentially an extension of MDG in the sense that it wants to widen the scope of the entire programme by including even the rich world in its purview. If we view poverty as the lack of or the deprivation from a decent standard of living, even in some cases, the riches can be viewed as poor. Inequality is there in every aspects of life. Every one of us is, therefore, suffering from one or other kind of poverty. If different forms of poverty cannot be eliminated from every nook and corner of the world, development can never be sustainable. If a country rich in income context suffers from achieving a decent standard in human development context, she will bound to fail in sustaining development. A country deficient in education, health, green environment, socio-political equality, etc. can never retain her progress in the long run.

The elimination of every form of poverty requires a balanced, harmonized development strategy in every sector of an economy. It is essential because if we go for a thorough search of each of the 17 goals, we can find that they are all interwoven. So our first task would be to find out the sectoral gaps. This is not an easy task. The task is much more difficult in a federal country like India because each state in India is not equally developed. The sectoral gaps are different in magnitude in different states. So in India if we want to bridge the sectoral gaps in each state, we have to set state-specific priority. In this volume, the contributors, therefore, try to focus on an exhaustive state-wise performance analysis. It is quite natural that all the states may not have adequate strength to perform equally in achieving all the goals. What is the problem, where the gap is, would be clear to all of us, to the academicians, the researchers and the policymakers as well, from the state-wise analysis. In a nutshell, the objective of this volume is to offer the readers a comprehensive performance analysis of Indian economy in all the seventeen aspects of the SDG, so that they can themselves judge the ability and lacuna of ours in achieving these goals.

Kolkata, India Somnath Hazra
Kolkata, India Anindya Bhukta
September 19, 2019

The Book

In 2015, the UN launched a new set of goals, named Sustainable Development Goal (SDG), to further the Millennium Development Goal (MDG), it took in the year 2000. The new set of goals has been taken to make this world developed without compromising the future of this planet. Accordingly, each member country of the UN has set its own task to achieve these universal goals. This present book tries to take account of India's progress towards the achievement of these goals—how practicable are these goals to achieve and how is the progress so far.

Contents

Editors and Contributors

About the Editors

Somnath Hazra is a trained ecological economist, with an expertise in climate change economics, forest economics, sustainable development, ecosystem services, environmental valuation and policy analysis. He has vast research experience in different Indian research centres as well as teaching experiences at UG and PG level in different universities in West Bengal. He has experience in managing and executing development programmes in various capacities for government, international development organizations and private sector. He has visited different countries all over the world for performing different project assignments and also for the presentation of research articles on climate change and environmental and development economics. He has more than 50 research articles and 5 edited volumes related to environmental economics and sustainable development to his credit.

Anindya Bhukta is a scholar of reputation, teaches economics at undergraduate level over 30 years. He is a prolific writer, who has more than 60 books and 300 articles to his credit. He writes on Indian economic issues and environmental concerns. He has special interest on intellectual property rights. The main essence of his writings is lucidity of language with an in-depth analysis of the subject matter.

About the Contributors

R. Balasubramanian is currently a professor of agricultural economics at Tamil Nadu Agricultural University, Coimbatore, where he has been working for the past 31 years. He has handled more than 30 research projects and published a number of research papers and book chapters. He was a visiting scholar at the University of Wisconsin-Madison and taught at Konkuk University, Seoul.

Sumanta Banerjee is a research scholar at the Department of Economics, Ravenshaw University, Cuttack, India.

Vaibhav Bangar is currently pursuing M.Phil. in Environmental Science in School of Environmental Sciences, Jawaharlal Nehru University, New Delhi.

Soumya Bhowmick is a research fellow at the Observer Research Foundation, Kolkata. His area of research includes Indian economy, governance, sustainability and energy.

Anindya Bhukta is an associate professor at Netaji Mahavidyalaya, West Bengal, India, writes especially on issues in Indian economic problems and on environmental issues. However, he obtained his doctoral degree in Intellectual Property Right Law. He has authored a considerable number of books and articles.

Sanchita Daripa is a senior research fellow at the Department of Economics, Burdwan University, West Bengal, India.

Soumyananda Dinda is a professor of economics at the University of Burdwan, India, and has gained a lot of experiences as faculty member from various academic institutions like Bath University, UK, and Madras School of Economics, Chennai. His research area includes climate change, health, social capital and trade with special emphasis on environment and development economics.

Avani Dubey is an associate researcher of The Environment Law and Development Foundation firm India.

Nilanjan Ghosh is the Director of Observer Research Foundation, Kolkata. His research interests include ecological economics, international trade and developmental issues, water resources and financial markets. He is also working with World Wildlife Fund, India as chief economist. He has published a number of books, monographs and research papers.

Tuhin Ghosh currently works as a professor at the School of Oceanographic Studies, Jadavpur University, is a coordinating lead author for the IPCC Special Report on Ocean and Cryosphere, published in 2019.

Badri Narayanan Gopalakrishnan is Chief Economist, Mckinsey, and Professor at the University of Washington, USA, and is a renowned economist and interdisciplinary scientist with extensive experience in producing and managing global economic datasets and models. He has a good number of international publications in the areas of trade, energy, environment, agriculture, labour and manufacturing. He is peer reviewer and editorial board member for several international journals.

Rajat Goyal associated with Forest Research Institute, Deemed University, Dehradun. He has expertise in forestry, climate change, and socio-environmental issues.

Sushil Kr. Haldar is a professor at the Department of Economics, Jadavpur University, and is an expert in the field of demography and econometrics. His research interest centres on issues of economics of social sector with special reference to India.

Asha Hans is the Founding President of Sansristi, a gender research organization, and was Professor of Political Science and Director of School of Women's Studies, Utkal University, Odisha. She works on gender issues, peace and conflict.

Somnath Hazra is currently working as a senior fellow at the School of Oceanographic Studies, Jadavpur University, India, has more than 15 years of teaching, consulting and advisory experience with government and bi-lateral and multi-lateral agencies. An ex-visiting researcher at the Basque Centre for Climate Change (BC3), Bilbao, Spain. Somnath has worked as an international consultant for the Ministry of Finance, Government of Kenya.

Sulekha Hembram is an assistant professor at the Department of Economics, Muralidhar Girls' College, Kolkata.

Zubayer Hossen is Research Economist at the South Asian Network on Economic Modelling (SANEM), Dhaka, Bangladesh.

Sebak K. Jana is Professor of Economics at Vidyasagar University, West Bengal, and is a whole-hearted research worker in the fields of environmental economics, energy economics, agricultural economics and economics of education. Recently he was awarded Vivekananda Memorial Research Award in recognition to his contribution to qualitative research by the Vidyasagar University, West Bengal.

Bazlul Haque Khondker is Professor of Economics at the University of Dhaka, and holds the Chair of the South Asian Network on Economic Modelling. He is an expert in assessing resource gap implications of various investment interventions using consistent macroeconomic framework, in constructing social accounting matrices and analysing impacts of trade and tax policy reforms using computable general equilibrium models. He constructed technical framework for the Five-Year Plan of Bangladesh.

Suvra Majumder is Energy Expert in UNDP, India. He is also certified as an Energy Auditor by BEE, Government of India.

Umanath Malaiarasan is an assistant professor at Madras Institute of Development Studies, Chennai, Tamil Nadu, India. His area of specialization is agricultural economics and interested in econometric modelling of agricultural crisis and interlinked phenomenon. He is currently working on food consumption pattern, dietary transition, policy intervention and food and nutrition security.

Arup Mitra is currently engaged as professor in South Asian University, Delhi, and has served a number of international organizations in different times as a consultant or as a researcher. His research interest includes development economics, labour and welfare, industrial growth and productivity and gender inequality. Glimpses of his writings can be found, for example, in Encyclopedia of Life Support Systems (UNESCO). He was also the Director General of National Institute of Labour Economics Research and Development, GOI.

Hippu Salk Kristle Nathan is an associate professor at IRMA, Gujarat. Prior to joining IRMA, he was with National Institute of Advanced Studies.

Shounkie Nawani Advisor, Infinite Sum Modelling LLC, USA. She is a trade policy consultant and her research focuses on the design and analysis of trade policies, regional integration, and the world trading system using the lens of interdisciplinary research. She has experience in working on trade policy issues and is conversant with econometric modeling using suitable econometric software packages.

Rajiv Pandey is a senior scientist in Indian Council of Forestry Research and Education, Dehradun, and is a coordinating lead author of Chapter 8: Poverty, livelihoods and sustainable development of the WG II Contribution to the IPCC, AR6. He is the member of editorial boards of different journals of Elsevier, Springer and Nature Publishing group.

Amrita Patel is a researcher, trainer and teacher on Women's Studies and has worked on issues of gender budgeting, land rights of women, women empowerment, climate change and declining child sex ratio.

Selim Raihan is a professor at the Department of Economics, University of Dhaka, Bangladesh, and the Executive Director of the South Asian Network on Economic Modeling. He is the Honorary Senior Research Fellow at the University of Manchester, UK, and an alumni of the Harvard University's program on "Cutting Edge of Development Thinking".

Roshan Saha is a research assistant at Observer Research Foundation, Kolkata. His primary interest is in international and development economics.

V. Saravanakumar is an associate professor at Tamil Nadu Agricultural University. He has long research experience in agricultural productivity, environmental economics and nutritional security issues. He has so far received four best researcher awards. He has written books, many book chapters and published research articles. He handled many externally funded research projects from various international and national organizations.

Jitender Singh is Joint Director in Indian Economic Service.

Clement A. Tisdell is Professor Emeritus at the School of Economics at The University of Queensland, and is one of the forerunners in the field of environmental and ecological economics. He writes extensively on sustainability issues. He has given particular attention to Indian environmental and inequality topics. His latest book, *Gender Inequality*, is published by World Scientific.

Sanjay Upadhyay is an India Visiting Fellow at the Boalt Hall School of Law, University of California; a legal intern to the Earth Justice Legal Defence Fund, San Francisco; a global fellow in Marine Policy at the Duke University; and the Founder of the India's first environmental law firm. He has been in the drafting Committees of several laws in India and abroad including Wildlife Act, Forest Rights Act and Indian Forest Act. He practices in the Supreme Court of India.

No Poverty: How Much, How Far

1

Anindya Bhukta

1.1 Introduction

The Earth Summit of 1992 brought a path-breaking change in any discussion on economy and environment. The world since then has learnt to believe in sustainable development, instead of only present development. Beforehand, there was a contradiction between development and distress. In Indian economy, for example, our planners emphasized on development measured in terms of per-capita income. With regard to poverty alleviation, they believed in trickle-down theory. Eventually, they learnt that growth does not automatically translate into betterment of the standard of living of the poor unless and until it ensures elimination of inequality in all its forms from the society. So long this cannot be ensured, development can be earned only at the cost of distress to a section of people. At this juncture, the concept of sustainable development keeps its eye to an end to this contradiction and teaches us to think of development in a wider perspective. The process of development will be sustainable only when it raises income, lowers poverty and inequality, eliminates hunger, ensures healthy life, establishes gender equality, and conserves energy and environment. In the year 2000, the UNDP set a global agenda toward achieving such a development. It was termed Millennium Development Goals (MDGs).

The Millennium Declaration was a major departure from the previous efforts of global development in the sense that instead of monolithic focus on macroeconomic growth the MDGs emphasized on social development like health and sanitation, poverty and hunger, education and empowerment, etc. (Mickey Chopra 2015). The MDGs also set different monitorable targets to achieve their goals over a 15-year time frame, ranging from 2000 to 2015. In terms of these monitorable targets, the MDG achieved considerable success which prompted the world leaders to step out for the second round of this programme. The sustainable development goals (SDGs) are, in fact, set to launch this second-round world development programme.

1.2 Poverty and Development

Poverty is basically an abstract concept at least in the perception of the person concerned. Initially, when it is believed that poverty is alleviated automatically as a result of economic growth, there was no problem with this perception, although attempts were made to measure poverty in numerical terms. In fact, when we like to view poverty from macroeconomic angle, measurement of poverty becomes essential. It becomes a

A. Bhukta (✉)
Netaji Mahavidyalaya,
Arambagh, Hooghly, West Bengal, India

S. Hazra, A. Bhukta (eds.), *Sustainable Development Goals*, Sustainable Development Goals Series,
https://doi.org/10.1007/978-3-030-42488-6_1

necessity when it is realized that benefits of development do not automatically translate into benefitting the poor and a direct government intervention is required to pass on the benefits of development to the poor as well. This realization, in fact, led to the declaration of MDGs.

Development cannot be sustainable until and unless initiatives are taken to alleviate poverty. The poor always drag development initiatives backward, and development always leads to inequality. To phase out this contradiction between development and distress, governments must take initiatives to redistribute with justice the benefits of development among the poor and those who already enjoy the benefits. In a globalized world, these initiatives should be taken on international basis. Otherwise, the poor countries will stand in the way of the development of developing countries. Primarily, the MDGs and then the SDGs were set by the UNDP from this very realization.

1.3 Toward Poverty Eradication

To reach at the objective of complete removal of poverty, what is essential first is to define poverty. However, this is quite a difficult task. Who is a poor and who is not depends to a great extent on the socio-economic character of a country. Based on the socio-economic needs, a threshold level of income is first set to identify the poor. Who are unable to earn even this threshold level of income are identified as poor. This threshold level of income varies from one country to another depending on their perception of poverty. For example, in major part of Europe, a family with a net income, net of taxes, of less than 60 percent of the 'median net disposable income' of the country is considered as 'poor'. In USA, on the other hand, the basic cost of food required by a family is multiplied by three to get this threshold income level (Choudhury and Chauhan 2013).

In contrast, in India, only the amount of money which is required to purchase the minimum necessity of life is identified as the poverty line. The head count of population who, with their income, cannot purchase even this subsistence requirement, in relation to total population, gives us the poverty ratio of the country. By the minimum necessity only, the requirement of food was considered for a considerable period of time. However, since 2009, with the launching of Multidimensional Poverty Index (MPI), the concept of minimum necessity is being defined alternatively.

With this brief introduction on the definition of poverty, we can now turn to the first and also the foremost goal of SDG, which is complete eradication of extreme poverty in all its forms by 2030. This goal is the foremost, because the UNDP submitted that without eliminating poverty completely the other goals cannot be reached in its true sense.

The first among the 17 goals of the SDG reads as 'end poverty in all its forms everywhere'. There are five targets under this goal, in addition to two subtargets referred to as resources and cooperation policies. All these targets and subtargets try to cover different dimensions of poverty. Unlike MDGs, the SDGs emphasize on maximization of local participation. It asks for 'each government to set its own national SDG targets, guided by global ambitions and national circumstances'. Accordingly, NITI Aayog prepared baseline report 2019 in which 19 indicators were developed to achieve these targets. In the present section, our discussion will be centred around these indicators, and suggestions will be given for some other indicators, if it is felt.

The first two targets of Goal 1 are to reduce poverty as it is defined in terms of national and international definition. The measure of poverty, \$1.25 per person per day, is an internationally defined measure of poverty. This poverty line is set by the World Bank in 2008 and is based on the poverty lines of 75 countries which were adjusted on the 2005 Purchasing Power Parity (PPP) dollars. The set of data had shown that the average poverty line of 15 poor countries were \$1.25 a day, which was then accepted as the international poverty line by the World Bank and the United Nations Development Programme (UNDP). In a similar fashion, a poverty line of \$1 a day was set in the year 1990, and on the basis of that particular line, the Millennium Development Goal

Table 1.1 Poverty in India by International Definition (percent of population)

	Poverty estimates					
Poverty standard	1981	2005	2009	2010	2011	2015
$1.25 (2005 PPP)	60	41.6	–	32.7	–	–
$1.90 (2011 PPP)	–	–	31.1	–	21.2	13.4

Source: http://povertydata.worldbank.org/poverty/country/IND

(MDG) sets the target to halve the number of people below the target line by 2015 (Qian-Qian 2015).

A question may arise at this point. By the time when SDGs were declared, the World Bank already updated the international poverty line on the basis of 2011 PPP dollars. According to the new estimate, the poverty line was set at $1.90 per day. If so, then why the target was set on the basis of a line which was determined by the use of a decade-old price line? It is, therefore, being prescribed from different corner to set the poverty line at $1.90 a day. Otherwise, comparable figure will not be available in the future. This fact would be clear from Table 1.1.

The data represented in Table 1.1 clearly demonstrate a declining trend of poverty in India over time by the international poverty standard. If this trend continues, we can at least approximate to the goal of eradicating extreme poverty from our country, if not completely be reached, by 2030.

The national poverty indicator of India, on the other hand, was basically based on calorie norm. By this norm, the minimum monetary requirement of a person is set on the basis of the minimum nutritional requirement of the person. This minimum monetary requirement is then called the poverty line, and this line is considered as the cut-off line between the poor and the non-poor.

Planning Commission of India started estimating poverty line since the early 1960s. In 1962, a study group appointed by the Commission defined a consumption basket worth Rs. 20 per capita per month as the poverty line, without mentioning any logic behind the magic number. A poverty line, supported by a definite logic, was first constituted by the 'Task Force on Projections of Minimum Needs and Effective Consumption Demand' in 1977. This poverty line, which was monetarily equivalent to a consumption basket providing 2400 calories per capita per day in rural areas and 2100 calories per capita per day in urban areas, was based on the recommendation by the Indian Council of Medical Research. The monetary equivalence of this calorie intake standard, based on 1973–1974 prices, was Rs.49.09 in rural areas and Rs. 56.64 in urban areas. The Task Force suggested updating this line time to time by adjusting for price changes, keeping the consumption basket unaltered (Table 1.2).

In 1993, another expert group under the chairmanship of D. T. Lakdawala recommended certain changes in the estimation of poverty, although it did not opine against calorie-based norm of poverty estimation. It was Tendulkar Committee (2009) who first recommended a shift away from calorie-based norm to a consumption expenditure-based norm in order to emphasize on expenditure on health and education.[1] Moreover, this expert group recommended a uniform poverty line basket for both the rural and urban areas. Another major recommendation was to estimate the number of poor on the basis of Mixed Reference Period[2] (MRP) (Table 1.3).

On the basis of Tendulkar Committee report, the Planning Commission computed a new pov-

[1]Prior to this period, expenditure on health and education was not considered for inclusion in poverty line calculation because it was assumed that expenditure on these two would be borne by the State.

[2]The consumer expenditure data collected by the NSSO, on which poverty ratios are estimated, was based on 30 days recall period prior to the 61st round survey. In contrast to this practice of Uniform Recall Period (URP), the NSSO, in their 61st round survey, introduced another method, called Mixed Recall Period (MRP). Under the MRP method for five infrequently used items (clothing, footwear, durables, education, and institutional health expenditure), consumers are surveyed on the basis of 365 days recall period, and for all other items, the 30 days recall period is used as usual.

Table 1.2 Incidence of poverty by national definition

Reference year	Poverty ratio (by Lakdawala methodology)			Poverty ratio (by Tendulkar methodology)			Poverty ratio (by Rangarajan methodology)		
	Rural	Urban	Total	Rural	Urban	Total	Rural	Urban	Total
1973–1974	56.4	49.0	54.9						
1977–1978	53.1	45.2	51.3						
1983–1984	45.7	40.8	44.5						
1987–1988	39.1	38.2	38.9						
1993–1994	37.3	32.4	36.0	50.1	31.8	45.3			
1999–2000[a]	27.1	23.6	26.1	33.8	20.9	29.8			
2004–2005	28.7	25.9	27.9	41.8	25.7	37.2			
2009–2010	–	–	–	33.8	20.9	29.8	39.6	35.1	38.2
2011–2012	–	–	–	25.7	13.7	21.9	30.9	26.4	29.5

Source: Planning Commission (2012, 2013), Expert Group Report (2014)
[a]Lakdawala estimates for 1999–2000 are based on the mixed recall period (MRP) method and are not comparable with his estimates for other years, which are based on the uniform recall period (URP) method. In Tendulkar methodology, MRP method is used all through

Table 1.3 Poverty line

Reference year	Rural	Urban
1973–1974	49.63	56.76
1977–1978	56.84	70.33
1983–1984	89.50	115.65
1987–1988	115.2	162.16
1993–1994	205.84	281.35
1999–2000	327.56	454.11
2004–2005	446.68	578.80
2009–2010	672.80	859.60
2011–2012	816	1000

Source: Planning Commission (1997, 2013); Press Information Bureau (2001, 2007)

erty line which was Rs. 446.68 per capita per month in the rural areas and Rs. 578.80 in the urban areas in 2004–2005. On the basis of this poverty line, poverty ratio came to be 41.8% for rural areas, 25.7% for urban areas, and 37.2% for the country as a whole. However, these figures for 2004–2005 are, in fact, revised estimates. Beforehand, on the basis of Lakdawala methodology, the Planning Commission published the numbers of below-the-poverty-line (BPL) people for 2004–2005 which were 28.3%, 25.7%, and 27.5% for rural, urban, and the country as a whole, respectively. A movement was started after the publication of these figures by a civil society organization, the Right to Food Campaign, against the statistical manipulation that produced such low figures (Jain 2015). The organization even dragged the Government up to the Supreme Court on this issue. In response, the Tendulkar Committee was constituted by the Government. Later on, poverty figures based on Tendulkar methodology, however, also raised severe debates, and the government constituted another committee, the Rangarajan Committee, to review the recommendations of Tendulkar Committee. In fact, the lack of reliability of government data is a major drawback in measuring the progress of poverty alleviation in our country. The success or failure of the SDGs with regard to poverty eradication in terms of national indicator on poverty has to be judged in the light of these debates and statistical juggleries.

With regard to national indicator of poverty, the target of the SDG is to halve the proportion of men, women, and children of all ages living in poverty in all its dimensions by 2030. As per the latest data available with regard to proportion of population living below the national poverty line, the base value for the reference period of the SDG is set at 21.92%. The poverty line in India is revised every occasion on the basis of National Sample Survey Organization (NSSO) data. Since no such data become available after 2015, progress toward poverty eradication cannot be measured. However, if we take into account the rate at which poverty in India has declined during the last two decades,[3] it can well be expected that India can reach the goal set by the SDG in due course of time.

Let us now turn to the most pertinent question at this point. Does eradication of poverty mean mere achievement of a minimum level of income? If a government wishes to eradicate poverty to gain some political mileage, it can reach its goal by simply providing some financial doles on a short-term basis to the poor who lie in close approximation to the poverty line. In fact, this is a common practice in India as well as in many other developing countries. But to reach at a permanent solution to this problem, what is essential is to arrange for employment to all. Employment to all, however, has some prerequisites like education for all, health for all, etc. In fact, the international poverty line is a Multidimensional Poverty Index (MPI) which covers various dimension of poverty. The MPI[4] defines that a 'poor' is one who is deprived of basic necessities of life like health and education and led a poor standard of living (Table 1.4).

In fact, similar outlook can be found in measuring Human Poverty Index (HPI). This index also views poverty as deprivations in three basic dimensions of human life, namely, health, education, and a decent standard of living. This simply implies that the world of today does not see poverty from the view point of income only. Rather, the availability and access to health and education facilities are being emphasized on. Although India has made considerable progress in these two fields also, the achievement is not sufficient. As a result, while the number of population below the poverty line as per national indicator is 21.9% to total population in 2011–2012, as per MPI it was 41.3% in the same year. Thus, if we have to have an idea of India's success in eradicating poverty by international standard, we have to take into account of her achievement in the expansion of public health service, success in the extension of educational opportunity, etc. In fact, in 2018 the UNDP has released a new, revised estimation procedure for calculating MPI. In this revision, the UNDP sets ten indicators to cover three dimensions of poverty. They are nutrition and child mortality for health, years of schooling and school attendance for education, and for standard of living the six other indicators are cooking fuel, sanitation, drinking water, electricity, housing, and assets. In the remaining part of this section, we shall try to explore the importance of education and good health in removing poverty in its every form.

The relationship between poverty and health is a cyclical relation, where poverty leads to ill health and ill health ensures deepening of poverty (Peters and Garg 2008). Financial constraints restrict poor from accessing health services, if not provided by the government at free of cost or at a minimal cost. Inability to accessing healthcare services often directly affects the productivity of poor, because it is seen that they often suffer from communicable diseases which are mainly associated with poor environment, malnutrition, and genetic problems. These diseases have long-run effects on their health and thereby affect their productivity. Moreover, ill health of the earning members of the family may have some intergenerational impact since loss of income opportunity stands on the way of creating educational opportunity of their offspring.

The most unfortunate reality is that in spite of these well-known facts and figures government expenditure on public health services is very low

[3]The MDG targeted the population that was under BPL in1990 to be halved by 2015. In 1990, the estimated BPL population of India was 47.8 per cent to its total population. India managed to reduce the number to 21.9 by 2011–2012. The corresponding figures for rural and urban areas were 52.6 and 30.5 in 1990 and 25.7 and 13.7 in 2011–2012 (Ahmed 2016).

[4]The Multidimensional Poverty Index (MPI) is an international index of poverty which, while defining poverty, considers its multidimensional facets. A 'poor' is one who is deprived of two basic necessities of life, namely, health and education, and led a poor standard of living owing to poor earning. To capture the poor earning, poor health, and poor education status of a poor, the MPI depends on ten indicators, namely, years of schooling, school attendance, child mortality, nutritional status, asset ownership, access to electricity, access to improved sanitation, access to safe drinking water, flooring, and cooking fuel. The concept of MPI was developed by the Oxford Poverty and Human Development Initiative and the United Nations Development Programme.

Table 1.4 Multidimensional poverty picture in India

Year of survey	Headcount ratio: population in multidimensional poverty	Percentage of people who are poor and deprived in…									
		Health		Education		Living standard					
		Nutrition	Child mortality	Years of schooling	School attendance	Cooking fuel	Sanitation	Drinking water	Electricity	Housing	Assets
2015–2016	27.5	20.53	2.39	11.59	5.50	25.75	24.25	6.14	8.52	23.27	9.42
2011–2012	41.3	35.44	15.45	9.41	5.75	39.72	36.95	6.92	14.23	30.70	17.31
2005–2006	53.7	38.20	22.50	17.60	19.50	51.10	48.20	11.90	28.30	39.40	37.50

Source: Compiled from various reports on MPI

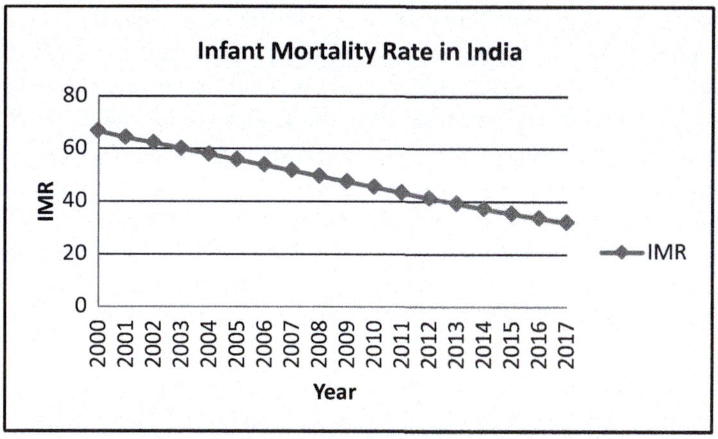

Source: Global Health Observatory data repository, WHO, updated as on 05.02.2019

in all the poor and developing countries in comparison to rich and developed countries. The Indian scenario is much more abysmal. Against the world average of 6%, the public health expenditure (sum of central and state governments) as a percentage of GDP in India was 1.4% in 2016–2017. It hovered around 1.3% during the period 2008–2015. Moreover, what is alarming is that centre's share in total health expenditure has shown steady declining trend over the years in recent time, except 2017–2018 (CBHI 2018). However, even with this meagre amount of public health expenditure, India managed to achieve some success in the betterment of various health indicators, although these successes are far away from some of the best performers in this area.

Infant mortality rate levelled down more than half during the period 2000–2017. In 2000, the IMR was 66.7%, while it is only 32% in 2017. However, if we remember that the first 10 countries in the world have reduced this figure below 3 only,[5] we find no reason to be contended.

Another major health indicator is maternal mortality rate. In the context of poverty eradication, this indicator is much more important in the sense that this rate cannot be made better off without improving the overall health infrastructure and solving the basic nutritional problem of the poor. The reduction of maternal mortality drew special attention in the list of MDGs, but is somehow neglected in the SDGs. The target of the MDG was to reduce maternal mortality ratio by three quarters from the 1990 level. In 1990, India's MMR was 560 per 100,000 live births. The target was, therefore, to reduce it to 140. However, the rate of decline in MMR between 2006 and 2012 was only around 16%. In 2013, the MMR was 190 (WHO 2014). It was then expected that India will certainly miss the MDG target. Then, the Union health minister asserted Rajya Sabha members that India will hit the MDG target in due course of time (Pandey 2015). The final country report on MDG, published by the CSO, Government of India, however, acknowledged that success could not finally be achieved in reducing MMR as per MDG target. Thereafter, in May 2018, Office of the Registrar General of India published a 'Special Bulletin on Maternal Mortality 2014-16', in which surprisingly we discovered that India made magical performance in this field and the MMR came down to 130 in 2016 (Office of Registrar General 2018). In response, by a press release, Regional office of WHO for South-East Asia acclaimed India for its ground-breaking progress in reducing the maternal mortality ratio. Yet, this claim and counterclaim by the different offices of the Government of India reveal that somewhere and somehow in these statistics there is some statistical manipulation. However, the practice of resorting to statistical manipulation may be a good practice for achieving political mileage, but in ultimate

[5]In 2017, the top three countries in this regard were Monaco (1.8), Japan (2.0), and Iceland (2.1).

course, such a practice of self-deception may be a suicidal effort for future course of development.

The health status of the poor and the vulnerable, as we have already argued, cannot be improved without government assistance. For the target of implementation of nationally appropriate social protection systems for the poor and the vulnerable, an indicator that has been set by the NITI Aayog is the percentage of households covered by any health insurance scheme. One such scheme, Rashtriya Swasthya Bima Yojana, was launched in 2007 to bring BPL families under health insurance coverage. However, government expenditure in this scheme has decreased continuously since 2012–2013. In 2018, however, this Yojana was subsumed to a new scheme, Ayushman Bharat. The new scheme has been targeted to bring ten crore poor and deprived rural families under the health insurance net. The success of SDG target in extending social protection in the form of health insurance will depend a lot on the success of this scheme.

In the educational front, success of India is not negligible. Attendance in the school, measured both in terms of mean years of schooling and expected years of schooling, has increased significantly particularly due to the introduction of mid-day meal programme. Poor parents regularly send their children to schools just to ensure a major meal at least once in a day. Mid-day meal programme at present covers students up to class eight. It should be extended to the students up to class 10, and compulsory skill development curriculum should be tied up with this extension. This will help not only to pull some people out of poverty level but also to enable them to get rid of poverty permanently after completion of their education. Bangladesh, for example, has adopted one such programme of education and training for the enhancement of skill of her labour force.

Along with better education and improved health, living a decent standard of living requires something more like improved sanitation facilities, roof over the head, adequate and safe drinking water, etc. Presently, almost all of these are considered to be the basic necessities of life. The government also has taken different initiatives at different times. However, in some cases, progress is remarkable and in some cases not.

India's achievement in improving sanitation facility, especially in terms of a toilet for each household, is quite astonishing during the last 5 years after the launch of Swachh Bharat Mission (SBM) in October 2014. Swachh Bharat Mission (SBM) was started to make India open defecation free by October 2019. In October 2014, only 38.7% household was reported to be a toilet at home. This percentile figure has reached 99.09% as displayed in the Ministry of Drinking Water and Sanitation (MDWS) website on April 18, 2019. Complete success in this programme is expected to be achieved well before the targeted date. However, it should also be noted here that this statistics is based entirely on the self-declaration by the local bodies. The real picture is not so bright. First of all, a part of this statistics has either been manufactured or has been exaggerated at the local bodies. For example, Gujrat declared her to be open defecation free under SBM. On the contrary, a Comptroller and Auditor General of India (CAG) report categorically described this claim as false (Sengupta and Verma 2018). The report said data had been fudged. The audit report of CAG covering eight districts of the state pointed out that toilets had not been constructed at all for 30% of the households. The second reality is that in some cases where toilets were built up finally, the toilets are not being used due to various reasons including lack of water supply.

A similar doubt can be raised against the government demand regarding electrification. To reach electricity connection to every villages and towns, two schemes were launched by the government, namely, Deendayal Upadhyaya Gram Jyoti Yojana and Pradhan Mantri Sahaj Bijli Har Ghar Yojana. The government is now claiming that electrification has been completed in 99.25% villages and towns (Central Electricity Authority 2018). However, a bare eye observation does not support this claim if we do not accept the view that electrifying a village or town does mean reaching an area by establishing only an electric pole there.

It can now be drawn from the above discussion that India's prospect toward achieving the goal of 'no poverty' on the basis of past experience, especially on the basis of experience regarding the achievement of MDGs, is more or less satisfactory. It is very likely that India can hit the target in due course of time. However, it should also be noted here that we can reach at this conclusion, given that governement- figures are reliable in most of the cases.

1.4 A State-Level Picture

In the state level also poverty ratios have declined steadily over time. But what is most alarming is that there are wide disparities among the states in reducing poverty ratio. The range of variation between the states with the highest and lowest poverty ratio was 48.20 in 2004–2005. However, this range value came down to 34.84 in 2011–2012. The standard deviation of all the states and union territories taken together shows a similar declining trend. This means that although significantly higher till now, disparities among the states in alleviating poverty is becoming narrowing down (Table 1.5).

The states that were able to reduce the poverty ratio well below the national level as per 2011–2012 estimates are Andhra Pradesh, Goa, Himachal Pradesh, Jammu & Kashmir, Kerala, Punjab, and Sikkim. The reasons behind their success may not definitely be the same. However, some of these states have achieved considerable success in achieving the target of different base level indicators mentioned in Table 1.6.

Table 1.5 Percentage of population below poverty line in Indian states

States/union territories	2004–2005	2009–2010	2011–2012
Andhra Pradesh	29.90	21.10	9.20
Arunachal Pradesh	31.10	25.90	34.67
Assam	34.40	37.90	31.98
Bihar	54.40	53.50	33.74
Chhattisgarh	49.40	48.70	39.93
Goa	25.00	8.70	5.09
Gujarat	31.80	23.00	16.63
Haryana	24.10	20.10	11.16
Himachal Pradesh	22.90	9.50	8.06
Jammu and Kashmir	13.20	9.40	10.35
Jharkhand	45.30	39.10	36.96
Karnataka	33.40	23.60	20.91
Kerala	19.70	12.00	7.05
Madhya Pradesh	48.60	36.70	31.65
Maharashtra	38.10	24.50	17.35
Manipur	38.00	47.10	36.89
Meghalaya	16.10	17.10	11.87
Mizoram	15.30	21.10	20.40
Nagaland	9.00	20.90	18.88
Odisha	57.20	37.00	32.59
Punjab	20.90	15.90	8.26
Rajasthan	34.40	24.80	14.71
Sikkim	31.10	13.10	8.19
Tamil Nadu	28.90	17.10	11.28
Tripura	40.60	17.40	14.05
Uttarakhand	32.70	18.00	11.26
Uttar Pradesh	40.90	37.70	29.43
West Bengal	34.30	26.70	19.98
Andaman and Nicobar Islands	3.00	0.40	1.00
Chandigarh	11.60	9.20	21.81
Dadra and Nagar Haveli	49.30	39.10	39.31
Delhi	13.10	14.20	9.91
Daman and Diu	8.80	33.30	9.86
Lakshadweep	6.40	6.80	2.77
Puducherry	14.10	1.20	9.69
Standard Deviation	14.45	13.48	12.24

Source: Handbook of Statistics on Indian States, 2019 (Table 14), RBI

Note: Computed as per Tendulkar method on Mixed Reference Period (MRP)

1.5 Concluding Remarks

The success of India in eradicating poverty over the last two decades is not insignificant. Even if we admit that much of her success is statistically manipulated, this success cannot be denied altogether. However, the target of the UNDP is not only to remove poverty by mere numbers but also to remove it from every sphere of life and in its every form. Moreover, removal of poverty must be permanent. This all-round removal of poverty requires three 'e' to be ensured. These three 'e' are employment, equality, and empowerment.

With regard to combat against poverty, NITI Aayog acknowledged that a holistic anti-poverty strategy must be growth oriented, because growth creates job (NITI Aayog 2016). The Aayog also prescribed for an employment-intensive growth strategy by drawing the example of South Korea and China. Unfortunately, the same NITI Aayog did not consider, while constructing the SDG index for India, any indicator directly relating to employment.

Table 1.6 Achievements of different base level indicators at the subnational level

Indicators	Base value (2015–2016)	Target value (2030)	Lowest achiever	Highest achiever
Households with at least one member covered under any health scheme	28.70	100	3.60 (Manipur)	74.60 (Andhra Pradesh)
Infant mortality rate (2016)	34 (2015)	Not set	44 (Assam, Odisha)	8 (Goa)
Maternal mortality ratio (2014–2016)	139 (2015)	Not set	237 (Assam)	46 (Kerala)
4. Proportion of the population (out of total eligible population) receiving social protection benefits under maternity benefit	36.40	100	No data available	No data available
Percentage of rural household with access to electricity (2017)		Not set	39.89 (Jharkhand)	100 (Andhra Pradesh, Goa, Gujrat, Punjab, Tamil Nadu)
Proportion of homeless population to total population	0.15	Not set	No data available	No data available
Person days generated (in lakhs) under MGNREGA (2017–2018)	23513.13	Not set	0.99 (Goa)	3125.57 (West Bengal)

Source: Compiled
Notes: (1) Figures corresponding to highest and lowest achievements are based on latest available data (corresponding year is mentioned in the first column within parentheses) from various sources. (2) Target values are mentioned where it is set by the NITI Aayog

In the last few decades, India's GDP growth rate is moderately high. However, this growth rate is scantly employment generating. It has achieved a good shape mainly due to better performance of the service sector. The service sector generates high income, but only with a few employments. On the other hand, the growth rate of agriculture, on which most of our poor people depend on, is abysmally low, hovers around 1–2% per annum. As a result, agricultural sector continuously generated surplus labour over this entire period. However, the manufacturing sector, growth rate of which was more or less stagnant over a long period of time,[6] could not absorb this surplus labour. This labour force, therefore, moved mainly to the construction sector, as unskilled labour with a very poor wage rate. This nature of employment, in turn, has deepened the problem of poverty in the country. These low wage earners might not scale the number of poor up, but with a scanty income, they live just above the BPL level. Owing to this reason when Rangarajan Committee revised the poverty line to set it a little bit upper, the head-count ratio of poverty increased up to several points.

Reduction of inequality is another major criterion of complete and permanent alleviation of poverty. It is widely accepted that growth will be pro-poor only if it is backed by appropriate redistributive policy. A study on poverty and inequality in India asserted that sharp increase in economic inequality during the 1990s resulted in deteriorated poverty picture despite higher growth (Sen and Himanshu 2004). However, India is no exception. Similar outcome has been found in case of other countries also with higher degree of inequal-

[6]A report on employment strategy for India was jointly prepared by the International Labour Organization (ILO) and Ministry of Labour and Employment, Government of India in 2009. In this report, it was acknowledged that although since the 1980s India achieved remarkable success in accelerating GDP growth, its performance in employment generation is disappointing. Agriculture, till the first half of the first decade of the present millennium contributed less than a quarter of GDP. However, more than half of the total employed labour force rests on agriculture. The share of manufacturing sector, on the other hand, rose marginally from 11:24 per cent in 1983 to only 12–20 per cent in 2004–2005. Growth in employment concentrated mainly in construction and other services (ILO 2009). This trend in employment pattern also reveals how inequality is gradually increasing in India. When more than half of the people earn less than one fourth of the GDP, the inequality is bound to increase.

ity. In India, adequate attention was not given to the aspect of inequality mainly out of the belief that inequality is an obvious by-product of rapid growth. This misbelief, along with another misconception that inequality in India is comparatively lower than other developing countries, prevented the government to take any sincere initiative to reduce inequality. In 2017, Oxfam International and Development Finance International developed a 'Commitment to Reducing Inequality Index' to measure commitments of the government toward reducing inequality.[7] Out of 152 countries, India ranked 132 in this index. This poor commitment of the government placed India among the most unequal countries in the world. Now, evidences from both primary and secondary sources are also asserting that this inequality has been rising over the last three decades (Ibid, 2018). If this trend continues, achievement of the SDG regarding poverty elimination will bound to fail. However, recent shifting of government policies towards inclusive growth is expected to make the government much more committed in reducing inequality and thereby sustaining development in an uninterrupted manner.

Political empowerment of the poor is also essential for the removal of poverty at the root. Poverty is seen to be prominent in case of marginalized groups. These people know the root cause of poverty and problems relating to poverty better than the others by the experience of their own lives. If these people can be brought into the arena of policy-making framework, then poverty alleviation programmes may give better outcomes. What is required for this is to make space for them in the political arena. Political empowerment of the poor is thus an important issue, at least in Indian context, in any discussion on poverty alleviation. The poor should not only be involved in the policy-making process, but they should also be involved in the implementation of the poverty eradication programmes. This can be made possible by the reservation of some administrative posts in the local bodies for the representatives from economically backward classes, not only from socially backward classes as it is presently arranged by the constitutional amendments. An index with regard to political empowerment of representatives of BPL people may give us a better idea of removal of poverty in its every form.

[7]This index measures government commitment in reducing the gap between the rich and the poor on the basis of three indicators, namely, social spending, tax, and labour rights.

References

Ahmed S (2016) End poverty in all its forms everywhere. In: India and sustainable development goals: the way; research and information system for developing countries, UN

Central Bureau of Health Intelligence, GOI (2018) National Health Profile 2018

Central Electricity Authority (2018) All India electricity statistics: general review 2018, Government of India

Choudhury G, Chauhan C (2013) Measuring poverty: how the world counts its poor. Hindustan Times, 28 July 2013

Peters DH, Garg A (2008) Poverty and access to health care in developing countries. Ann N Y Acad Sci 1136:161–171

ILO (2009) Towards an employment strategy for India, ILO

Jain SK (2015) Reading between poverty lines. https://www.downtoearth.org.in/blog/reading-between-poverty-lines-361

Mickey Chopra EM (2015) Millennium development goals: background. Arch Dis Child 100(suppl 1):s2–s4

NITI Aayog (2016) Eliminating poverty: creating jobs and strengthening social programs, Government of India

Office of Registrar General (2018) Special bulletin on maternal mortality in India 2014-16, Government of India

Pandey K (2015) Maternal mortality: India likely to miss MDG target. https://www.downtoearth.org.in/news/maternal-mortality-india-likely-to-miss-mdg-target%2D%2D49036

Qian-Qian LY (2015) Poverty reduction within the framework of SDGs and post-2015 development agenda, Ke Ai

Sen A, Himanshu H (2004) Poverty and inequality in India: II: widening disparities during the 1990s. Economic and Political Weekly, vol 39, no. 39

Sengupta S, Verma R (2018) Sanitation will remain a mirage even after India stops defecting in the open. https://www.downtoearth.org.in/news/maternal-mortality-india-likely-to-miss-mdg-target%2D%2D49036. Last updated 1 Oct 2018

WHO (2014) Trends in maternal mortality: 1990 to 2013, WHO

Sustainable Agriculture, Poverty, Food Security and Improved Nutrition

2

V. Saravanakumar, Umanath Malaiarasan, and R. Balasubramanian

2.1 Introduction

Despite the reduction in proportion of people under malnutrition by half from 23.3% in 1992 to 12.9% in 2016, yet the absolute number of undernourished people increases to 821 million, i.e. every 1/9th person in the world is undernourished (UN 2017). Of this, 515 million were undernourished in Asia during 2017. Unless appropriate measures are taken to improve/sustain the global food and agricultural systems, it is estimated that these figures are likely to get worse and thereby endangering future food and nutritional security. Historically, poverty and malnutrition continue to be the major challenges faced by the human society. Therefore, Millennium Development Goals (MDGs) were set in 2001 to ensure reducing extreme poverty and all forms of malnutrition with eight time-bound targets to be met by 2015. Following the MDGs, Sustainable Development Goals (SDGs) were formulated in 2015 to cover all the dimensions of social, economic and environmental issues comprising 17 different goals covering both developed and developing countries (Steffen et al. 2015). Out of 17 SDGs, the second goal (SDG-2) seeks to eliminate hunger and malnutrition by ensuring food and nutritional security through sustainable agriculture. SDG-2 is interrelated to the society, economy and environment, which is crucial to the success of the entire SDG (FAO 2016). Since all the developing countries heavily relied on agriculture and allied sectors, production and consumption of food is fundamental to their economic and social stability. Hence, achieving SDG-2 is expected to establish significant trade-offs with other SDGs across space and over time. Food production, distribution and access, as well as nutritional status of people of all age groups, gender and social categories, supported by sustainable agriculture are intrinsically linked to other SDGs. Besides, SDG-2 emphasizes equitable access to nutritious food for all along with improving sanitation and hygiene. Therefore, it is imperative to understand the current status of malnutrition, performance of MDGs and efficacy of government schemes, which would provide a powerful advocacy to implement and monitor SDG-2 and achieve its target by 2030.

V. Saravanakumar (✉) · R. Balasubramanian
Tamil Nadu Agricultural University,
Coimbatore, Tamil Nadu, India

U. Malaiarasan
Madras Institute of Development Studies (MIDS),
Chennai, Tamil Nadu, India

S. Hazra, A. Bhukta (eds.), *Sustainable Development Goals*, Sustainable Development Goals Series,
https://doi.org/10.1007/978-3-030-42488-6_2

This chapter examines the unfinished targets of MDG; current targets and performance of SDGs; current status of food production, consumption, nutrient availability and the problems thereof; and different nutritional and agricultural development programmes to sustain the food production and their impact on food and nutritional security outcomes. Further, this chapter discusses the projection of food supply and demand to realize the targets under SDG-2 and also explores the roadmap and possible pathways to achieve SDG 2 by 2030.

2.1.1 India's Unfinished Agenda of MDG Target 2

Although substantial progress has been made in halving the extreme poverty rate and people suffering from hunger under MDGs by 2015, the achievements have been uneven. For instance, the proportion of underweight children who are under 5 years is one of the composite indicators for child malnutrition. The results of three different rounds of National Family Health Survey (NFHS) show uneven performance, i.e. the percentage of underweight children has decreased over the years but not to the desired extent (Table 2.1). The trends in nutritional status in rural and urban areas are presented in Fig. 2.1. The underweight children have been considerably reduced by around 18% over the last 30 years; however, 35% of children less than 5 years were reported as underweight at the end of 2016.[1] Similarly, the stunted and wasted children declined to 38.4% from 48% in the country as per the NFHS-4. The decrease in stunting over time was greater in rural areas than in urban areas. Underweight children decreased slightly more in rural areas than in urban areas, but there was very little improvement in the percentage of children who were severely underweight in both rural and urban areas.

Table 2.1 Percentage of different categories of malnourish children under 5 years

Survey	Underweight (weight for age)	Stunted (height for age)	Wasted (weight for height)
NFHS-1 (1992–1993)	53.4	52.0	17.5
NFHS-2 (1998–1999)	47.0	45.5	15.5
NFHS-3 (2005–2006)	42.5	48.0	19.8
NFHS-4 (2015–2016)	35.7	38.4	21.0

The status of underweight children aged below 5 years across different states of the country is presented in Fig. 2.2. As per the NFHS-4, ten states, namely, Mizoram (11.9%), Manipur (13.8%), Sikkim (14.2%), Kerala (16.1%), Jammu and Kashmir (16.6%), Nagaland (16.8%), Arunachal Pradesh (19.5%), Himachal Pradesh (21.2%), Punjab (21.6%), Goa (23.8%) and Tamil Nadu (23.8%) are the top ten states that achieved the all India MDG target, and its average is far lesser than the all India average for prevalence of underweight children. However, the prevalence of underweight children was higher than national average in the states of Jharkhand (47.8%), Bihar (43.9%), Madhya Pradesh (42.8%), Gujarat (39.3%), Chhattisgarh (37.7%) and Rajasthan (36.7%). The figures for Odisha, Karnataka and Maharashtra states are also very close to the national average. As compared to NFHS-3, Mizoram, Sikkim, Arunachal Pradesh, Kerala, Himachal Pradesh and Tamil Nadu states have performed better in reducing the underweight children.

The percentage of children aged less than five years, whose height for age is below from the median height is defined as stunting (WHO 2018). The number of stunted children has decreased from 48% (NFHS-3) to 38.4% (NFHS-4) in India. Among the states, Kerala (19.7%), Goa (20.1%), Tripura (24.3%), Punjab (25.7%), Himachal Pradesh (26.3%), Tamil Nadu

[1] NFHS-1 round assessed the number of underweight children was less than age of 4; the numbers are not exactly comparable. However, it provides the indication about reduction in underweight children.

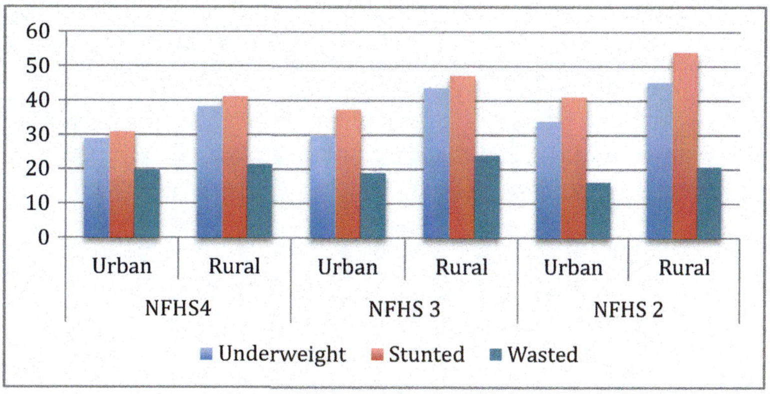

Fig. 2.1 Trend in nutritional status of children under 5 years in rural and urban regions

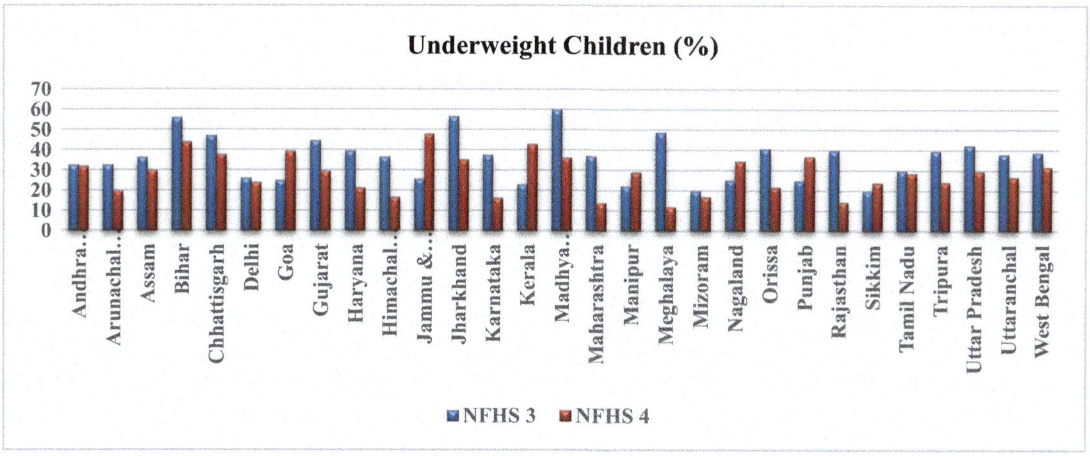

Fig. 2.2 Percentage distribution of underweight children across different states

(27.1%), Jammu and Kashmir (27.4%), Telangana (28.1%) and majority of the northeast states (Mizoram, Manipur, Nagaland, Arunachal Pradesh, Sikkim) were the top performing states with respect to reduction in number of stunted children. Bihar (48.3%), Uttar Pradesh (46.3%), Jharkhand (45.3%), Meghalaya (43.8%) and Madhya Pradesh (42%) were the states in which stunting is the most prevalent form of malnutrition. The status of stunted children across the states under NHFS-3 and NHFS-4 is presented in Fig. 2.3.

Wasting is defined as deficits in weight for height, which is usually used to express a more severe form of malnutrition (WHO 2018). Comparison of different NFHS data revealed that the reduction in the level of wasting of children aged less than 5 years has much deteriorated; it has further increased by 3.5% from 17.5% (NFHS-1) to 21% (NFHS-4) over the 30 year period. There were 19 states including all north-eastern states that have performed better than the national average. Of these, Manipur (6.8%) and Mizoram (6.1%) states performed extremely well followed by Nagaland (11.2%), Jammu and Kashmir (12.1%), Himachal Pradesh (13.7%), Sikkim (14.2%), Meghalaya (15.3), Punjab (15.6%), Kerala (15.7%), Tripura (16.8%), Assam (17%), Andhra Pradesh (17.2%), Arunachal Pradesh (17.3%), Uttar Pradesh (17.9%), Telangana (18%), Uttarkhand (19.5%), Tamil Nadu (19.7%) and West Bengal (20.3%). Again, the states like Jharkhand (29%), Gujarat (26.4%), Karnataka (26.1%), Madhya Pradesh

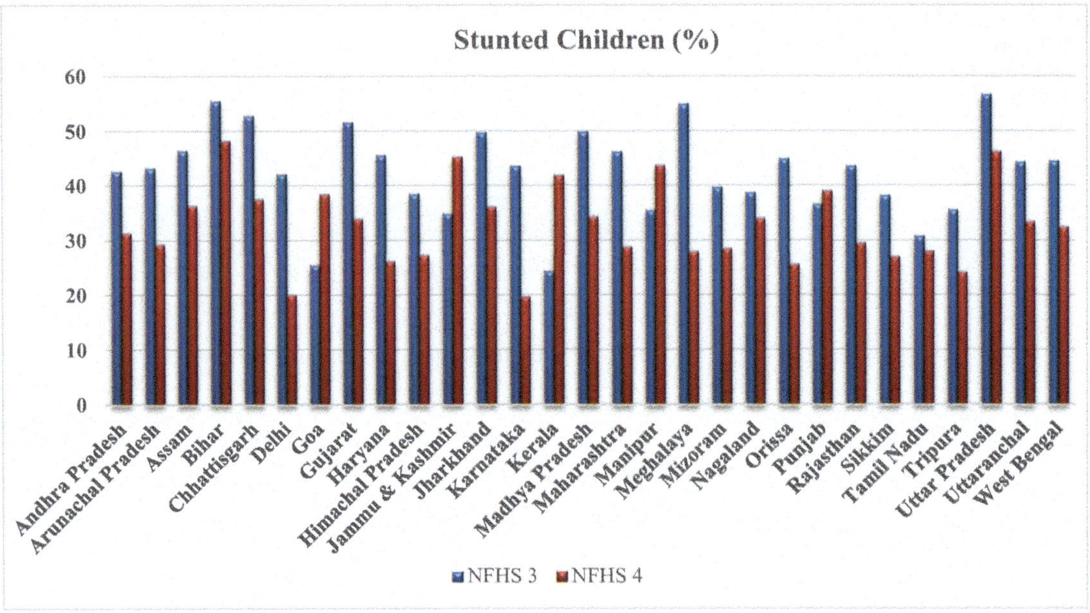

Fig. 2.3 Percentage distribution of stunted children across different states

(25.8%), Maharastra (25.6%), Chhattisgarh (23.1%) and Rajasthan (23%) have performed above national average (21%). As per the NFHS data, the interstate inequalities in the prevalence of wasting were observed to be very high. Mizoram state recorded the lowest wasted children (6%) in India, whereas Jharkhand had the highest prevalence of wasting (29%), which was almost five times higher than Mizoram (Fig. 2.4).

India has experienced remarkable economic growth in recent years and remains one of the fastest growing economies in the world. However, poverty and food insecurity continue to be the areas of concern. In 2016, India figured 97th rank among 118 developing countries in Global Hunger Index (GHI). About 15% of under 5 years children still suffer from wasting and 39% from stunting. According to NHFS-4 data, though 58.4% of children remain anaemic in 2015–2016, there has been a substantial improvement from 2005 to 2006 levels by 11% points. Incidence of anaemia among pregnant women has been reduced by about 8% over 10 years from 2005–2006 to 2015–2016. About 53% of nonpregnant women (15–49 years) suffered from anaemia in 2016 as compared to 55.2% in 2005–2006, showing a marginal reduction over a period of 10 years. On the other side, incidence of obesity has almost doubled over the decade ending 2015–2016, i.e. there is an increment from 9.3 to 18.6% among men and from 12.6 to 20.7% among women. Thus, while the food security situation is progressively improving, access to balanced food is problematic for the vulnerable population. Despite many efforts have been taken to halve hunger by 2015 in India, this target has not been met under MDG (Saxena 2018).

2.1.2 A Transformative Agenda from MDG 2 to SDG 2

The MDGs were focused on developing countries to reduce poverty and malnutrition from 2001 to 2015. SDGs build upon the MDGs by including sustainability, holistic development and uniform treatment to all the countries. While MDGs had only 8 goals, 21 targets and 63 indicators, SDGs include 17 goals with 169 targets. SDGs focus mainly on environmental sustainability, economic development and social inclusion in an integrated manner. It further addresses

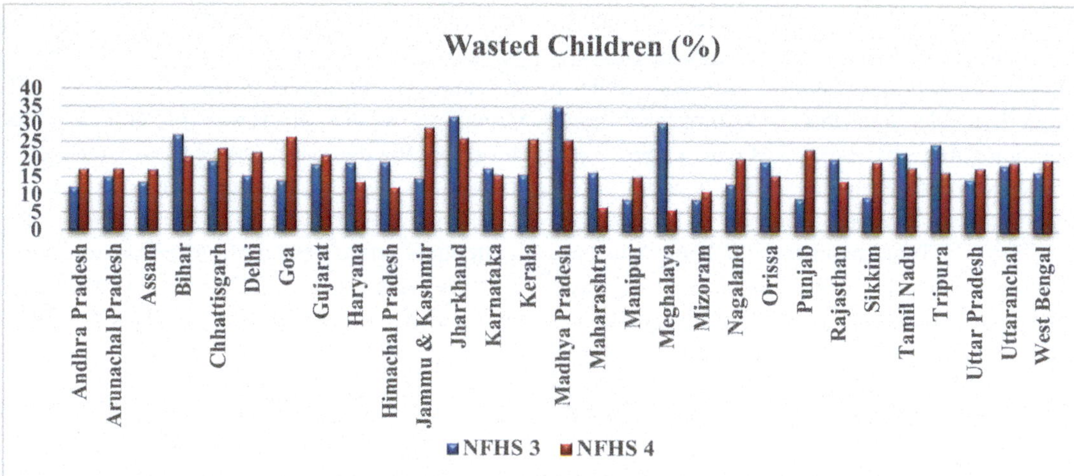

Fig. 2.4 Percentage distribution of wasted children across different states

all the interrelated challenges in a comprehensive way involving multiple actors.

In spite of the progress made in meeting the MDGs, scepticism has surfaced regarding its implementation and the actual achievements of its targets. A study by Friedman (2013) suggests that there was no statistically significant trend in accelerations of the MDG indicators after 2000, when the MDGs were published. Any of the successes in eradicating poverty as well as hunger may be indicative of the impact of long-term and broader economic trends. To paraphrase William Easterly, 'the MDGs communicated a very wrong idea about how development happens: technocratic, patronizing, and magically free of politics…it's not about western saviours, but home-grown efforts linked to a gradual extension of political freedom'.

MDGs serve as the foundation for establishing SDGs to realize improved and better goals based on critical evaluation of the different goals. Unlike the newly developed Sustainable Development Goals in 2015, there were not only fewer goals but targets set for these goals applied primarily to least developed/poor countries. The new SDG goals set targets that call all countries for action, no matter how developed, and they 'challenge the west lectures rest' (Caballero 2016). In order to achieve hunger eradication in sustainable manner, the specific goal under SDG has been set as 'End hunger, achieve food security and improved nutrition, and promote sustainable agriculture', unlike in MDG where it was clubbed with eradication of poverty.

2.1.3 SDG-2: Targets and Indicators

SDG-2 is composed of eight targets (i.e. specific, measurable and time-bound outcomes that directly contribute to the achievement of a goal) and 15 indicators (i.e. metrics used to measure focus of this study are directly related to food security and agricultural sustainability). The last three (2a–2c) are market-related measures aimed at increasing agricultural investments and reducing market restriction, distortions and volatility. The global targets and indicators for SDG-2 to be achieved by 2030 are given in Annexure 1.

2.1.4 How the SDG-2 is Interrelated with other SDG Indicators to Achieve the Goal of SDG?

SDG-2 aims to promote research and extension in agricultural sector, develop infrastructural facilities in the rural areas, develop agricultural technologies and establish gene banks for crops and livestock to increase agricultural productivity in developing countries. FAO Report (2016) on

SDGs indicates that 'these objectives are materially connected, *inter alia*, to ending poverty (SDGs 1), child and maternal care, and (SDG 3 target 3.2) diet and food system-related non-communicable diseases (SDG 3 target 3.4), women's empowerment, especially target 5.a—ensure women's right to own land (SDG 5 target 5.a), water and sanitation for all (SDG 6), access to modern energy—a prerequisite for inclusive rural poverty reduction (SDG 7), sustainable consumption and production, especially to reduce food losses and waste (SDG 12 target 12.3), peaceful and inclusive societies and effective, accountable and inclusive institutions, especially targets to reduce all forms of violence, and develop effective, accountable and transparent institutions (SDG 16 target 16.1 and 16.6)'. It would be useful to take cognizance of this interrelatedness while devising a response for realizing the targets under SDG-2. The interrelationship and linkages among different SDGs with SDG-2 are given in Fig. 2.5.

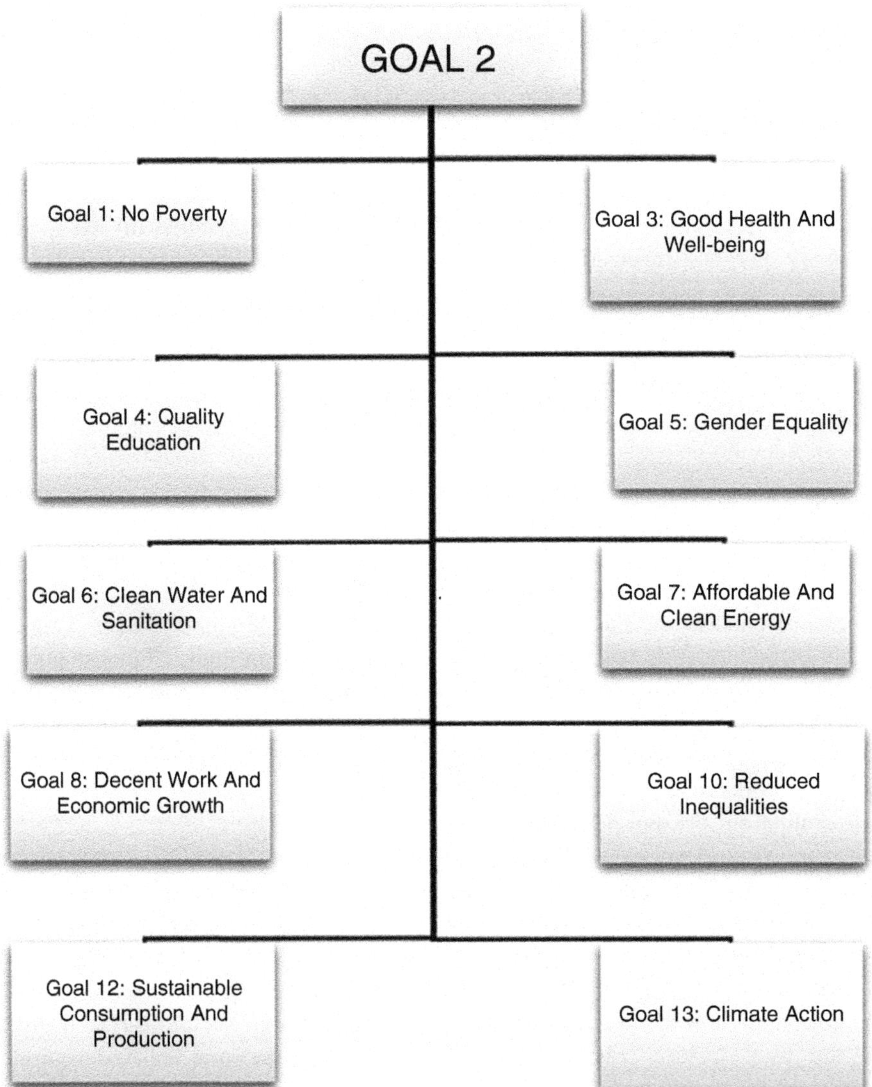

Fig. 2.5 Linkages of SDG-2 with other Sustainable Development Goals

2.1.5 Current Status and Performance of SDG-2 in India

The composite scores of the SDG-2 among the different states of India are presented in Fig. 2.6. SDG-2 score for 0 hunger ranges between 35 (Jharkhand) and 80 (Goa). This composite score was greater than (or equal to) 65 for seven states and three union territories categorized as 'front runners'; while the score was less than 50 in ten states and four union territories categorized as 'aspirant category'.

The performance of states and UTs are classified into four different categories based on the composite scores of all indicators such as (1) achiever (100), (2) front runner (65–99), (3) performer (50–64) and (4) aspirants (0–49). The different states and UTs classified based on its composite score are given in Table 2.2.

SDG-2 aims to end hunger and malnutrition by ensuring food and nutritional security to all, which demands food production systems to be resilient and sustainable. Further, equitable access and nutritious food for all and reducing vulnerability to climate vagaries and natural disasters have to be ensured under this goal. Although considerable progress has been made in poverty alleviation, higher level of malnutrition problems, viz. underweight, stunting, wasting and anaemia, are considerably prevalent in India. As per the NFHS-4, almost 50% of the pregnant women are anaemic, around 33% of women have a low body mass index and a substantial number of children under age 5 are stunted (38%), wasted (21%) and underweight (35.7%). Food and nutritional insecurity due to lack of basic amenities (water and sanitation), low agricultural productivity and social inequalities (caste, gender and illiteracy) are the root causes of the above problems.

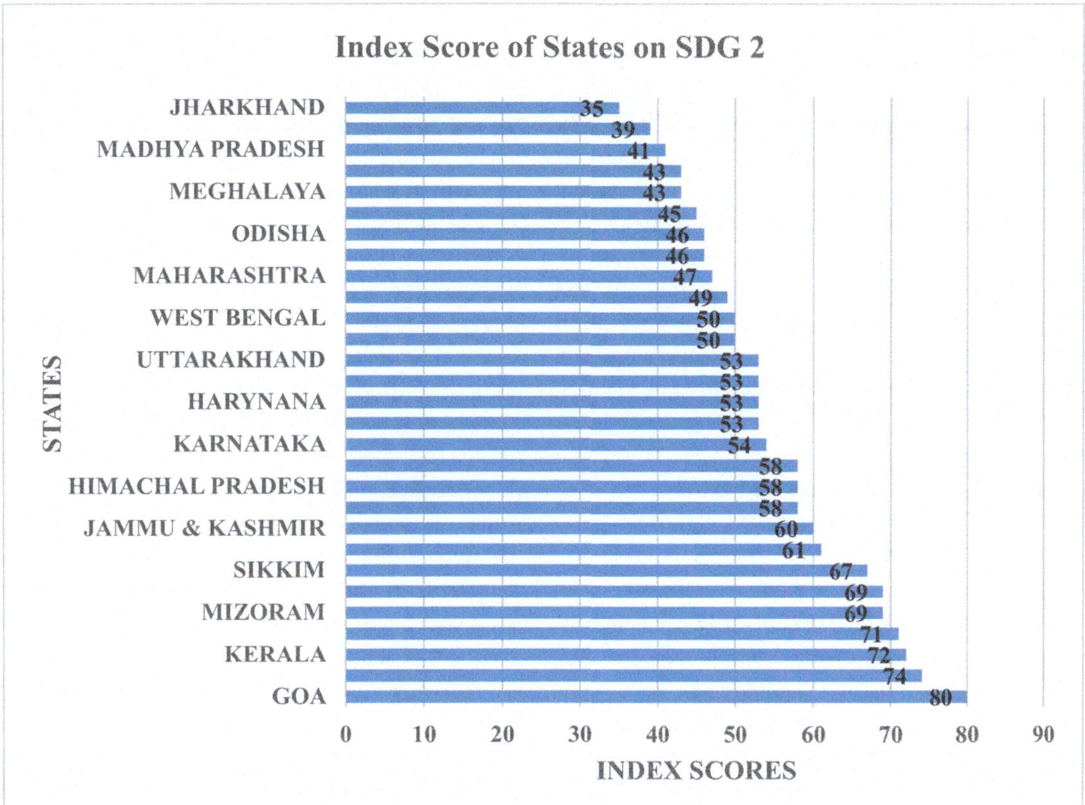

Fig. 2.6 Composite index score of Indian states on SDG-2

Table 2.2 Classification states and UTs based on SDG composite scores

Front-runner	Performer	Aspirants
Goa	Tamil Nadu	Gujarat
Manipur	Jammu & Kashmir	Maharastra
Kerala	Arunachal Pradesh	Chhattisgarh
Punjab	Himachal Pradesh	Odisha
Mizoram	Tripura	Rajasthan
Nagaland	Karnataka	Meghalaya
Sikkim	Assam	Uttar Pradesh
Delhi	Haryana	Madhya Pradesh
Puducherry	Telangana	Bihar
	Uttarakhand	Jharkhand
	Andhra Pradesh	Lakshadweep
	West Bengal	Daman & Diu
		Dadra & Nagar Haveli
		Andaman & Nicobar Islands

Source: NITI Ayog (2018)

Enhancing and sustaining agricultural productivity remains to be a major challenge in achieving food and nutritional security, especially in the wake of dwindling water resources and increased demand for water for farm and non-farm sectors. Currently, average cereal (wheat, rice and coarse cereals) productivity is 2509 kg per ha, which has to be increased by 200% to achieve the SDG-2 by 2030. Recent government schemes such as Rashtriya Krishi Vikash Yojana (RKVY), National Food Security Mission (NFSM), Integrated Watershed Management Program, Minimum Support Price, Micro-irrigation programmes and Prime Minister's Fasal Bima Yojana have been implemented to enhance the productivity and growth of agriculture to meet the target of SDG-2.

2.2 Current Status of Food Consumption and Dietary Pattern

Adequate nutrition throughout early life is fundamental for proper growth, development and survival and poor nutrition continues to be a major public health problem in developing countries (Latham 1997). However, the food items, which are predominantly consumed, may not contain all the nutrients. Specifically, cereals are the staple food items rich in carbohydrate and energy. All other nutrients may be available either in small proportion or nothing. The major policy breakthroughs such as liberalization and globalization, associated improvements in transportation and storage facilities and rise of supermarkets have made easy accessibility of all kinds of food products from any parts of the world. Therefore, identifying food consumption transition as well as dietary diversity would be imperative to avoid the problem of malnutrition.

The dietary changes that characterize the 'nutrition transition' include both quantitative and qualitative changes in the diet (Vasileska and Rechkoska 2012). Nutrition transition has been varying over the period from age of hunting and collecting food items which are rich in fat with unsaturated fatty acid, carbohydrate, protein etc., to age of consuming food items which cause degenerative diseases and to the age of behavioural changes concerning health (Popkin 2002). Though global per capita calorie intake has increased by more than 17% over four decades and expected to increase to 3040 and 3130 kcal in 2030 and 2050, respectively (Kearney 2010), the entire world, indeed, has experienced transition in food consumption pattern and nutrient intake from diversified food articles (Chengappa et al. 2017). It is apparent that there is a shift in the dietary pattern from traditional fibre-rich, grain-based food to fat-rich, processed, packaged and animal-based food items across the world. However, this transition is not uniform across income classes and geographies. For instance, the diet of the poor is based on limited number of foods, specifically with cereals in countries like India.

Several authors argued that there is a declining trend in food grain demand and increasing trend in the consumption of high-value food substances, which are rich in fat and protein (Kumar and Joshi 2016), but not towards the healthy dietary pattern. As far as India is concerned, evidences show that cereals, including rice and wheat, have occupied a major share of Indian diet, irrespective of rural and urban areas, for a longer period. However, sluggishness can be observed in rice and wheat consumption; specifically, wheat consumption has declined more than that of rice. Between 1983–1984 and 2011–2012, coarse cereal consumption has decreased substantially, and pulse consumption has decreased as well by 16%. In contrast, consumption of edible oil, fruits, vegetables and livestock products has increased significantly over the years. The rate of increase in consumption was found to be more for fruits followed by edible oils. Fruit consumption increased by 268% in rural areas and 241% in urban areas and edible oils by 107% in rural areas and 45% in urban areas during the same period. Consumption of livestock products such as milk and meat has been increasing consistently over the years (Table 2.3). Similarly, per capita consumption of processed food and food away from home is expected to increase over the years (Chengappa et al. 2017).

Expenditure share of food commodities is shown in Fig. 2.7. The monthly percapita consumption expenditure (MPCE) shares of high-value commodities such as fruits and vegetables, egg, meat and fish showed increasing trend, whereas there is a negative trend for traditional food items like cereals and pulses, and it was reported that this decline in consumption expenditure on cereals and pulse was mainly due to low cost and increase in income levels of the household (Ramachandran 2007). It is found that the expenditure share of food grains is higher in rural areas, while urban households consumed more of high-value commodities such as fruits, vegetables, milk and meat items. Although the share of food grains in food expenditure has declined significantly, these still constitute major portion of the Indian diet (Chengappa et al. 2017).

Across income groups, rice consumption has increased by 18% between 1983–1984 and 2011–2012 for low-income group. This may be due to supply of rice at subsidized prices through Public Distribution System (PDS) in India. It is interesting to note that percentage change for the commodities like fruits, vegetables, meat, fish and eggs was found to be lower for low-income people, compared to middle- and high-income people, indicating that as income increases, the changes in the consumption of all these food commodities are highly responsive (Table 2.4). Few studies in India indicated that cereal consumption was least responsive to income changes, whereas fruit and vegetables (Umanath et al. 2015) and livestock products (Chengappa et al. 2016) showed higher-income elasticity. It was found that the consumption of cereals and millets by children has declined from 1975–1976 to 2011–2012 and no change was observed for pulses and legumes. Nevertheless, children's intake of vegetables, milk and milk products has considerably increased during the same period (Table 2.5). Thus, it is clearly evident that irrespective of residential location and income groups, there is a change in consumption pattern in India.

All these dynamics are the results of changes in both composition of meals of household and its structure associated with intensive consumption of snacks and junk foods, fast foods, pre-packaged meals and carbonated drinks—mostly observed in the countries of Asia and Latin America. Particularly, youngsters are observed as the profound consumers of these fast foods and soft drinks (Tiwari and Sankhala 2007), and almost 30% of youngsters (under age of 18) are affected by obesity problem in India (FAO 2017). This change in food consumption pattern is attributed to numerous factors such as increased per capita income, urbanization, development in technologies and marketing, women empowerment, changing tastes and preferences, diversified occupations, health consciousness etc. (Kumar and Joshi 2016).

Table 2.3 Changes in monthly per capita quantity consumption of food commodities in rural and urban regions of India

	Rural (in kg)					Urban (in kg)				
Food items	1983–1984	1999–2000	2004–2005	2011–2012	% Change between 1983–1984 and 2011–2012	1983–1984	1999–2000	2004–2005	2011–2012	% Change over which period?
Rice	7.25	6.59	6.38	7.12	−1.80	5.85	5.10	4.71	5.96	1.92
Wheat	4.46	4.45	4.19	3.92	−12.05	4.86	4.72	4.36	4.07	−16.33
Course cereals	3.49	1.35	1.14	0.59	−83.11	1.09	0.31	0.44	0.26	−76.26
Total cereals	15.20	12.39	11.71	11.63	−23.49	11.80	10.42	9.94	10.29	−12.83
Pulses	0.96	0.84	0.71	0.80	−16.20	1.04	1.00	0.82	0.88	−15.49
Edible oils	0.30	0.50	0.48	0.63	107.14	0.51	0.72	0.66	0.74	45.28
Vegetables	3.86	5.36	4.68	4.60	19.18	4.25	5.60	5.76	4.82	13.56
Fruits	0.24	0.47	0.60	0.88	268.77	0.35	1.01	1.04	1.18	241.59
Milk and milk products	3.27	3.93	4.08	5.08	55.45	4.63	5.64	5.42	5.93	28.11
Sugar	0.93	0.77	0.74	0.82	−12.34	0.98	1.00	0.87	0.86	−12.38
Meat, fish and eggs	0.39	0.32	0.38	0.58	47.55	0.55	0.54	0.99	0.70	27.62

Source: NSSO's consumer survey data published in 1983–1984 and 2011–2012

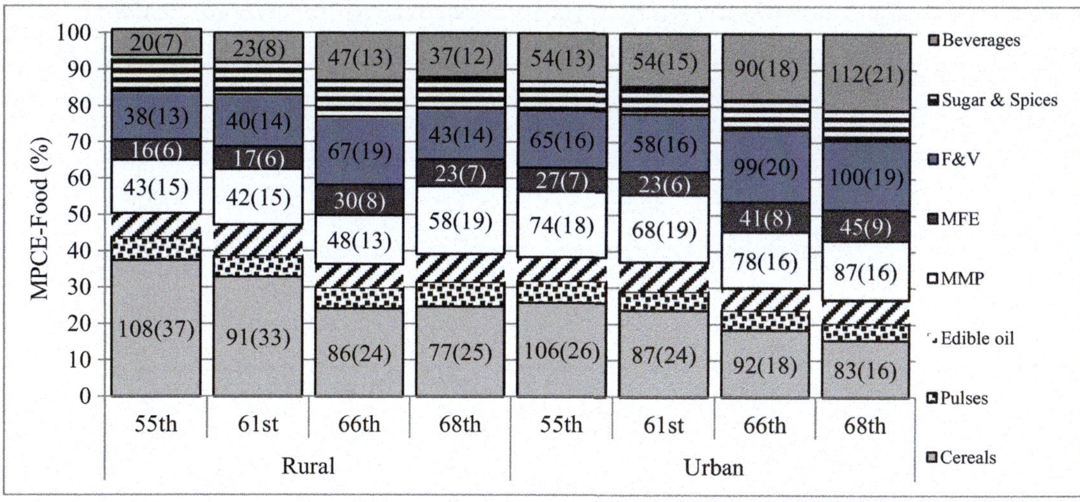

Fig. 2.7 Changes in the food expenditure composition in rural and urban regions

Table 2.4 Income group-wise changes in monthly per capita quantity consumption of food commodities in India

	Low income			Middle income			Higher income		
Food items	1983–1984	2011–2012	% change	1983–1984	2011–2012	% change	1983–1984	2011–2012	% change
Rice	5.40	6.39	18.41	7.50	6.82	−8.98	7.84	6.69	−14.66
Wheat	3.77	3.46	−8.2	4.46	3.77	−15.31	5.92	4.30	−27.47
Corse cereals	2.95	0.42	−85.88	2.52	0.53	−78.97	2.35	0.44	−81.4
Total cereals	12.12	10.27	−15.26	14.48	11.13	−23.14	16.11	11.42	−29.11
Pulses	0.66	0.59	−10.65	0.96	0.72	−24.67	1.48	0.99	−33.03
Edible oils	0.23	0.45	100	0.36	0.59	63.81	0.61	0.81	33.01
Vegetables	2.98	3.52	17.89	3.99	4.22	5.57	5.42	5.41	−0.17
Fruits	0.13	0.34	161.29	0.23	0.61	171.96	0.54	1.47	174.07
Milk	1.39	1.89	35.99	3.36	3.73	11.01	7.47	7.74	3.64
Sugar	0.56	0.52	−6.42	0.89	0.72	−19.59	1.56	1.02	−34.59
Meat, fish and eggs	0.26	0.29	11.04	0.41	0.48	16.16	0.74	0.84	12.71

Source: NSSO's consumer survey data published in 1983–1984 and 2011–2012

2.2.1 Changes in Dietary Pattern

Dynamics in the intake of different nutrients by the Indian households are presented in Tables 2.6 and 2.7. In general, under Indian context, energy requirements for average working men and women are 2730 and 2230 kcal/day, respectively, and heavy working people require more than these. From Table 2.6, it could be observed that both the rural and urban household consumed less than the requirement in all the periods. In particular, rural energy intake has decreased from 2243 kcal in 1983–1984 to 2131 kcal in 2011–2012, whereas urban households show increasing trend in energy intake. With respect to protein

Table 2.5 Time trends in selected food intakes (g/day) from the NNMB surveys 1975–1979 and 2011–2012

	Period of survey	Children		Adult men	Adult women
Type of foodstuff (g/day)		1–3 years	4–6 years	18 years	18 years
Cereals and millets	1975–1979	158	228	474	410
	1989–1990	176	263	–	–
	1996–1997	152	243	474	414
	2011–2012	131	209	380	341
Pulses and legumes	1975–1979	14	20	36	29
	1989–1990	14	20	–	–
	1996–1997	13	20	36	29
	2011–2012	15	20	32	28
Vegetables	1975–1979	35	52	146	129
	1989–1990	31	51	–	–
	1996–1997	35	64	145	124
	2011–2012	41	70	143	138
Milk and milk products	1975–1979	74	57	101	92
	1989–1990	68	62	–	–
	1996–1997	66	59	101	81
	2011–2012	86	67	91	82

Source: NNMB baseline and repeat surveys (data for adults are not available for the year 1989–1990)

Table 2.6 Region-wise changes in nutrient intake in India

	Rural			Urban		
Nutrients	1983–1984	2011–2012	% Change	1983–1984	2011–2012	% Change
Calories (kcal/capita/day)	2243.00	2131.00	−4.99	1987.00	2059.00	3.62
Protein (g/capita/day)	63.40	57.00	−10.09	56.20	55.60	−1.07
Fat (g/capita/day)	26.50	42.10	58.87	34.50	48.30	40.00
Calcium (mg/capita/day)	474.00	556.00	17.30	518.00	617.00	19.11
Iron (mg/capita/day)	41.10	35.70	−13.14	39.80	36.10	−9.30
Zn (mg/capita/day)	8.80	10.00	13.64	7.80	9.70	24.36
Vitamin A equivalent (μg/capita/day)	161.00	201.00	24.84	186.00	223.00	19.89

Source: Joshi et al. (2016)

and fat, Indian households consumed more than enough in all the periods, but intake of protein decreased considerably in 2011–2012. It is interesting to note that fat intake has increased manifold over the years. Among the micronutrients, Ca and Fe consumption was closer to the recommended daily allowance (RDA), and Zn and vitamin A consumption was less than the RDA. In addition, except iron, the intake of all the micronutrients has increased by about 20% (Table 2.7).

Across income groups, both low- and middle-income groups have consumed lesser energy than the higher-income people, even lesser than the RDA. Further, negative trends in energy intake could be found in all income groups. Specifically, high-income group has reduced their energy intake by 15% from 1983–1984 to 2011–2012. Intake of protein, fat, Ca, Zn and vitamin A by low-income people is very less, while high-income households have consumed all these nutrients more than the RDA and middle-income people consumed closer to the RDA (Table 2.7). Shankar et al. (2017) found an interesting feature with respect to the dietary intake that all these

Table 2.7 Income group wise changes in nutrient intake in India

	Low income			Middle income			High income		
Nutrients	1983–1984	2011–2012	% change	1983–1984	2011–2012	%change	1983–1984	2011–2012	% change
Calories (kcal/capita/day)	1703.00	1690.00	−0.76	2176.00	1965.00	−9.70	2755.00	2342.00	−14.99
Protein (g/capita/day)	48.60	44.10	−9.26	61.00	52.00	−14.75	77.70	63.80	−17.89
Fat (g/capita/day)	17.50	26.10	49.14	27.60	36.50	32.25	47.70	56.20	17.82
Calcium (mg/capita/day)	303.00	314.00	3.63	460.00	457.00	−0.65	781.00	752.00	−3.71
Iron (mg/capita/day)	31.60	28.70	−9.18	39.90	33.20	−16.79	54.00	40.20	−25.56
Zn (mg/capita/day)	6.80	8.00	17.65	8.50	9.20	8.24	10.60	11.00	3.77
Vitamin A equivalent (µg/capita/day)	109.00	126.00	15.60	163.00	172.00	5.52	262.00	264.00	0.76

Source: Joshi et al. (2016)

Table 2.8 Time trends in selected median nutrient intakes (per day/person) from the NNMB surveys 1975–1979

	Period of survey	Children		Adult men	Adult women
Type of nutrients		1–3 years	4–6 years	18 years	18 years
Energy (kcal)	1975–1979	834	1118	2217	1824
	1989–1990	908	1260	–	–
	1996–1997	807	1213	2402	2070
	2011–2012	767	1082	1875	1709
Protein (g)	1975–1979	22.8	30.2	61.0	49.1
	1989–1990	23.7	33.9	–	–
	1996–1997	20.9	31.2	32.4	53.4
	2011–2012	21.3	30.3	52.7	46.5
Iron (mg)	1975–1979	10.2	15.0	30.4	24.9
	1989–1990	10.2	15.3	–	–
	1996–1997	8.7	14.3	28.5	24.1
	2011–2012	5.8	8.9	15.4	13.7

Source: Shankar et al. (2017)

nutrients, except fat, show an increasing trend over the years, indicating that there is scope for improving the dietary diversity among the low-income household in the future (Table 2.8). It was observed that children under age 6 were deficient in energy intake and moderately deficient in iron intake, while intake of energy, protein and Fe has declined among the children over the period (Table 2.9).

2.2.2 Food Production Dynamics

Production changes in agricultural and allied sectors in India over the years are presented in Table 2.9 and Fig. 2.8. It could be observed that production of all the food items has increased manifold over the last four decades. Specifically, high-value commodities, viz. oilseeds, fruits, vegetables, milk, meat and fish, have increased

Table 2.9 Production of food items in India (in million tonnes)

Year	Cereals	Rice	Coarse cereals	Pulses	Oilseeds	Sugarcane	Vegetables	Fruits	Milk	Egg	Meat	Fish
1970–1971	96.60	42.22	30.55	11.82	9.02	126.37	NA	NA	NA	NA	NA	1.76
1975–1976	108.00	48.74	30.41	13.04	9.87	140.6	NA	NA	NA	NA	NA	NA
1980–1981	118.96	53.63	29.02	10.63	8.74	154.25	NA	NA	31.6	10060	NA	2.44
1985–1986	137.08	63.83	26.20	13.36	10.15	170.65	NA	NA	44	16128	NA	2.87
1990–1991	162.13	74.29	32.70	14.26	17.56	241.05	58532	28632	53.90	21101	NA	3.84
1995–1996	168.11	76.98	29.03	12.31	21.04	281.10	75074	40458	66.20	27198	NA	4.95
2000–2001	185.74	84.98	31.08	11.08	17.36	295.96	88622	43001	80.60	36632	1.90	5.66
2005–2006	195.22	91.79	34.07	13.39	26.81	281.17	114993	59563	97.10	46235	2.30	6.57
2010–2011	226.26	95.98	43.40	18.24	30.98	342.38	156325	76424	121.80	63024	4.90	8.23
2011–2012	242.2	105.30	42.01	17.09	27.35	361.04	162187	81285	127.90	66450	5.50	8.67
2014–2015	234.87	105.48	42.86	17.15	25.49	362.33	169064	90183	146.30	78484	6.69	10.25
%change in 2014–2015 over 1970–1971	143.14	149.83	40.29	45.09	182.59	186.72	188.84	214.97	362.97	680.16	252.11	483.77

Source: Handbook of Statistics on Indian Economy 2017–2018

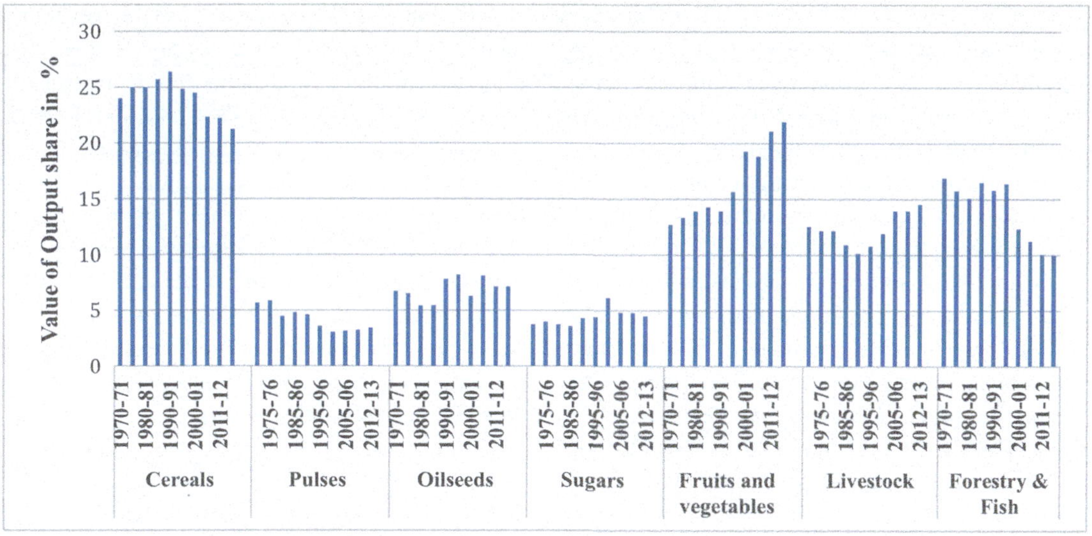

Fig. 2.8 Output share of agriculture and allied products over the years. Data source: (National Accounts Statistics, CSO, Government of India)

more than 150–250%, individually, which are significantly high compared to cereals and pulses (Table 2.9). Although food grains occupied more than 25% of total value of agricultural products, there is a decreasing trend in their share over the years, while the share of fruits, vegetables and livestock products continues to increase over the years (Fig. 2.8). Similar findings were reported by several authors (Birthal and Negi 2012; Chengappa et al. 2016), and both crop and livestock sectors have witnessed interesting developments in recent decades. Food grain production exhibited downward trend against other crop groups. With the increase in the share of livestock and horticulture sectors continuously, there has been a concomitant decrease in the share of food grains. Interestingly, the share of livestock sector surpassed the share of food grains in the total value of output in the mid-2000s with a share of around 28%. Thus, horticulture sector is emerging in a big way as its contribution to Indian agriculture is increasing over the years.

2.3 Nutrient Availability from Different Sources of Foods

Most of the literature emphasized that crop diversification is one way for achieving dietary diversity and consequently to reduce nutritional imbalances (Meenakshi et al. 2010; Ruel 2003). Diversifying food production has increased nutritional outcomes in farm households, who primarily consume from own production (Ecker and Qaim 2011). It is argued that intensification of agriculture towards energy dense food commodities such as paddy and wheat can contribute to food security, but this may have negative impact on food quality and dietary diversity. Consequently, reliance on too much of starchy foods and inadequate consumption of nutrient-rich diet causes micronutrient deficiencies and hidden hunger (Muthayya et al. 2013; Kennedy et al. 2003). Agriculture can be linked to food security and improved nutritional status in multiple and bidirectional pathways as agriculture not only supplies food for nutrition but also offers employment and income through which majority of population can improve their nutritional status. Hence, understanding the interplay between farm production diversification and dietary diversity is a policy relevant in developing countries, where more than 50% rely on agriculture for food and income (Ayenew et al. 2018).

Source-wise nutrient supply and its dynamics over the years are presented in Fig. 2.9. Accordingly, cereals continue to be the major supplier of energy (around 60%), though the energy supplied by cereals has decreased from 73% in 1983–1984 to 60% in 2011–2012. It could be noted that energy supply by edible oil, milk and other food items has increased significantly. In the case of protein supply, food grain share has decreased considerably, while the share of fruits, milk, meat and other food products showed increasing trend. Edible oils and milk not only were the major contributors for fat supply but also show increasing trend over the years. Milk was the prime supplier of calcium followed by cereals, vegetables and pulses in all the periods. Other than cereals, vegetable and pulses were the major suppliers of iron (Fe) and zinc (Zn). However, it is interesting to note that the supply of these two minerals by other food items has increased tremendously from 1983–1984 to 2011–2012. Vegetables and milk were the major suppliers of vitamin A in all the period, and the share of the other food items has increased significantly in 2011–2012. From these results, it can be discerned that contribution of cereals to total nutrient supply is decreasing over the years, while milk, vegetables, fruits and other food items continue to increase their share, specifically with respect to protein, minerals and vitamins.

2.3.1 Problem of Malnutrition

The world has not witnessed the expected reduction in the absolute number of chronically hungry people over the years. Although economic growth has supported the reduction in undernutrition problems, there exist regional and social disparities in executing the programs for alleviation of undernourishment and malnutrition problems. It could be found that malnutrition persists in term

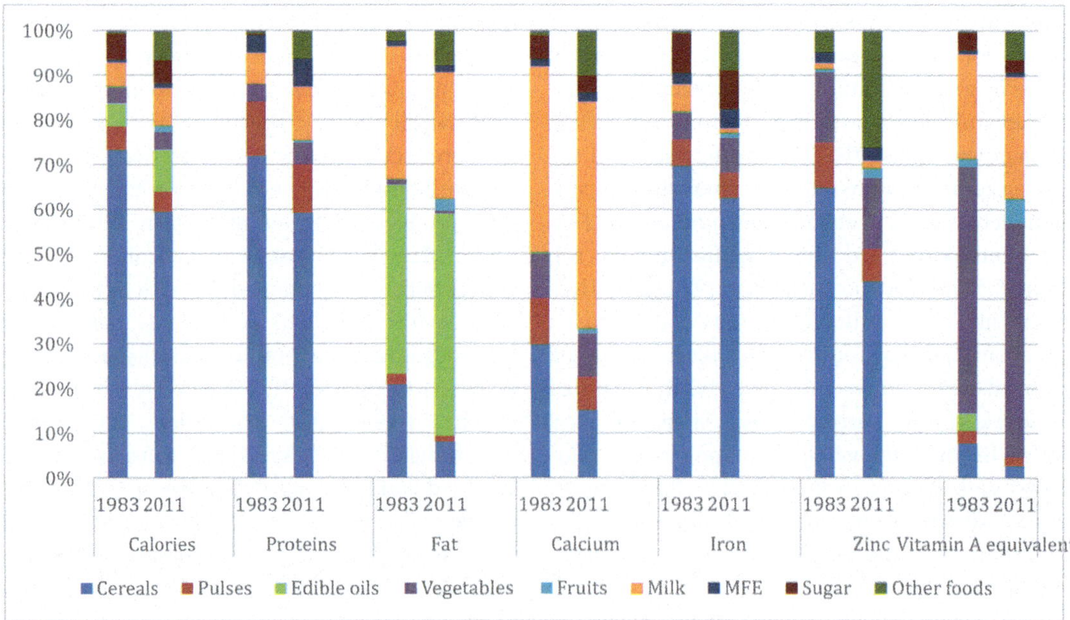

Fig. 2.9 Source-wise nutrient supply and change over the years (Joshi et al. 2016)

of micronutrient deficiencies in the countries, where food security problem has been mitigated by larger supply of calorie through cereals. In addition, the persistence of undernutrition and micronutrient deficiencies along with the prevalence of excess weight is found in some areas, while economic transitions (Popkin 2006), growing income inequalities, and rapid changes in the structure of food systems are noticed in most regions.

Dietary diversity is one among the several factors which often affects nutritional status of households (Taren, & Chen 1993; Hatløy et al. 2000; Arimond & Ruel 2002; Ekesa et al. 2008). The lack of diversity in terms of amount and composition in the food basket is considered to be suboptimal growth leading to poor development and long-term health outcome (Charmarbagwala et al. 2004). As far as India is concerned, burden of undernutrition, micronutrient deficiency and overnutrition remains high (Keshari and Mishra 2017); specifically, children and women are the most vulnerable people who are frequently affected by malnutrition problems (Kehoe et al. 2014; Aurino 2017). Moreover, children in low-income group and rural areas are heavily affected than those in high-income groups and urban areas. Overnutrition problems are mostly associated with high income and urbanization. Moreover, while there is a decreased level of undernutrition problem among adolescent, the overweight problem has increased significantly from 1.8% in 1999 to 7.5% in 2015 for boys and from 1.9 to 6.1% for girls during the same period. Similarly, the problem of obesity, diabetes and overnutrition shows increasing trend for all categories of people. While significant progress has been made over the last 50 years in improving food production to achieve sufficiency (Table 2.9 and Fig. 2.8), still majority of rural households and communities have to face uncertainty in food and nutrition security. On the whole, over one-fifth of India's population suffers from chronic hunger, and India ranks 100th among 119 countries in the 2017 Global Hunger Index (Von Grebmer et al. 2017).

It is noted that dietary diversity is positively linked with overall diet quality, micronutrient intake, and better nutritional status of young children, household food security (Sawadogo et al. 2006; Steyn et al. 2006; Kennedy et al. 2007). Balanced diet with right quantities of protein, carbohydrate, fat, vitamins and low cholesterol content in childhood stage support better health outcomes for the remaining life period of a child, thus help better cognitive development of children. Gausman et al. (2018) found that individual and contextual socioeconomic factors are associated with children's diet with variety of food items. Codjoe et al. (2016) stated that children in affluent homes grow better for a number of reasons—improved nutrient adequacy is expected to be one important mechanism, where wealth and resources of households translate into better outcomes for children.

People in countries like India have limited access to adequate and healthy food due to many reasons, and this has created wider space between food and people. This space can be filled in by ensuring food availability, accessibility and affordability. Science and technological developments have ensured food availability manifold over the last 50 years in many developing countries including India. Also, major policy changes of recent decades such as liberalisation and globalization and associated developments in transportation and storage facilities and rise of supermarkets have contributed for better accessibility to most of the food products from any parts of the world. However, the challenge of malnutrition remains a major development concern hampering health and human development, thus affecting the achievement of Sustainable Development Goals in India. The main reason behind malnutrition problems is the inability of households to pay for healthy food items due to high cost and less purchasing power. This keeps the household members including children nutritionally insecure.

Public policies in favour of cereal production and distribution have largely ensured the availability of energy-dense food items at cheaper prices and reduced the food insecurity problem significantly, whereas accessibility of high-value commodities such as fruits and vegetables and livestock products is challenging for the poor section of the population due to unexpected price hike and fluctuations. Hence, policy measures to keep food prices especially fruit and vegetable prices in control would help in achieving Sustainable Development Goals. Specifically, availability and affordability for pulses, fruits and vegetables and livestock products should be ensured as they are not only less supplier of energy and fat but also possess protein and other micronutrients such as calcium, iron, vitamins etc., in large volume, compared to cereals and junk foods. A healthy diet including fruits and vegetables is necessary for optimal child growth, weight management and prevention of chronic diseases (USDA 2010) including some types of cancer and cardiovascular disease (WHO 2005). Livestock products, particularly fish and milk, are rich in protein (Adeniyi 2010) and possess richer content of amino acids, unsaturated fatty acids, vitamins and trace metals that are easily digestible (Kızılaslan and Nalıncı 2013). Consumption of fish helps prevent cardiovascular diseases, high blood pressure, cholesterol, Alzheimer's disease and cancer to humans (McNaughton et al. 2008). Moreover, fish has low cholesterol over red meat and is recommended for both the young and the old people (Eyo 2002). Low intake of nutrient-rich foods affects children's future productivity negatively (Strauss and Thomas 1995). Similarly, intake of adequate millets can possibly protect from diabetes as they contain high level of proteins, minerals, vitamins, antioxidants, non-glutinous and non-acid-forming low glycaemic index (GI), non-starch polysaccharides and dietary fibres (Muthamilarasan and Prasad 2015).

2.4 Nutritional Programmes in India

Nutrition programmes are an integral component of a comprehensive nutritional policy. Nutritional improvements require a multifaceted approach since malnutrition, poverty and inequality, educational status, water supply and sanitation, food production, gender issues and health are intricately interwoven issues that need to be tackled through collective efforts of multiple agencies and ministries. The National Nutrition Policy of India recognizes this fact and lists out a wide range of nutrition policy instruments covering direct, short-term instruments and indirect, long-term structural and institutional measures. In what follows, we provide a comprehensive overview of major nutritional intervention programmes as well as indirect food security measures implemented by the governments at various levels.

2.4.1 Summary of Nutritional Programmes in India

Over the years, the government has introduced several policy measures and nutrition programmes to combat malnutrition problems among its people. Such interventions include the Integrated Child Development Services (ICDS) scheme launched in 1975 targeting early childhood care and development; Special Nutrition Programme (SNP) in 1970–1971 for the same target group; bal (children) wadi (home or centre) Nutrition Programme in 1970–1971 aiming to supply about one-third of the calorie and half of the protein requirements of the preschool child—as measure to improve nutritional and health status; National Goitre Control Programme (NGCP) in 1962 and renamed as National Iodine Deficiency Disorders Control Program (NIDDCP) with a view to cover a wide spectrum of Iodine Deficiency Disorders like mental and physical retardation, deaf mutinous, cretinism, still births, abortion squint and various types of goitre; National Nutritional Anaemia Prophylaxis Programme (NNAPP) in 1970 as a centrally sponsored scheme targeting children of 6–10 years old by providing 30 mg elemental iron and 250 µg folic acid per child per day for 100 days in a year; Tamil Nadu's Midday Meal Scheme/Nutritious Meal Programme (NMP) to ensure an atmosphere in which children can grow with good health and good education; Rajiv Gandhi Scheme for Empowerment of Adolescent Girls (RGSEAG)—scheme for empowerment of adolescent girls; Weekly Iron and Folic Acid Supplementation (WIFS) Programme in 2012–2013 to meet the challenge of high prevalence and incidence of iron deficiency anaemia among adolescent girls and boys; Mother's Absolute Affection (MAA) in 2016 to bring undiluted focus on promotion of breastfeeding and to revitalize efforts towards promotion, protection and support of breastfeeding practices through health systems to achieve higher breastfeeding rates; Pradhan Mantri Matru Vandana Yojana in 2017 in all the districts of the country to fulfil specific conditions relating to Maternal and Child Health; and POSHAN Abhiyaan in 2017–2018 aiming to improve the nutritional status of children from 0 to 6 years, adolescent girls, pregnant women and lactating mothers.

The various nutritional programmes implemented in India along with their objectives, target group and key components are summarized in Table 2.10.

2.4.2 Leveraging Agriculture for Increasing Productivity and Nutritional Security

Likewise, to leverage agriculture for increasing productivity and nutritional security, the government has introduced some flagship programmes. These include the National Agriculture Development Programme (Rashtriya Krishi Vikash Yojana or RKVY), National Food Security Mission (NFSM) and programmes related to irrigation Pradhan Mantri Krishi Sinchayee Yojana (PMKSY). These geared toward providing much needed boost to enhance agricultural productivity and, thereby, higher agricultural growth. Besides, a pilot scheme on Nutri Farms has been launched in 2013–2014 with an outlay of Rs. 200 crores to promote cultivation of biofortified food crops enriched with critical micronutrients like iron-rich bajra, protein-rich maize and zinc-rich wheat, to improve the nutrition status of the most vulnerable sections of population of the country, and it will add nutritional dimension to farming sector. The National Mission on Agricultural Extension and Technology (NMAET) was introduced in 2010 to make the extension system farmer-driven and farmer-accountable by way of new institutional arrangements for technology dissemination. This mission was expected to take-up and deliver new technologies to farmers with the help of mass media support, agriclinic and agribusiness centres, National e-Governance Plan in Agriculture, Kisan Call Centres (KCC), Farmers' Portal and MKisan App.

2.4.2.1 National Food Security Mission (NFSM)

The National Food Security Act (NFSA) in 2013 was enacted by the Government of India as food and nutritional security are expected to be basic right of all human being born in India by ensuring access to adequate quantity of quality food at affordable prices to people to live a life with dignity. NFSA-2013 created legal entitlements to existing governmental food and nutrition security programmes. NFSM targeted additional production of 25 million tonnes of food grains comprising 10 million tonnes of rice, 8 million tonnes of wheat, 4 million tonnes of pulses and 3 million tonnes of coarse cereals by the end of XII Plan. By the Act, it is ensured that there is a free daily meal for children and maternity benefits, including cash for pregnant women, which can combat rampant undernutrition (calorie deficiency) and malnutrition (protein deficiency) across India. These steps are expected to complement the current nutritional programs such as midday meals and Integrated Child Development Services. Now, the availability of food grains, specifically rice and wheat, exceeds consumption requirement of food grains in the country. However, the rate of malnutrition has not declined to the desired level, and still prevalence of micronutrient deficiency, for instance, iron, zinc and vitamin A, can be found among Indian households. The promotion of cultivation of micronutrient-rich cultivars and crops and development of their effective supply chain would help in the reduction of malnutrition problems.

Table 2.10 Summary of nutritional programmes implemented in India

Programme	Objective	Target group	Components
ICDS	Provide early childhood care	Children age 0–6 years, pregnant women and lactating mothers	Provide supplementary nutrition, preschool non-formal education, nutrition and health education, immunization, health check-up and referral services
Balwadi nutrition programme	To improve nutritional and health status of the preschool child	Beneficiaries of this scheme are preschool children between the age 3 and 5 years, with priority to low-income group	Provide supplementary nutrition consisting of 300 calories and 10 g of protein per child per day, activities for social and emotional development
National Iodine Deficiency Disorders Control Programme (NIDDCP)	To bring the prevalence of IDD to below 5% in the country by ensuring 100% consumption of adequately iodized salt (15 ppm) at the household level	Entire country	Production and supply of iodized salt, monitor iodine deficiency and health and education publicity
National Nutritional Anaemia Prophylaxis Programme (NNAPP)	To prevent nutritional anaemia in mothers and children	Children age 6–10 years, expectant and nursing mothers	Provide elemental iron and folic acid
Special Nutrition Programme	Same as ICDS	Children age 0–6 years and expectant and lactating mothers in tribal areas and slums	Provide supplementary feeding of about 300 calories and 10 g of protein to preschool children and about 500 calories and 25 g of protein to expectant and nursing mothers for 6 days a week
Tamil Nadu Midday Meal Scheme/ Nutritious Meal Programme (NMP)	Improve the nutritional status of school-going children	Children studying in government schools, government-aided schools, special training centres and Madrasas	Provide hot, nutritive noon meal for school-going children of all ages

Rajiv Gandhi Scheme for Empowerment of Adolescent Girls (RGSEAG)—'SABLA'	Empowering adolescent girls along through nutrition and non-nutrition interventions	Adolescent girls	Out-of-school adolescent girls in the age group of 11–14 years attending *Anganwadi* centres (AWC) and all girls in the age group of 14–18 years are provided supplementary nutrition containing 600 calories, 18–20 g of protein and micronutrients, per day for 300 days in a year Upgrading various skills like home skills, life skills and vocational skills
Weekly Iron and Folic Acid Supplementation (WIFS) Program	To reduce the prevalence and severity of anaemia in adolescent population (10–19 years)	School going adolescent girls and boys in 6th to 12th class enrolled in government/ government aided/municipal schools. Out-of-school adolescent girls	• Administration of supervised weekly iron-folic acid supplements of 100 mg elemental iron and 500 µg folic acid using a fixed day approach. • Biannual deworming for control of helminthic infestation
Pradhan Mantri Matru Vandana Yojana	To provide cash benefits for maternity expenses	All eligible pregnant women and lactating mothers throughout the country	Monetary sum of Rs. 5000 in three instalments. In addition, maternity benefit of Rs. 1000 provided under Janani Suraksha Yojana (JSY) after institutional delivery. Supplemental nutrition under *Anganwadi* services
POSHAN Abhiyaan (National Nutrition Mission- NNM)	To reduce the level of stunting, undernutrition, anaemia and low birth weight babies and to ensure holistic development and adequate nutrition for pregnant women, mothers and children	Improving the nutritional status of children from 0 to 6 years, adolescent girls, pregnant women and lactating mothers in all districts in three phases with distinct milestones for each phase	Infant and young child feeding, food and nutrition, immunization, institutional delivery, WASH (water, sanitation and hygiene), deworming, ORS-zinc, food fortification, dietary diversification, adolescent nutrition, maternal health and nutrition, ECD (early childhood development)/ECCE (early childhood care and education), convergence, ICT-RTM (information and communication technology-enabled real-time monitoring) and capacity building

2.4.3 Performance and Current Status of Nutritional Security Programmes

The government of India in the year 2017–2018 allocated Rs. 16,745 crore and Rs. 10,000 crore, respectively, for the Integrated Child Development Services (ICDS) and the Mid-day Meal Scheme (MDMS). Each year, the coverage of the ICDS increases with the opening of new centres, but the budget allocation remains the same as the previous year allocation. The scenario can be seen in the case of Mid-day Meal Scheme. Thus, all nutrition-related schemes including the food subsidy shared only about 1.8% of the GDP in 2017–2018. Excluding the food subsidy, the nutrition budget is less than 1% of GDP of India. Under the XI Five Year Plan, universalization of ICDS coupled with setting up of mini Anganwadi Centres (AWC) was undertaken in deprived areas. However, there is a need to further strengthen ICDS in poorly performing states based on the lessons learnt from various successful models. After overhauling the existing problems of ICDS, the Ministry of WCD suggested to improve nutrition scenario of the country. The Food Security Bill ensuring 5 kg food grain/person/month at highly subsidized rates was passed in 2013. It is commendable that food and nutrition security is being promoted through several national level programmes like TPDS, MGNREGA, ICDS and MDMS. Further, programmes like Swachh Bharat and 'Beti Bachao, Beti Padao' address critical nutritionally sensitive issues. Fortification of food items for the people availing PDS food items is taken up to address the issue of malnutrition in the country. Thus, all programmes under the umbrella ICDS are expected to contribute to the attainment of the Millennium Development Goals (MDGs) 1 (reduction in severe to moderate malnutrition among children), 4 and 5 (reduction in infant mortality rate, child mortality rate, maternal mortality rate) and 2 (increase in enrolment, retention rates, and reduction in dropout) as well as SDG 2 (end hunger) by laying foundation at AWC and NNM. ICDS has helped to reduce wasting under 3 years old children by 6%, but less than what was achieved in Thailand (from 12.5 to 1%).

The results of the impact studies on nutritional programmes are mixed. An early study on the impact of ICDS in India found that there has been a significant decline in malnutrition among children included in the programme (Tandon 1989). A more recent study in West Bengal found that though the ICDS has improved the institutional child delivery by about 12%, it did not improve the health outcomes of children (Dixit et al. 2018), while another study in the same state revealed that mere inclusion of children in the ICDS programme did not result in reduced stunting (Dutta and Ghosh 2016). There are also several gaps in the implementation of ICDS programme such as infrastructure gap, shortcomings in training programmes and coverage, supply and provision of supplemental nutrition and preschool education (Chudasama et al. 2016). A study from rural Bihar by Mittal and Meenakshi (2016) reported that cooked meals provided to children in the age-group 3–6 years under the ICDS scheme has increased calorie intake by 135, protein intake by 6 grams and iron intake by 2 g though there was no change in the net intake of vitamin A. However, a pan-India study based on National Family Health Survey-3 reveals that large long-run benefits are expected from the ICDS (Jain 2015). Using data from National Family Health Surveys conducted during 2005–2006 and 2015–2016 covering large samples, it was found that the coverage of ICDS scheme has improved significantly between 2006 and 2016 on multiple dimensions. The mean proportion of respondents receiving supplementary food increased from 9.6 to 37.9% between 2006 and 2016, and the corresponding figure for health and nutrition education increased from 3.2 to 21.0%, health check-ups from 4.5 to 28%, and those availing child-specific services such as immunization, growth monitoring etc. have increased from 10.4 to 24.2%. In spite of the progress made at national level, gaps continue to remain at subnational level with the most disadvantaged households from the lowest socio-economic strata, and women with low educational status are still more excluded from the scheme (Chakrabarti et al. 2019). The evolution of nutritional policy in India is plagued by several structural weaknesses such as involvement of multiple stakeholders with entrenched views about problems and solutions, serious institutional weaknesses in the ministries implementing nutrition and health programmes such as Ministry of Women and Child Development and Ministry of Health and Family Welfare (Balarajan and Reich 2016).

2.5 Conclusions

The projected world population is 8.6 billion in 2030, and it is around 10% more than the year 2017. It has greater implication on world food demand (United Nations 2017). Similarly, India's current population is projected to increase by around 15% (from 1.3 to 1.5 billion) by 2030. On the other hand, the demand-supply gap for different food commodities under different scenarios indicated that India is not likely to remain rice surplus and may even become deficit in rice production to the extent of 3–5 Mt in the coming years and the wheat demand will continue to be met from the domestic production and there may even be marginal surplus of about 14.2 Mt by the year 2030 (Kumar et al. 2016). It is notified that there would be a shift in consumption from rice to wheat even in major rice-consuming states in India. Thus, it is expected that the surplus wheat production is likely to substitute rice, which might lead to lower availability of wheat in predominantly wheat-consuming states. Moreover, supply and demand balance of food grains in India suggests that the future domestic demand should be met with national production, but likelihood of surplus production is marginal with 28 Mt in 2030 if the current efforts on doubling the agricultural productivity may continue in the future.

In this situation, increasing food production is indispensable to achieve the SDG 2. To increase the current productivity level of 2500–5000 kg/ha by 2030, research investments have to be increased to ensure output growth and more efficient use of resources (Samantara and Badatya 2012). Technological innovations by the Agricultural Research Institutions have been the backbone of productive and resilient farms, fisheries and livestock operations ensuring safe and wholesome food supply in India. Access to these innovations by the farmers and producers along the value chain is essential to meet the rising global demand for agriculture in the face of climate change. The thrust areas should be identified for immediate attention to augment agricultural output, specifically in the areas of irrigation, rainfed farming, wasteland development, soil and water conservation, animal husbandry/dairy development and fisheries. More focus and investment are required on the already existing food security programmes to address these issues holistically to manage sustainable agriculture and food security in India. Approximately 55% of India's croplands are rainfed, and it has vast coastal lands under agriculture. Food production in the country is, therefore, vulnerable to climate change. The National Mission on Sustainable Agriculture (NMSA), in collaboration with other Missions under the National Action Plan on Climate Change, is striving towards mitigating the impact of climate change and sustaining agricultural productivity. Further, a comprehensive plan is also being implemented for doubling productivity and farmers' income by 2022 including sustainable and climate-adaptive agriculture promoting organic farming and issuing soil health cards to farmers, and increasing water use efficiency to increase the availability of food grains by 2030.

Though significant progress has been made in improving food and nutrition security India is facing the dual problem of undernutrition and overnutrition. Inequitable distribution of food among different segments of the population is one of the major factors responsible for such inequality in nutrition problems in India. Hence, it is necessary to reduce the gap between food and people by ensuring availability, accessibility and affordability to achieve SDG2 by 2030.

To accelerate the progress on both reducing all forms of malnutrition and ending hunger by 2030, a holistic and integrated approach among all stakeholders', viz. governments, donors and nutrition and non-nutrition communities, should prepare targets and deliver actionable commitments and programmes to end malnutrition (the road map to achieve the targets under SDG2 is given in Fig. 2.10 in the Annexure 2). All stakeholders need to prioritize, allocate resources and scale up finance based on assessment of the different forms of malnutrition by involving scientific data collection and analysis. Besides, it is imperative to stimulate public-private partnership for effective realization of nutritional programmes. The progress should be assessed periodically, and necessary efforts have to be made to correct the leakages and keep track in the right direction to achieve the targets under SDG2 by 2030.

Annexure 1: Target and Indicators of Global SDG2

Target no.	Global SDG	Indicator no.	SDG global indicators
2.1	'By 2030 end hunger and ensure access by all people, in particular the poor and people in vulnerable situations including infants, to safe, nutritious and sufficient food all year round'	2.1.1	'Prevalence of undernourishment'
		2.1.2	Prevalence of moderate or severe food insecurity in the population, based on the Food Insecurity Experience Scale (FIES)
2.2	'By 2030 end all forms of malnutrition, including achieving by 2025 the internationally agreed targets on stunting and wasting in children under 5 years of age, and address the nutritional needs of adolescent girls, pregnant and lactating women, and older persons'	2.2.1	'Prevalence of stunting (height for age–2 standard deviation from the median of the World Health Organization (WHO) Child Growth Standards) among children under 5 years of age'
		2.2.2	'Prevalence of malnutrition among children under 5, disaggregated by type viz., wasting and overweight'
2.3	'By 2030 double the agricultural productivity and the incomes of small-scale food producers, particularly women, indigenous peoples, family farmers, pastoralists and fishers, including through secure and equal access to land, other productive resources and inputs, knowledge, financial services, markets and opportunities for value addition and non-farm employment'	2.3.1	'Volume of production per labor unit by classes of farming/pastoral/forestry enterprise size'
		2.3.2	'Average income of small-scale food producers, by sex and indigenous status'
2.4	'By 2030 ensure sustainable food production systems and implement resilient agricultural practices that increase productivity and production, that help maintain ecosystems, that strengthen capacity for adaptation to climate change, extreme weather, drought, flooding and other disasters, and that progressively improve land and soil quality'	2.4.1	'Percentage of agricultural area under sustainable agricultural practices'
		2.4.2	'Percentage of agricultural households using irrigation systems compared to all agricultural households'
		2.4.3	'Percentage of agricultural households using eco-friendly fertilizers compared to all agricultural households using fertilizers'
2.5	'By 2020 maintain genetic diversity of seeds, cultivated plants, farmed and domesticated animals and their related wild species, including through soundly managed and diversified seed and plant banks at national, regional and international levels, and ensure access to and fair and equitable sharing of benefits arising from the utilization of genetic resources and associated traditional knowledge as internationally agreed'	2.5.1	'Number of plant and animal genetic resources for food and agriculture secured in either medium or long-term conservation facilities'
		2.5.2	'Proportion of local breeds classified as being at risk, not-at-risk or at unknown level of risk of extinction'
2.a	'Increase investment, including through enhanced international cooperation, in rural infrastructure, agricultural research and extension services, technology development, and plant and livestock gene banks to enhance agricultural productive capacity in developing countries, in particular in least developed countries'	2.a.1	'The agriculture orientation index for government expenditures'
		2.a.2	'Total official flows (official development assistance plus other official flows) to the agriculture sector'
2.b	'Correct and prevent trade restrictions and distortions in world agricultural markets including by the parallel elimination of all forms of agricultural export subsidies and all export measures with equivalent effect, in accordance with the mandate of the Doha Development Round'	2.b.1	'Agricultural export subsidies'
2.c	'Adopt measures to ensure the proper functioning of food commodity markets and their derivatives, and facilitate timely access to market information, including on food reserves, in order to help limit extreme food price volatility'	2.c.1	'Indicator of food price anomalies'

Source: United Nations (2015)

Annexure 2: The Road Map and the Way Forward to Achieve the Targets Under SDG 2

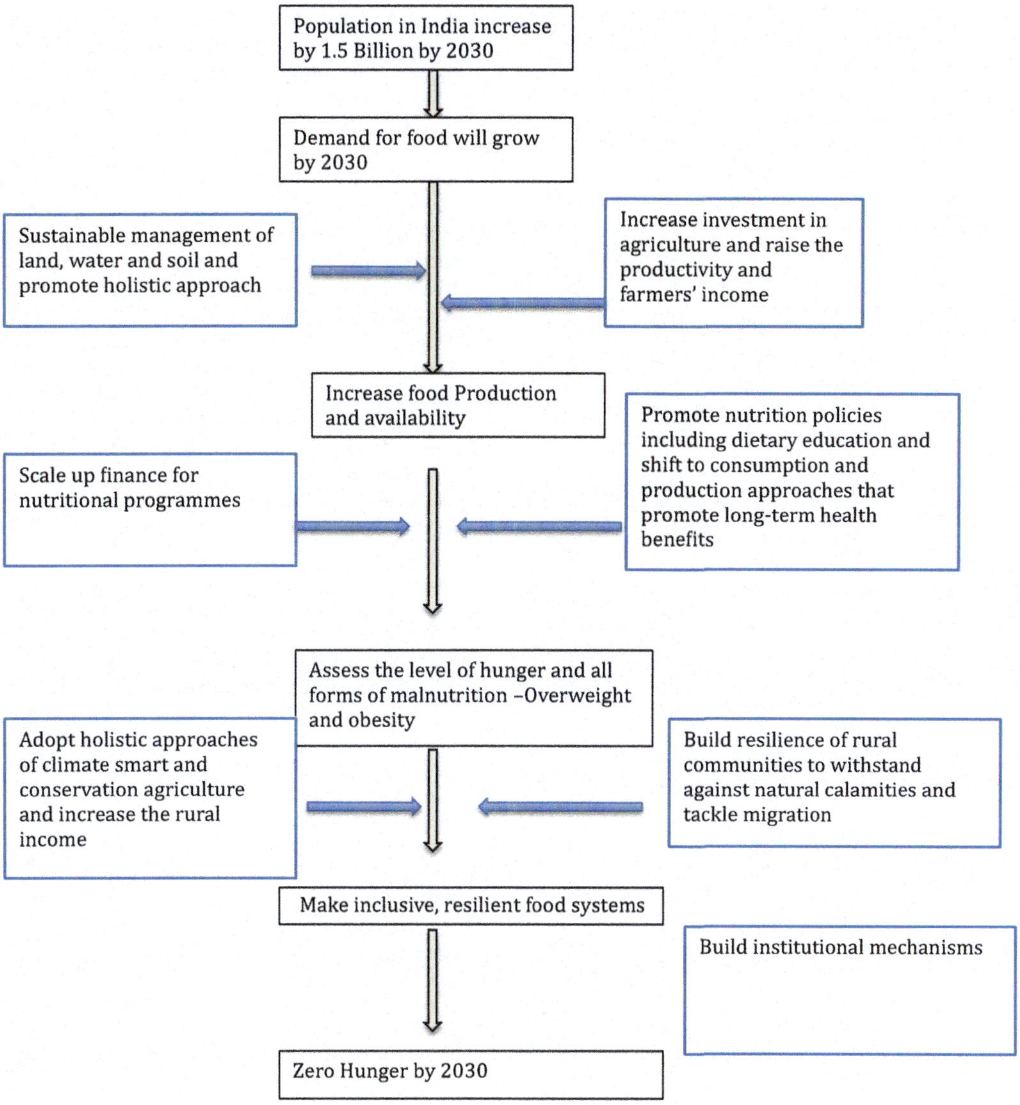

Source: Authors (2019)

Fig. 2.10 The road map and the way forward to achieve the targets under SDG 2

References

Adeniyi AS (2010) Livestock keepers' rights–an important concept for food security. FSN forum discussion no. 53

Arimond M, Ruel M (2002) Summary indicators for infant and child feeding practices: An example from the Ethiopia Demographic and Health Survey 2000. FANTA, AED Washington DC

Aurino E (2017) Do boys eat better than girls in India? Longitudinal evidence on dietary diversity and food consumption disparities among children and adolescents. Economics & Human Biology. 25:99–111

Ayenew HY, Biadigilign S, Schickramm L, Abate-Kassa G, Sauer J (2018) Production diversification, dietary diversity and consumption seasonality: panel data evidence from Nigeria. BMC Public Health 18(1):988

Balarajan Y, Reich MR (2016) Political economy of child nutrition policy: a qualitative study of India's Integrated Child Development Services (integrated child development services) scheme. Food Policy 62:88–98. https://doi.org/10.1016/j.foodpol.2016.05.001

Birthal PS, Negi DS (2012) Livestock for higher, sustainable and inclusive agricultural growth. Econ Polit Wkly 47:89–99

Caballero P (2016) One year on, the SDGs provide reason for hope. The World Bank, 1 Aug 2016

Chakrabarti S, Raghunathan K, Alderman H, Menon P, Nguyen PH (2019) India's Integrated Child Development Services programme; equity and extent of coverage in 2006 and 2016. Bull World Health Organ 97(4):270–282. https://doi.org/10.2471/BLT.18.221135

Charmarbagwala R et al (2004) The determinants of child health and nutrition: a meta-analysis. World Bank, Washington, DC

Chengappa PG, Umanath M, Vijayasarathy K, Manjunatha PA (2016) Changing demand for livestock food products: an evidence from Indian households. Indian J Anim Sci 86(9):1055–1060

Chengappa PG, Umanath M, Vijayasarathy K, Babu P, Devika CM (2017) Demand for Agro Processed Food Products: An Evidence from Indian Households. Asian J Agri Ext Econom & Socio 16(1): 1–1

Chudasama RK, Patel UV, Kadri AM, Mitra A, Thakkar D, Oza J (2016) Evaluation of integrated Child Development Services program in Gujarat, India for the years 2012 to 2015. Indian J Public Health 60:124–130

Codjoe SN, David O, Abu M (2016) Urban household characteristics and dietary diversity: an analysis of food security in Accra, Ghana. Food Nutr Bull 37(2):202–218

Dixit P, Gupta A, Dwivedi LK, Coomar D (2018) Impact evaluation of integrated child development services in rural India: Propensity Score Matching Analysis (2018), Sage Open. https://doi.org/10.1177/2158244018785713

Dutta A, Ghosh S (2016) Impact of integrated child development scheme on child malnutrition in West Bengal, India. Matern Child Nutr 1–8. https://doi.org/10.1111/mcn.12385

Ecker O, Qaim M (2011) Analyzing nutritional impacts of policies: an empirical study for Malawi. World Devel 39(3):412–428

Ekesa BN, Walingo MK, Abukutsa-Onyango MO (2008) Influence of agricultural biodiversity on dietary diversity of preschool children in Matungu division, Western Kenya. Afr J Food Agric Nutr Devel 8(4):390–404

Eyo AA (2002) Fish processing in the tropics. National Institute for Fresh Water Fisheries Research (NIFER), New Bussa, pp 1–4

FAO, IFAD and UNICEF, WFP, WHO (2017) The state of food security and nutrition in the world 2017. Building Resilience for Peace and Food Security. Food and Agriculture Organization, Rome

FAO, IFAD, WFP (2016) The state of food insecurity in the World 2015. In: Meeting the 2015 international hunger targets: taking stock of uneven progress. Food and Agriculture Organization, Rome

Friedman HS (2013) Causal Inference and the Millennium Development Goals (MDGs): Assessing Whether There Was an Acceleration in MDG Development Indicators Following the MDG Declaration. MPRA Paper no. 48973, 3rd August 2013. Available at https://mpra.ub.uni-muenchen.de/48793/

Gausman J, Perkins JM, Lee HY, Mejia-Guevara I, Nam YS, Lee JK, Oh J, Subramanian SV (2018) Ecological and social patterns of child dietary diversity in India: a population-based study. Nutrition 53:77–84

Hatløy A et al (2000) Food variety, socioeconomic status and nutritional status in urban and rural areas in Koutiala (Mali). Public Health Nutr 3(1):57–65

Jain M (2015) India's struggle against malnutrition—is the ICDS program the answer? World Dev 67:72–89. https://doi.org/10.1016/j.worlddev.2014.10.006

Kearney J (2010) Food consumption trends and drivers. Philos Trans R Soc B Biol Sci 365(1554):2793–2807

Kehoe SH, Krishnaveni GV, Veena SR, Guntupalli AM, Margetts BM, Fall CH, Robinson SM (2014) Diet patterns are associated with demographic factors and nutritional status in South Indian children. Matern Child Nutr 10(1):145–158

Kennedy G, Nantel G, Shetty P (2003) The scourge of "hidden hunger": global dimensions of micronutrient deficiencies. Food Nutr Agric 32:8–16

Keshari P, Mishra CP (2017) Growing menace of fast food consumption in India: time to act. Int J Community Med Public Health 3(6):1355–1362

Kızılaslan H, Nalıncı S (2013) The fish meat consumption habits of households and the factors affecting their fish meat consumption in the province of amasya. Gaziosmanpaşa Bilimsel Araştırma Dergisi 5:61–75

Kumar P, Joshi PK (2016) Food demand and supply projections to 2030: India. In: International Trade and Food Security: The Future of Indian Agriculture, Eds: Floor Brouwer and PK Joshi. CABI, UK. 29–63

Kumar P, Joshi PK, Mittal S (2016) Demand vs supply of food in India - futuristic projection. Proc Indian Natl Sci Acad 82(5):1579–1586. https://doi.org/10.16943/ptinsa/2016/48889

Kennedy GL, Pedro MR, Seghieri C, Nantel G, Brouwer I (2007) Dietary Diversity Score Is a Useful Indicator of Micronutrient Intake in Non-Breast-Feeding Filipino Children. The Journal of Nutrition 137(2):472–477

Latham MC (1997) Human nutrition in the developing world (No. 29). Food & Agriculture Organization, Geneva

McNaughton SA et al (2008) Dietary patterns of adolescents and risk of obesity and hypertension. J Nutr 138(2):364–370

Meenakshi JV, Johnson NL, Manyong VM, DeGroote H, Javelosa J, Yanggen DR, Naher F, Gonzalez C, Garcia J, Meng E (2010) How cost-effective is biofortification in combating micronutrient malnutrition? An ex ante assessment. World Dev 38(1):64–75

Michele G, Meera S, Monica DG, Caryn B, Lee YK (2005) India's undernourished children: a call for reform and action Health Nutrition and Population Discussion Paper. The World Bank, Washington, DC

Mittal N, Meenakshi JV (2016) Does the ICDS improve the quantity and quality of children's diets? Some Evidence from Rural Bihar, working paper no. 257

Muthamilarasan M, Prasad M (2015) Advances in Setaria genomics for genetic improvement of cereals and bioenergy grasses. Theor Appl Genet 128(1):1–1

Muthayya S, Rah JH, Sugimoto JD, Roos FF, Kraemer K, Black RE (2013) The global hidden hunger indices and maps: an advocacy tool for action. PLoS One 8(6):e67860

NITI Aayog (2018) SDG India Intex-Baseline Report. NITI Aayog, Government of India

Popkin BM (2006) Global nutrition dynamics: the world is shifting rapidly toward a diet linked with noncommunicable diseases. Am J Clin Nutr 84(2):289–298

Popkin BM (2002) The dynamics of the dietary transition in the developing world. In The Nutrition Transition. pp. 111–128. Academic Press

Ramachandran P (2007) Poverty nutrition linkages. Indian J Med Res 126(4):249

Ruel MT (2003) Operationalizing dietary diversity: a review of measurement issues and research priorities. J Nutr 133(11):3911S–3926S

Samantara S, Badatya KC (2012). A Perspective on Agricultural Credit for 2020. Available at https://www.igidr.ac.in/conf/money/mfc-13/Credit_Perspectives_for_2020_samir%20samantara.pdf

Sawadogo PS et al (2006) An infant and child feeding index is associated with the nutritional status of 6-to 23-month-old children in rural Burkina Faso. J Nutr 136(3):656–663

Saxena (2018) Hunger, under-nutrition and food security in India. In: Poverty, chronic poverty and poverty dynamics. Planning Commission, New Delhi. https://doi.org/10.1007/978-981-13-0677-8_4

Shankar B, Agrawal S, Beaudreault AR, Avula L, Martorell R, Osendarp S, Prabhakaran D, Mclean MS (2017) Dietary and nutritional change in India: implications for strategies, policies, and interventions. Ann N Y Acad Sci 1395(1):49–59

Steffen W et al (2015) Planetary boundaries: Guiding human development on a changing planet. Science 347(6223):1259855

Steyn NP et al (2006) Food variety and dietary diversity scores in children: are they good indicators of dietary adequacy? Public Health Nutr 9(5):644–650

Strauss J, Thomas D (1995) Human resources: Empirical modeling of household and family decisions. Handb Dev Econ 3:1883–2023

Tandon BN (1989) Nutritional interventions through primary health care: impact of the ICDS projects in India. World Health Organ Bull 87(1):77–80

Taren D, Chen J (1993) A positive association between extended breast-feeding and nutritional status in rural Hubei Province, People's Republic of China. Am J Clin Nutr 58(6):862–867

Tiwari P, Sankhala A (2007) Prevalence of obesity, weight perception and dietary behaviour of urban college going girls. J Hum Ecol 21(3):181–183

United Nations (2015). Transforming our world: the 2030 agenda for sustainable development, Resolution A/Res/70/1. New York, NY: UN General Assembly. Available at https://stg-wedocs.unep.org/bitstream/handle/20.500.11822/11125/unep_swio_sm1_inf7_sdg.pdf?sequence=1

United Nations (2017) United Nations Statistical Commission – 48th session (2017). In: UNSD - United Nations Statistical Commission [online], New York. Available from: https://unstats.un.org/unsd/statcom/48th-session

USDA, USDHHS (2010) Dietary guidelines for Americans, 2010. US Government Printing Office

Umanath M, Vijayasarathi K, Babu BP, Baskar M. (2015) Food consumption pattern and nutrient intake in rural and urban Karnataka. Indian J Agri Econom 70(4):487–501

Vasileska A, Rechkoska G (2012) Global and regional food consumption patterns and trends. Procedia Soc Behav Sci 44:363–369

Von Grebmer K et al (2017) Global Hunger Index: the inequalities of hunger. International Food Policy Research Institute, Washington, DC

WHO (2005) Mental health atlas 2005. Department of Mental Health, World Health Organization, Geneva

WHO (2018) Child growth standards: The WHO Multicentre Growth Reference Study (MGRS). World Health Organization. Available from: http://www.who.int/childgrowth/mgrs/en/

3 Health Progress in India with Respect to Millennium Development Goals: Are Health Targets of SDGs Achievable? An Empirical Study at Sub-National Level

Sushil Kr. Haldar and Sulekha Hembram

3.1 Introduction

India accounts for 16.5% of the world's population, but it contributes to one-fifth of the world's share of diseases. A major health challenge posed by the specific phase of the demographic transition that India is going through is related to infant mortality, child malnutrition, mothers suffering from anaemia and poor reproductive health of the mothers (Bhattacharya and Haldar 2015). This is inevitable given the fact that the proportion of women in the reproductive age group is projected to rise to a peak of nearly 55% in 2016 and remains around that level for the next decade. However, the increase in this proportion requires increased efforts to reduce not only maternal mortality but also the number of infant and child deaths Recent findings support absolute beta convergece of health across major states (Goli et al 2013); but a comprehensive study on health convergence is carried out by Hembram et al (2020) and their study support club convergnce along with low level health trap found in few major states in India. The recent national health policy-2017 (NHP 2017) recognizes the pivotal importance of health and well-being in sustainable development goals (SDGs). An indicative list of time-bound quantitative goals aligned to ongoing national efforts as well as the global strategic directions has some impact on the principles of NHP-2017.

The Millennium Development Goals (MDGs) was ended in 2015, and a more comprehensive development goal has been proposed by United Nations (UN). The UN has adopted the development agenda of 2030 towards achieving sustainable development. The sustainable development goals (SDG) are comprised of 17 global targets proposed by UN in 2015 for the year 2030. A critical assessment regarding performance of MDGs at subnational level is needed towards achieving the targets for SDGs. Health is an important component of human capital, and it plays a significant role in human development and future economic growth. This article tries to assess the comparative performance of states with respect to health indicators of MDG. The major constraints of achieving health targets of SDG are also highlighted so that proper policy intervention can be made. The present paper is organized as follows:

S. Kr. Haldar (✉)
Centre for Advanced Studies, Department of Economics, Jadavpur University, Kolkata, India

S. Hembram
Centre for Advanced Studies, Department of Economics, Jadavpur University, Kolkata, India

Department of Economics, Muralidhar Girls College, Kolkata, India

S. Hazra, A. Bhukta (eds.), *Sustainable Development Goals*, Sustainable Development Goals Series,
https://doi.org/10.1007/978-3-030-42488-6_3

Section 3.2 deals with the construction of health deprivation index (DIH) across 26 states over four time points corresponding to NFHS (1–4). How far and to what extent the states are dispersed over mean DIH is also explained in this section. The underlying reasons of DIH are explored in a panel data regression in Section 3.3. Concluding observations and policy options are highlighted in Section 3.4.

The Millennium Summit of the United Nations (UN) was held in 2000 in which 8 international development goals (having 18 targets) were set for the year 2015. These eight development goals are known as the Millennium Development Goals (MDG). All the member states of UN are committed to achieve the target by 2015. The following goals were set up:

1. To eradicate extreme poverty
2. To achieve universal primary education
3. To achieve gender equality and empowering women
4. To reduce child mortality
5. To improve maternal health
6. To control HIV/AIDS, malaria and other communicable diseases
7. To ensure environmental sustainability
8. To develop a global partnership for development

India's MDG framework is based on 2003 framework[1] comprising 8 goals, 12 targets (out of 18) and 35 indicators. The Ministry of Statistics and Programme Implementation, Government of India, has released the report of the MDGs in 2015. The progress of MDG based on health issues [viz. goals (4), (5) and (6)] is shown in the Table 3.1.

Keeping in mind the limitations of data on health indicators (suggested in MDG), we try to explore the progress and failure of states of India up to 2015–2016.

Now, we try to evaluate the state-specific performance with respect to health of MDG using NFHS (I–IV) data. This is reported in Tables 3.2, 3.3 and 3.4. We find that most of the states are approaching to the target with respect to goal-4 of MDG (under 5 mortality rate). The MDG has set a target level, to reduce the child mortality by two-thirds between 1990 and 2015. Some major states like Bihar, Rajasthan, Madhya Pradesh and Uttar Pradesh fail to achieve the target with respect to reducing under five mortality rate; the same picture is noticed in some smaller states like Delhi, Himachal Pradesh, Mizoram and Nagaland. Only the states like Goa, Karnataka and Kerala have successfully achieved the target.

The target of Goal 5 (of MDG) is to reduce the maternal mortality ratio (MMR) by three quarter. From the following table (Table 3.3), we can observe the status (viz. achievement and failure) of states. Most of the states have failed to achieve the target level; about 80% states fail to achieve their target level. The most surprising fact is that Bihar who had a high MMR (initially) could manage to reduce the MMR drastically and has ultimately achieved the target. Two states like Kerala and West Bengal have achieved their target in reducing the MMR.

We are not reporting the state-wise variation of incidence of AIDS; almost all the states have achieved their target. The sixth goal of MDG was to control HIV/AIDS, malaria and other communicable diseases. Most of the states are experiencing an increasing trend in percentage of women age 15–49 who has heard of AIDS from NFHS-1 to NFHS-4. Only one state, Rajasthan, fails to achieve the target level, but the difference between NFHS-1 and NFHS-4 is very low.

The MDG Goal 6b states that malaria and other communicable disease should halt by 2015 and begin to reverse the spread of malaria and major communicable diseases. The number of persons suffering from malaria in 1992–1993

[1]The United Nations Development Group (UNDG) in 2003 has provided a framework of 18 targets and 53 indicators for measuring the progress towards individual targets. A revised indicator framework drawn up by the Inter-agency and Expert Group (IAEG) on MDGs came into effect in 2008 which is comprised of 8 goals, 21 targets and 60 indicators. However, India has not accepted this revised framework of MDG.

Table 3.1 Millennium Development Goals (MDG) of health: Targets and achievement in India

MDG	Target	Achievement	Status
4. Reduce child mortality	Reduce by two-thirds between 1990 and 2015, under five mortality rate (U5MR)	Under 5 mortality rate (U5MR) was estimated at 125 in 1990 per 1000 births. As per SRS (RG Government of India), the U5MR was 49 in 2013	Very close to the target
5. Improve maternal health	Reduce by three quarters between 1990 and 2015 the maternal mortality ratio (MMR)	In 1990, the MMR was estimated at 437 (per 100,000 live births). As per the estimates during 2011–2013, the MMR is 167. The trend suggests that it will reach 140 by the year 2015	Fell short of target
6. Control HIV/AIDS, malaria and other communicable diseases	Halt by 2015 and begin to reverse the spread of HIV/AIDS	The prevalence of AIDS among pregnant women (15–24) is showing a declining trend from 0.89 percent in 2005 to 0.32 percent in 2012–2013	Fully achieved
	Halt by 2015 and begin to reverse the spread of Malaria and Major communicable diseases	Malaria consistently has come down from 2.12 in 2001 to 0.72 in 2013 [but slightly increased to 0.88 in 2014] per thousand. The TB prevalence per lakh population has reduced from 465 in 1990 to 211 in 2013. The TB incidence (new cases) per lakh population has reduced from 216 in 1990 to 171 in 2013; TB mortality has reduced from 38 in 1990 to 19 in 2013 per lakh population	Very close to the target

Source: The Ministry of Statistics & Program Implementation, New Delhi, Government of India (2015)

Table 3.2 State-wise variation of U5MR (Goal 4)

States	NFHS 1	Targeted	NFHS 4	Status
Andhra Pradesh	91.2	33	41	Close to the target
Arunachal Pradesh	72	25	33	Close to the target
Assam	142.2	47	56	Close to the target
Bihar	127.5	46	58	Fails to achieve
Delhi	83.1	29	42	Fails to achieve
Goa	38.9	17	13	Achieved
Gujarat	104	38	43	Close to the target
Haryana	98.7	37	41	Close to the target
Himachal Pradesh	69.1	23	38	Fails to achieve
Jammu & Kashmir	59.1	45	38	Close to the target
Karnataka	87.3	31	31	Achieved
Kerala	32	11	7	Achieved
Madhya Pradesh	130.3	49	65	Fails to achieve
Maharashtra	70.3	25	29	Close to the target
Manipur	61.7	23	26	Close to the target
Meghalaya	86.9	35	40	Close to the target
Mizoram	29.3	10	46	Fails to achieve
Nagaland	20.7	7	37	Fails to achieve
Orissa	131	45	48	Close to the target
Punjab	68	25	33	Close to the target
Rajasthan	102.6	38	51	Fails to achieve
Tamil Nadu	86.5	34	27	Close to the target
Tripura	104.6	32	33	Close to the target
Uttar Pradesh	141.3	51	78	Fails to achieve
West Bengal	99.3	34	32	Close to the target

Source: Authors' estimation from NFHS (I–IV) State Fact Sheet

Table 3.3 State-wise variation of MMR (Goal 5)

States	NFHS 1	Targeted (based 1990)	NFHS 4	Status
Andhra Pradesh	297.8	74.4	85.18	Fails to achieve
Assam	544.2	136	289.71	Fails to achieve
Bihar	735.8	186.9	169.53	Achieved
Gujarat	308	77	101.89	Fails to achieve
Haryana	108.4	27.1	140.48	Fails to achieve
Karnataka	315.9	79	122.27	Fails to achieve
Kerala	279.2	69.8	50.21	Achieved
Madhya Pradesh	602.8	150.7	190.72	Fails to achieve
Maharashtra	235.5	58.6	64.51	Close to the target
Odisha	482	120.5	199.34	Fails to achieve
Punjab	333.4	83.4	131.57	Fails to achieve
Rajasthan	724.9	181.2	210.7	Fails to achieve
Tamil Nadu	196.7	49.2	72.58	Fails to achieve
Uttar Pradesh	855.1	213.2	241.96	Fails to achieve
West Bengal	666.9	166.7	88.69	Achieved

Source: Authors' estimation from NFHS (I–IV) State Fact Sheet

Table 3.4 State-wise variation of population suffering from malaria (Goal 6b)

States	NFHS1	NFHS4	Status
Andhra Pradesh	1944	25,042	Increasing
Arunachal Pradesh	4213	5088	Increasing
Assam	2707	15,557	Increasing
Bihar	1428	4006	Increasing
Delhi			
Goa	243	651	Increasing
Gujarat	3228	41,566	Increasing
Haryana	933	9308	Increasing
Himachal Pradesh	1141	60	Decreasing
Jammu & Kashmir	853	216	Decreasing
Karnataka	457	12,445	Increasing
Kerala	112	1549	Increasing
Madhya Pradesh	4728	100,597	Increasing
Maharashtra	3742	56,603	Increasing
Manipur	1641	216	Decreasing
Meghalaya	4723	48,603	Increasing
Mizoram	4636	28,593	Increasing
Nagaland	2778	1527	Decreasing
Odisha	5148	436,850	Increasing
Punjab	2546	596	Decreasing
Rajasthan	5103	11,796	Increasing
Tamil Nadu	576	5587	Increasing
Tripura	2619	32,525	Increasing
Uttar Pradesh	7395	42,767	Decreasing
West Bengal	678	24,208	Increasing

Source: Authors' estimation from NFHS (I–IV) State Fact Sheet

(NFHS-1) and 2015–2016 (NFHS-4) clearly reveals that we could not tackle the spread of this communicable disease (as shown in Table 3.4). The absolute number of persons suffering from malaria is found to be increasing in most the states except Himachal Pradesh, Jammu & Kashmir, Manipur, Nagaland, Punjab and Uttar Pradesh. Indeed, it is a great concern of our health sector.

3.2 Health Deprivation Index

Using different rounds of National Family Health Survey (NFHS) data, we estimate health deprivation index (DIH) across 26 states over four time points. The principal component analysis is employed in order to find out the weights of DIH. Following Anand and Sen[2] (1997), we consider α order average of DIH as follows:

$$P(\alpha)=\left[\frac{W_1P_1^\alpha+W_2P_2^\alpha+W_3P_3^\alpha+W_4P_4^\alpha+W_5P_5^\alpha+W_6P_6^\alpha}{W_1+W_2+W_3+\cdots+W_6}\right]^{\frac{1}{\alpha}} \quad (3.1)$$

Here, $P(\alpha)$ is the health deprivation index (DIH) of order α, W_i be the weight attached to ith health deprivation parameter and P_i be the ith health deprivation parameter. For empirical estimation we consider:

P_1 = under 5 mortality rate of children, P_2 = maternal mortality rate, P_3 = total fertility rate, P_4 = percentage of births attended by trained health professionals, P_5 = underweight children under 5 years of age and P_6 = percentage of mothers (aged between 15 and 49) suffering from moderate anaemia. Equation (3.1) does satisfy certain properties of a good index like monotonicity, convexity and substitutability, but it does not satisfy decomposability. Moreover, it cannot represent head count; rather, it gives a pure number. The UNDP (1997) has suggested equal weights of Ps towards estimation of human poverty index; this means that $W_1 = W_2 = \ldots = W_6 = 1$. However, equal weights are being criticized on the grounds of variance and importance of the indicators.

Moreover, the arbitrary weightage scheme is also unscientific. In order to overcome this problem, one can employ principal component analysis (PCA) through which one can reduce the number of variables in one hand and take care of the weights of the PCs on other hand (Johnson and Wichern 2007, Haldar and

[2]Human poverty index (HPI) was developed by Sen and Anand (1997), and it was used by UNDP (1997). HPI has been considered a good index for capturing human deprivation, and $P(\alpha)$ does satisfy the following important properties. Here, P_i stands for deprivation parameter (i = 1, 2 and 3).

1. $\min\{P_1,P_2,P_3\} \le P(\alpha) \le \max\{P_1,P_2,P_3\}$
2. As $\alpha \to \infty$, $P(\alpha) \to \max\{P_1, P_2, P_3\}$
3. $P(\alpha)$ is homogeneous of degree 1 in $\{P_1, P_2, P_3\}$
4. For each i = 1, 2, 3; $\frac{\partial P(\alpha)}{\partial P_i} > 0$
5. $P(\alpha)$ is convex with respect to P_i. For each i = 1, 2, 3; $\frac{\partial^2 P(\alpha)}{\partial P_i^2} > 0$
6. For any i, $\frac{\partial P(\alpha)}{\partial W_i} \ge 0$ as $P_i \ge P(\alpha)$, similarly, $\frac{\partial P(\alpha)}{\partial W_i} \le 0$ as $P_i \le P(\alpha)$
7. For given P_1, P_2 and P_3 that are not equal, if $\alpha > \gamma > 0$, then $P(\alpha) > P(\gamma)$
8. The HPI is not sub-group decomposable. For $\alpha \ge 1$, $\sum_{j=1}^{m} \frac{n_j}{n} P_j(\alpha) \ge P(\alpha)$ where n_j be the population in the jth group, $n = \sum_{j=1}^{m} n_j$, $P_j(\alpha)$ be the *HPI* of jth group.
9. The elasticity of substitution (σ) between any two poverty sub-indices of $P(\alpha)$, that is, between any two of P_1, P_2 and P_3, is constant and equal to $\frac{1}{\alpha-1}$.

Raychaudhuri 2009). Here, the weights are not arbitrarily given; it is determined endogenously from the data matrix.

The following steps are adopted to find the weights (W):

We calculate the correlation matrix of the six deprivation parameters from the state-level panel data so that correlation matrix becomes time and space invariant. Only the first PC is taken into account because it absorbs the maximum variance of the data and it consists of six original variables but with differential weights (Nunnally and Bernstein 1994). Since the factor loadings represent the correlation coefficients between PC and original variables (*P*'s), we go for pair-wise statistical testing using the following *t* test:

$$t = \frac{r\sqrt{(n-2)}}{\sqrt{(1-r^2)}} \quad (3.2)$$

It is revealed that all the factors (*P*s) have appeared to be statistically significant, and we retain all the six parameters. In final stage, we calculate DIH across 26 states in four time points using the above power mean formula (Eq. 3.1), where α is taken as 3. The *W*s are estimated as follows:

$$W_1 = 0.934, W_2 = 0.853, W_3 = 0.775,$$
$$W_4 = -0.079, W_5 = 0.827, W_6 = 0.595$$

Following the above-mentioned methodology, the health deprivation index (DIH) across states over four time points corresponding to NFHS survey is reported in Table 3.5.

The following radar diagrams show the dispersion of DIH over mean across states over four time points.

In Fig. 3.1, states like Assam, Bihar, Gujrat, Haryana, Madhya Pradesh, Orissa and Rajasthan lie outside the mean DIH; few states like Karnataka, Meghalaya and Sikkim lie near the mean. The Radar picture of DIH for the year 1998–1999 (corresponding to NFHS-2) is shown in Fig. 3.2.

Only one state like Arunachal Pradesh which lie below the mean DIH in 1992–1992 has moved outside the mean DIH in 1998–1999. The opposite happens in case of Karnataka. The status of DIH of other states remains the same as it was in 1992–1993.

In 2005–2006, the health deprivation of states is found to be more or less same as it was in 1998–1999 (NFHS-2).

Two major states like West Bengal and Uttar Pradesh which lie within the mean DIH during NFHS-1 (1992–1993) to NFHS-3 (2005–2006) have performed badly; as a result, these two states lie outside the mean DIH in 2015–2016 (NFHS-4). This is quite surprising, and it is a great concern because UP and West Bengal share about 25% of India's population.

3.3 Reasons of Health Deprivation: An Econometric Analysis

In order to get a clear picture of state-wise variations of DIH, we consider the following panel data regression model:

$$\text{DIH}_{it} = \beta' X_{it} + \eta_i + \mu_t + \varepsilon_{it} \quad (3.3)$$

X_{it} = vector of the explanatory variables, η_i = captures the space effect, μ_t = captures the time effect and ε_{it} stands for white noise term. Here, $i = 1, 2, 3,\ldots, 26$ and $t = 1$ (1992–1993), 2 (1998–1999), 3 (2005–2006) and 4 (2015–2016). Due to multicollinearity of the explanatory variables, we have formulated different models. Moreover, in order to overcome the problem of heteroscedasticity, robust standard error is taken in all the models.

The following explanatory variables are taken into account in regression models:

per capita net state domestic product (PCNSDP), female literacy rate (FLR), per capita education expenditure (PCEE), per capita health expenditure (PCHE) and infrastructural index (II). The infrastructural index is comprised of a set of variables like availability of primary and upper schools, primary health centre (PHC), health workers (like ANMs), road and rail connectivity etc. The details about the explanatory variables and the factors used in

Table 3.5 Health deprivation index (DIH) across states over four time points corresponding to NFHS (1992–1993, 1998–1999, 2005–2006 and 2015–2016)

	Health deprivation index (DIH) over time			
States	DIH (1992–1993)	DIH (1998–1999)	DIH (2005–2006)	DIH (2015–2016)
Andhra Pradesh	59.85974(18)	55.82677(18)	52.20129(20)	37.92852(19)
Arunachal Pradesh	52.5187(14)	63.62345(21)	56.86587(22)	27.92238(6)
Assam	91.39992(26)	60.87026(20)	58.59936(23)	39.26534(23)
Bihar	83.14987(24)	69.7478(23)	60.66033(25)	45.62369(25)
Delhi	54.09668(15)	38.67006(6)	34.3665(6)	35.53958(16)
Goa	31.12159(4)	32.93327(2)	23.63022(2)	19.85662(3)
Gujarat	67.40236(22)	56.11426(19)	45.66903(18)	38.23312(21)
Haryana	63.56509(20)	50.4997(16)	41.49137(16)	38.62347(22)
Himachal Pradesh	48.03253(12)	35.44302(3)	33.89185(5)	33.41895(14)
Jammu & Kashmir	44.5818(9)	53.98443(17)	38.02601(12)	31.58078(9)
Karnataka	58.59005(17)	47.53747(13)	41.07775(14)	30.78352(8)
Kerala	23.93061(2)	19.38744(1)	15.08872(1)	19.05465(1)
Madhya Pradesh	83.10427(23)	87.08812(26)	64.01751(26)	46.58598(26)
Maharashtra	50.6303(13)	44.37062(11)	37.14137(11)	31.61411(10)
Manipur	40.03015(6)	36.54192(5)	29.69218(3)	19.44929(2)
Meghalaya	58.12527(16)	77.68032(25)	49.03391(19)	35.84027(17)
Mizoram	26.16908(3)	38.85513(7)	35.39183(8)	29.32063(7)
Nagaland	23.20194(1)	41.4967(9)	41.75691(17)	25.16666(5)
Odisha	83.41796(25)	69.33148(22)	60.54889(24)	37.70595(18)
Punjab	47.25405(11)	46.7529(12)	35.2763(7)	32.09056(11)
Rajasthan	65.62683(21)	73.33813(24)	55.44142(21)	39.79626(24)
Sikkim	60.94681(19)	48.62293(15)	36.22651(10)	20.7966(4)
Tamil Nadu	39.57173(5)	36.36181(4)	33.07286(4)	32.94073(4)
Tripura	43.41427(8)	47.61785(14)	41.19036(15)	32.71532(12)
Uttar Pradesh	43.0338(7)	39.01082(8)	35.42344(9)	35.38726(15)
West Bengal	45.93092(10)	42.82196(10)	40.05327(13)	38.16297(20)

Source: Authors' estimation. Note: Values in parentheses represent rank; higher value means lower rank since DIH is deprivation index

the computation of infrastructural index (II) are given in Appendix 2. The data on PCNSDP, PCEE, PCHE and FLR are reported in Appendix 3 (Tables 3.7, 3.8, 3.9 and 3.10).

Due to collinearity among the infrastructural variables, we create an index, infrastructural stock index (II), using principal component method as outlined earlier. A detailed discussion about the computation of II is reported in Appendix 1. We have taken different kinds of social and physical infrastructures like primary health centre, both primary and upper primary school, health workers at PHC (ANMs), doctor's availability at PHC and road and rail connectivity. However, to construct the infrastructure index, we do not take all the variables. According to their factor loading here, we include only number of PHC per 150,000 population, number of primary and upper primary school per 30,000 population and number of ANMs and number of doctors (working at PHC) per 20,000 population.

The coefficients of all the models (in Table 3.6) represent elasticity except infrastructure index (II). The FLR appeared as a profound effect on health deprivation (DIH) in all the models. The income and expenditure (PCHE and PCEE) elasticities of health deprivation are found to be statistically significant. This econometric result is assumed to have some important policy relevance towards improving the health status of the population across states in India. All the models support fixed effect (FE); this

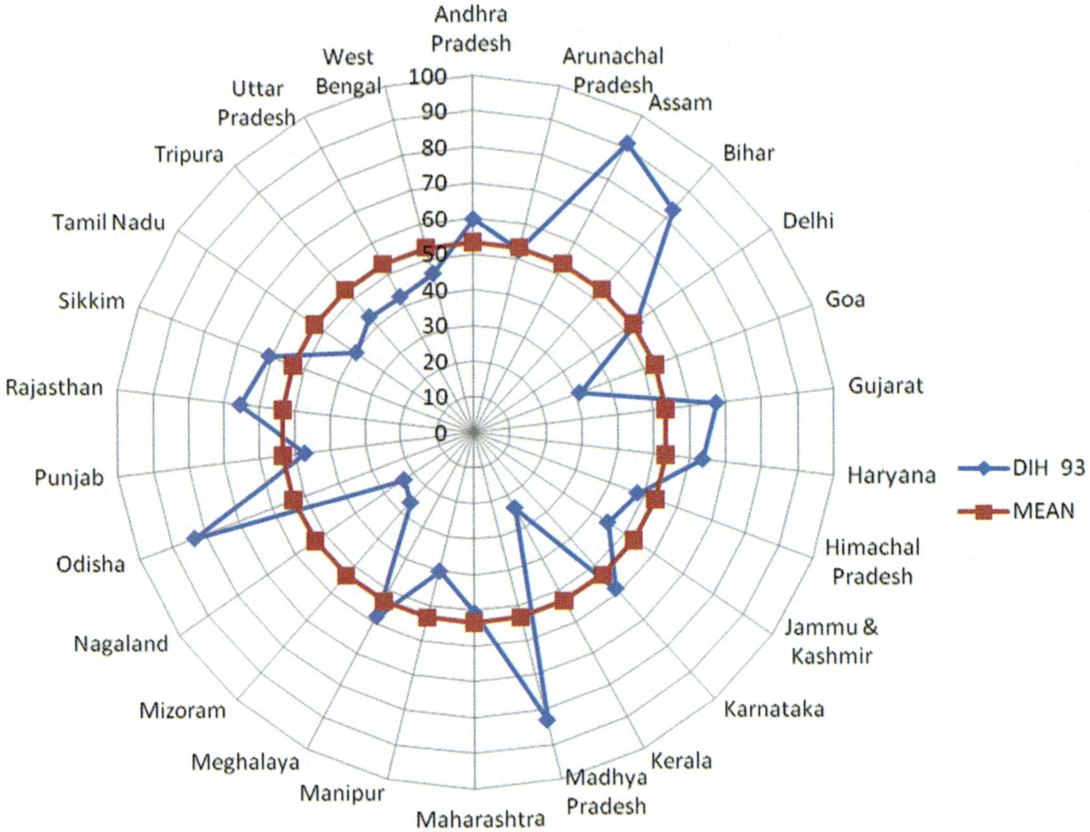

Fig. 3.1 Radar diagram showing dispersion of states with respect to DIH (1992–1993) over mean

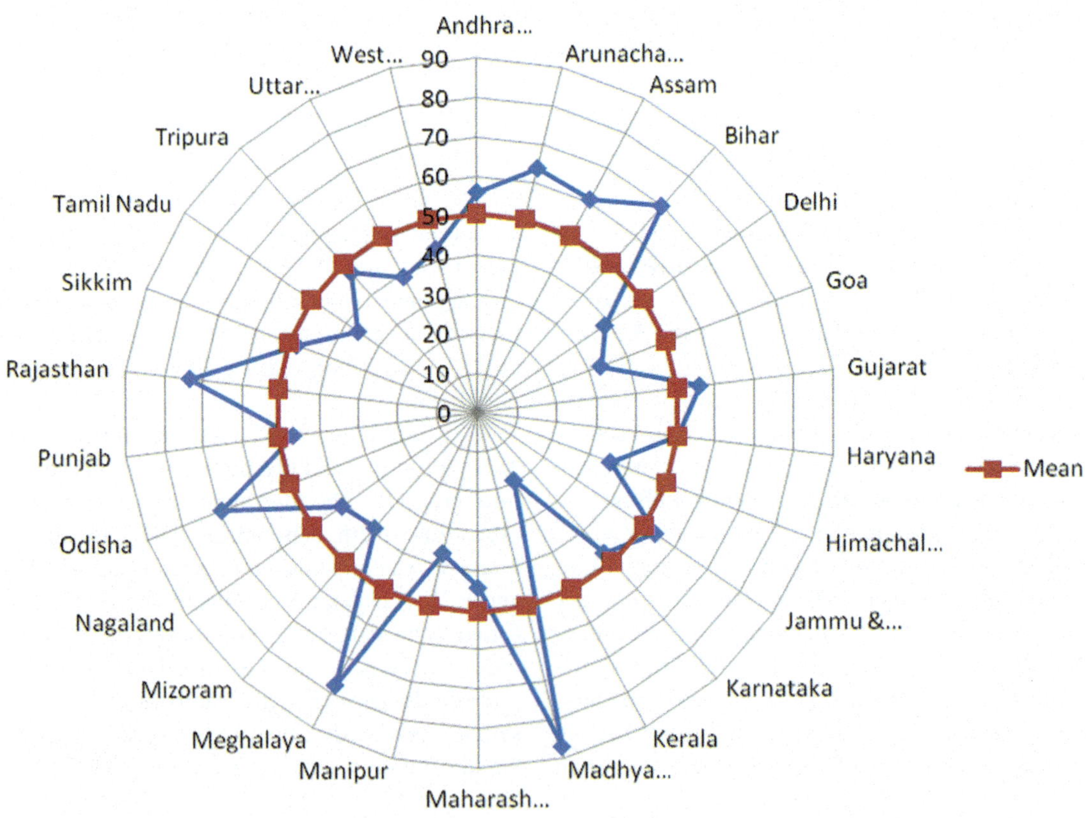

Fig. 3.2 Radar diagram showing dispersion of states with respect to DIH (1998–1999) over mean

Table 3.6 Panel data regression results [dependent variable = lnDIH]

Variables	Model-1	Model-2	Model-3
Constant	3.583***	2.773***	2.6304***
lnPCNSDP	−0.205***		
lnPCEE		−0.1354***	
lnPCHE			−0.1357***
lnFLR	−0.565***	−0.3376***	−0.294***
II	−0.00032	−0.00008	−0.00008
Overall *R*-Square	0.449	0.478	0.485
Within *R*-Square	0.59	0.65	0.681
Between *R*-Square	0.365	0.394	0.383
Huasman test	7.09**	6.34*	7.73*
N = 104			

Source: Authors' estimation, Note: Hausman Test clearly supports fixed effect (FE)
Note: ***, ** and * implies 1, 5 and 10 percent level of significance respectively

means state-specific characters of explanatory variables do play a significant role towards variations of DIH.

3.4 Concluding Observations and Policy Options

At the aggregate level, India has witnessed more or less satisfactory performance; however, the disaggregate performance (at the state level) of the MDG targets of health indicators is not so encouraging. Some major states like Bihar, Orissa, Madhya Pradesh, Rajasthan and Assam are found to be high in health deprivation captured by DIH, though all the states have witnessed a decline of DIH over time. In 2015–2016, two major states like Uttar Pradesh and West Bengal could not perform well in reducing health deprivation; therefore, these two states lie outside the mean DIH. It is clear from the above analysis that there has been a wide range of divergence in the health outcome variables measured by composite health deprivation index (DIH). Two distinct clubs (measured by above means and below means) have been appeared from the radar diagrams as shown in Figs. 3.1, 3.2, 3.3 and 3.4. It is interesting to note that in all four time points, states like Bihar, MP, Assam, Orissa and Rajasthan lie outside the mean. The performance of Uttar Pradesh and West Bengal is not satisfactory in reducing the DIH in the last round of NFHS.

The determinants of DIH clearly show that the economic development measured by PCNSDP has a strong impact on DIH. The female literacy rate (FLR) is found to be the most influencing variable in reducing DIH. States spending higher amount of social sector (like PCEE and PCHE) do experience a faster decline of DIH. The data on PCEE and PCHE of different states show significant variations as reported in Appendix 3 (Tables 3.7, 3.8, 3.9 and 3.10). It is revealed from the econometric analysis that public spending on health and education significantly reduces composite health deprivation index, DIH.

The composite physical infrastructure though shows an expected sign but does not appear significant. This may be due to the fact that we consider only the physical availability of different types of infrastructure, but the quality aspect of infrastructure is not taken into account. Government of India has initiated different affirmative actions like Backward Regions Grants Fund (BRGF), National Rural Health Mission (NRHM), Midday Meal Programme, MGNREGA and many others. We believe all these programmes directly or indirectly help in reducing deprivation of health, but these are inadequate. Health sector is state's responsibility; now, if income (PCNSDP) of any state is low, how do the state allocate more funds for health sector? Indeed, this is a critical issue that must be taken into account if we try to attain inclusive development especially with respect to health outcomes.

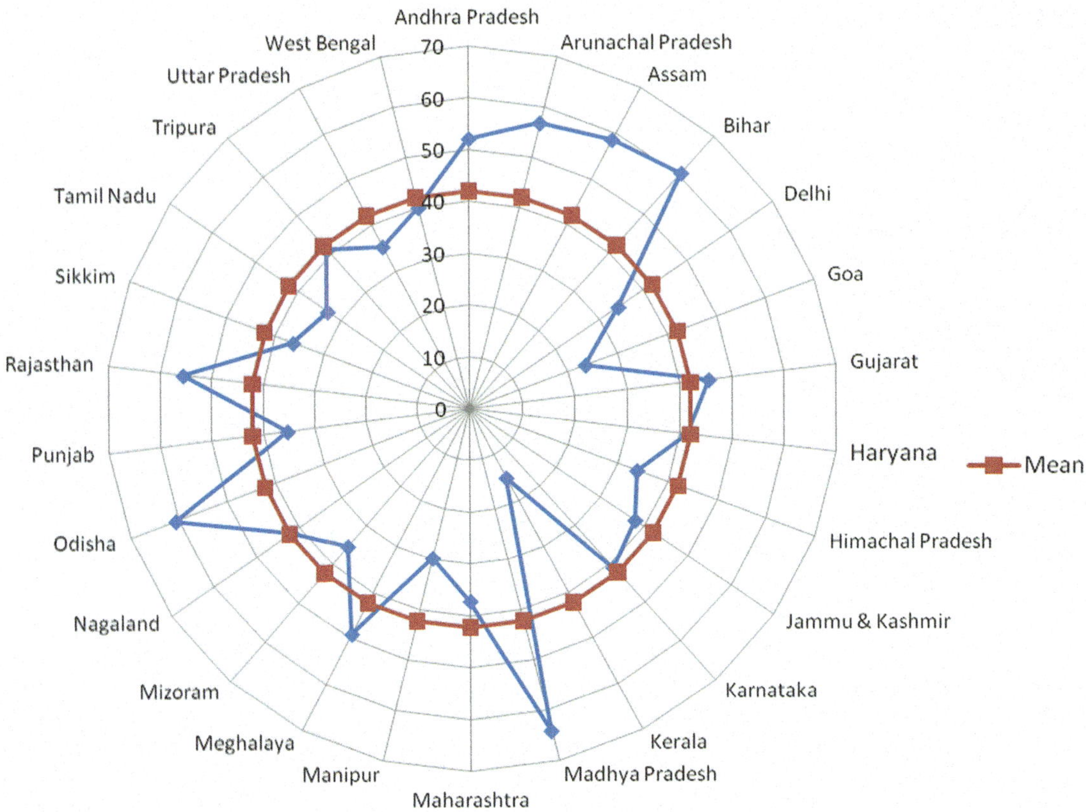

Fig. 3.3 Radar diagram showing dispersion of states with respect to DIH (2005–2006) over mean

3.4.1 Policy Options

The public health expenditure in India (total of centre and state governments) has remained constant at approximately 1.3% of the GDP between 2008 and 2015 and increased marginally to 1.4% in 2016–2017. This is less than the world average of 6%. India is set to increase its public health spending to 2.5% of its gross domestic product (GDP) by 2025. The NHP of 1983 and 2002 has served well in guiding the approach for the health sector in the 5-year plans. Now, after 14 years, a new health policy, NHP-2017 is initiated. The context, perspectives and macroeconomic picture of health and health status have changed in four major ways. First, the health priorities are changing as India is going to enjoy the benefit of demographic dividend around 2025; although the maternal and child mortality have rapidly declined, there is growing burden on account of non-communicable diseases and some infectious diseases. The second important change is the emergence of a robust health-care industry estimated to be growing at double digit. The third change is the growing incidences of catastrophic expenditure due to health-care costs, which are presently estimated to be one of the major contributors to poverty. Fourth, a rising economic growth enables enhanced fiscal capacity. Therefore, the new NHP-2017 should be responsive to these contextual changes in Indian backdrop. The SDG target of good health and well-being will remain unachievable if those considerations are ignored.

In a democratic set-up like India, we must address the issue of social justice, equity and balanced regional development especially with respect to public health outcomes. This is possible if more development grants aiming to eradicate poverty are allowed to the poorer states.

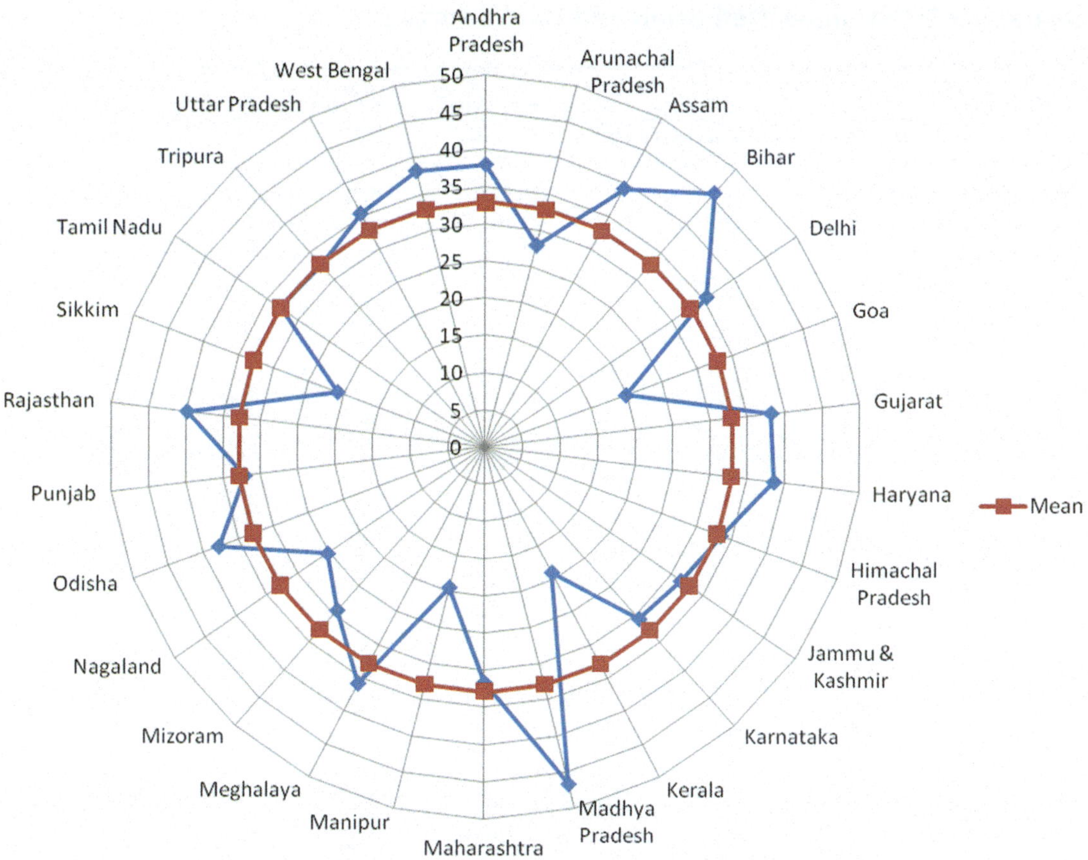

Fig. 3.4 Radar diagram showing dispersion of states with respect to DIH (2015–2016) over mean

Appendix 1

The six infrastructure variables are as follows: number of PHC, number of school (both primary and upper primary school), number of ANMs at PHC, number of doctors at PHC, road length and rail route.

The first PC captures the maximum variability of the data, and it is written as follows:

$$\begin{aligned} PC_1 = & a_{11}PHC + a_{12}School + a_{13}ANM + \\ & sa_{14}Doctors + a_{15}Road + a_{16}Rail \end{aligned} \quad (3.4)$$

The second PC is written as follows:

$$\begin{aligned} PC_2 = & a_{21}PHC + a_{22}School + a_{23}ANM + \\ & a_{24}Doctors + a_{25}Road + a_{26}Rail \end{aligned} \quad (3.5)$$

The third PC is written as follows:

$$\begin{aligned} PC_3 = & a_{31}PHC + a_{32}School + a_{33}ANM + \\ & a_{34}Doctors + a_{35}Road + a_{36}Rail \end{aligned} \quad (3.6)$$

We run a paired sample t test between PC_1 and the respective factor loadings of PC_1 in order to find the significant factor loadings, and we consider only the first PC since it captures the maximum variance of data.

Here, the first PC captures more than 50% of variation; so there is no need to consider the second and third PC. Using those significant factors, loading one can estimate the infrastructure index. The first PC can be written as follows:

$$\begin{aligned} PC_1 = & 0.851 + PHC + 0.758 + School \\ & + 0.827 + ANM + 0.767 + Doctors \end{aligned} \quad (3.7)$$

Appendix 2: Variables, Definition and Data Sources

Definitions	Abbreviation	Sources
Per capita net state domestic products at constant price 2004–2005	PCNSDP	RBI
Per capita education expenditure	PCEE	RBI, NITI Aayog
Per capita health expenditure	PCHE	RBI, NITI Aayog
Female literacy rate	FLR	National Family Health Survey
Number of primary health centre per 150,000 population	PHC	NITI Aayog, Rural Health Statistics 2014–2015
Number of primary school per 30,000 children age between 6 and 14 (primary and upper primary school)	School	DISE State Report
Number of ANMs at PHC per 20,000 population	ANM	NITI Aayog, Rural Health Statistics 2014–2015
Number of doctors (working at PHC) per 20,000 population	Doctor	NITI Aayog, Rural Health Statistics 2014–2015
Road length per 1000 km^2	RL	RBI
Rail route per 1000 km^2	RR	RBI

Appendix 3

Table 3.7 State-wise PCNSDP at constant price 2004–2005

State/UT	NFHS 1	NFHS 2	NFHS 3	NFHS 4
Andhra Pradesh	14,389.13	19,421.28	27,179	44,831
Arunachal Pradesh	16,998.61	18,943.16	26,870	37,683
Assam	14,360.97	14,521.58	17,050	23,968
Bihar	6244.685	6613.388	7588	16,801
Delhi	40,349.98	54,491.86	69,128	91,495.1
Goa	45,816.15	71,695.39	80,844	110,388.7
Gujarat	20,011.51	26,723.82	36,102	67,451
Haryana	23,393.07	27,479.8	40,627	70,629
Himachal Pradesh	18,437.5	24,237.01	35,806	57,405
Jammu & Kashmir	16,844.62	19,248.63	22,406	30,612
Karnataka	16,100.72	22,925.27	29,295	48,907
Kerala	17,360.26	23,242.86	34,837	62,477
Madhya Pradesh	11,703.53	14,685.54	15,927	29,218
Maharashtra	22,890.1	29,041.83	40,671	72,200
Manipur	14,241.6	15,552.32	19,478	24,990
Meghalaya	14,483.71	18,582.56	25,642	39,503
Mizoram	8400.429	16,443	25,826	95,317
Nagaland	26,842.86	25,984.29	33,072	51,888
Odisha	11,487.86	13,419.73	18,194	26,531
Punjab	25,480.03	29,430.15	34,096	51,403
Rajasthan	13,628.68	17,354.31	19,445	33,186
Sikkim	25,591.52	20,640.6	29,008	88,959
Tamil Nadu	17,382.35	24,224.49	34,126	66,635
Tripura	11,118.35	16,125.8	25,688	51,260
Uttar Pradesh	10,741.67	11,596.16	13,445	20,057
West Bengal	12,778.4	17,571.67	23,808	38,624

Table 3.8 State-wise PCEE at constant price

State/UT	NFHS 1	NFHS 2	NFHS 3	NFHS 4
Andhra Pradesh	1681.708	3681.179	6908.237	16,295.79
Arunachal Pradesh	1752.668	5664.32	19,829.76	10,0716
Assam	3249.543	4958.928	9133.292	51,599.8
Bihar	3019.475	3830.845	5190.907	23,382
Delhi	2817.839	5851.379	10,372.85	39,067.97
Goa	4129.665	13,041.26	24,432.68	99,184.8
Gujarat	2262.856	5177.64	8400.409	32,427.38
Haryana	2153.507	5882.993	9157.274	42,092.78
Himachal Pradesh	4118.856	11,074.24	19,400.21	62,639.7
Jammu & Kashmir	2523.943	6215.551	11,356.83	40,998.02
Karnataka	2833.269	5033.935	8990.158	31,055.23
Kerala	3116.821	6611.769	10,862.77	39,175.65
Madhya Pradesh	2182.556	3757.781	4762.246	25,215.41
Maharashtra	3637.619	6788.928	10,734.41	38,059.58
Manipur	9679.6	10091.05	17,033.15	53,396.53
Meghalaya	4656.435	7756.214	12,845.23	70,705
Mizoram	9345.839	15,363.98	29,325.79	10,9337.3
Nagaland	8445.185	11,538.6	16,390.54	49,907.95
Odisha	1886.663	3783.41	6186.289	28,416.32
Punjab	2343.139	5682.59	9439.589	28,975.38
Rajasthan	2743.801	5005.432	7886.4	29,712.33
Sikkim	6552.509	36,279.49	39,712.46	154,803.2
Tamil Nadu	2980.869	5449.584	8219.999	36,467.62
Tripura	4790.847	9393.263	15,379.2	51,904.88
Uttar Pradesh	2055.374	3299.974	5252.005	19,362.05
West Bengal	1491.304	3566.42	6852.701	24,367.34

Table 3.9 State-wise PCHE at constant price

State/UT	NFHS 1	NFHS 2	NFHS 3	NFHS 4
Andhra Pradesh	593.4922	1263.98	2116.036	5302.599
Arunachal Pradesh	1591.859	3525.666	6209.319	50,357.98
Assam	604.489	895.925	1492.942	8774.055
Bihar	777.3928	953.5719	1191.79	5077.234
Delhi	1171.423	2792.948	5708.549	21,561.1
Goa	1600.065	4999.369	9733.344	36,783.77
Gujarat	314.847	1076.082	2066.767	11,733.59
Haryana	469.2591	1319.745	2118.474	9962.789
Himachal Pradesh	1440.3	3726.501	6329.146	19,110.42
Jammu & Kashmir	1163.906	2847.097	5495.239	18,221.34
Karnataka	712.5881	1320.786	2119.109	9772.626
Kerala	741.7671	1700.137	3075.604	12,660.42
Madhya Pradesh	738.0198	1227.168	1587.415	7326.1

(continued)

Table 3.9 (continued)

State/UT	NFHS 1	NFHS 2	NFHS 3	NFHS 4
Maharashtra	554.3623	1159.774	2187.906	8523.76
Manipur	1110.71	1765.858	3318.145	24,791.25
Meghalaya	1222.328	2421.494	4557.986	31,010.97
Mizoram	3080.957	4860.148	7659.721	32,675.5
Nagaland	3188.948	4336.451	6075.805	18,578.87
Odisha	368.4042	844.3798	1262.508	8983.225
Punjab	824.2572	1910.052	2840.23	8915.501
Rajasthan	628.3279	1263.065	2017.451	9963.415
Sikkim	1155.803	8407.904	10,691.81	52,907.43
Tamil Nadu	804.2425	1516.329	2538.529	10,847.96
Tripura	661.5171	1850.365	4724.33	20,427.08
Uttar Pradesh	443.7927	824.698	1762.186	6583.097
West Bengal	490.4613	1138.188	1950.769	7366.871

Table 3.10 State-wise FLR (%)

State/UT	NFHS 1	NFHS 2	NFHS 3	NFHS 4
Andhra Pradesh	38.5	46	57.77942	66.4775
Arunachal Pradesh	42.1	57	53.43077	61.34286
Assam	50.7	59.1	62.23981	71.56964
Bihar	28.6	34.8	38.29288	58.54821
Delhi	70.8	78.3	75.25	81.17857
Goa	73.1	74.8	76.82115	87.05357
Gujarat	51.3	53.6	61.08654	75.495
Haryana	45.9	57.3	65.16519	73.16821
Himachal Pradesh	57.4	68.7	76.13462	81.85
Jammu & Kashmir	51.8	44.7	37.17308	54.35
Karnataka	46.5	55.5	64.80865	74.60036
Kerala	82.4	85.1	90.36038	94.90643
Madhya Pradesh	34.3	44.5	59.10038	68.62357
Maharashtra	55.9	61.4	72.66904	81.5725
Manipur	63	58.7	57.35192	72.81071
Meghalaya	60.2	66.8	63.83462	75.03571
Mizoram	88.9	89.4	86.84423	88.86786
Nagaland	71.8	68.3	57.825	73.88929
Odisha	41.2	51.3	58.88904	70.8325
Punjab	52	64.9	73.53942	77.91179
Rajasthan	25.4	37.1	54.53962	62.125
Sikkim	54.82581	64.4	67.41042	80.64516
Tamil Nadu	56.1	58.3	67.12596	78.6325
Tripura	64.4	57.52	54.45538	75.2
Uttar Pradesh	31.5	42.7	51.84846	66.89571
West Bengal	55.2	57.4	62.33423	74.85214

References

Anand S, Sen A (1997) Concepts of Human Development and Poverty: A Multidimensional Perspective, Background Papers for Human Development Report. New York, UNDP, New York

Bhattacharya G, Haldar SK (2015): Does demographic dividend yield economic dividend? India, a case study", Economics Bulletin, 2015, 35(2), 1274–1291

Goli S, Arokiasamy P (2013) Trends in health and health inequalities among major states of India: assessing progress through convergence models. Health Econ Policy Law 9(2):143–168

Hembram S, Haldar SK (2020): Is India experiencing health convergence? An empirical. Economic Change and Restructuring, https://doi.org/10.1007/s10644-020-09269-6analysis

Johnson RA, Wichern DW (2007) Applied multivariate statistical analysis, 6th edn. Pearson Prentice Hall, Upper Saddle River

National Health Policy (2017) Ministry of Health and Family Welfare. New Delhi, Government of India

Nunnally JC, Bernstein IH (1994) The assessment of reliability. Psychom Theory 3:248–292

Raychaudhuri, A. & S.K. Haldar (2009). An investigation into the inter-district disparity in West Bengal, 1991–2005. Economic and Political Weekly, 44(26 and 27), 258–63

The Ministry of Statistical & Program Implementation (2015) Government of India, New Delhi

United Nations Development Programme (1997) Human Development Report. Oxford University Press, New York

4 Education in India: Goals and Achievements

Sebak K. Jana

4.1 Introduction

The foundation of any society is education. The importance of education in economic and social development of a country has been extensively recognized since the days of Adam Smith. Education plays a significant role in transforming human being into 'human capital' thorough imparting knowledge and skills required for the development of both traditional and the modern sectors of the economy. It also opens up new opportunities and avenues leading to enhancement of individual potentials and group potentials. Education is found as the most crucial input for empowering people to access to productive employment opportunities. Improvement in education level of an individual not only enhances efficiency but also augments the overall quality of his life. The other function of education is that it performs sorting function to assess and classify students on the basis of ability and performance. The current research evidences reveal that investing in the improvement of the quality of education contributes to higher economic returns and social returns. Enhancing the quality of an expanding and diverse education system has become the major challenge facing most countries around the globe (Varghese et al. 2018). According to GOI (2018), the number of educational institutes in India in 2016 was as follows: schools, 1,522,346; universities, 799; and colleges, 39,071. The enrolment of students at different levels of education was as follows: 25.36 million in school education and 1.71 million in higher education. The major problem of education system in India is that a vast share of passed-out students fail to achieve the desired quality. An obsessed education system with examinations in India has been characterised by rote learning to pass examinations with very low level of practical skills (Muralidharan 2019).

The Sustainable Development Goals (SDGs) were formulated as a continuation of the Millennium Development Goals (MDGs) and are being implemented globally. The SDGs consist of 17 Goals with 169 targets set by the United Nations in 2015 for the year 2030 and form a comprehensive framework for global development. The major goals of SDG4 are ensuring inclusive, equitable and quality education (including technical and vocational training) for all. It also aims to build and improve educational facilities for ensuring that all learners are able to acquire knowledge and skills required to promote sustainable development.

Some of the targets under SDG4 to be achieved by 2030 are *that* all girls and boys getting access to free, equitable and quality primary and secondary education, access to quality early childhood development etc. All these are expected to achieve quality education which is one of the

S. K. Jana (✉)
Department of Economics, Vidyasagar University, Midnapore, West Bengal, India

S. Hazra, A. Bhukta (eds.), *Sustainable Development Goals*, Sustainable Development Goals Series,
https://doi.org/10.1007/978-3-030-42488-6_4

goals of sustainable development. Education as Sustainable Development Goal has been discussed by various authors (Didham and Paul 2015; GOI 2017; Griggs et al. 2013; Kopnina and Meijers 2014; NITI Aayog 2015; Pandey 2018; Sachs 2012; United Nations 2019; Waage et al. 2015; WNTA 2017; NUEPA 2014; CII 2018; Goel & Vijay 2017).

The present paper aims to address the current status of education sector in India with respect to SDG. It will assess the performance of India in education sector and state-wise status of education sector at different levels of education, namely, primary, secondary and higher. The paper also evaluates the educational status of the states on the combined ranking. The methodology for constructing the state-wise index of education quality for present study is proposed as principal component analysis. The relevant variables for this study have been collected from different reports like ASER, DISE, AISHE and various other government documents.

4.2 Progress in the Education Sector in India

The major goal of Sustainable Development Goal 4 (SDG4) is to ensure universal quality education and lifelong learning. The flagship government education scheme of India, *Sarva Shiksha Abhiyan*, is aimed at achieving universal quality education for all Indians and is complemented in this effort by targeted schemes on nutritional support, higher education and teacher training programme. Literacy is a basic human right, guaranteed under the right to education enshrined in the Universal Declaration of Human Rights. It is a fundamental tool to empower people and, through them, communities and countries. There is a growing consensus throughout the world that equitable and inclusive access to 'quality' learning must be ensured for all children, youth and adults (ESCAP 2015). The Government of India has expressed commitment for achieving the Sustainable Development Goal (SDG4) by 2030. The Right to Free and Compulsory Education (RTE) Act 2009 enacted in 2010 provides for children's right to an education of equitable quality, based on principles of equity and non-discrimination (Government of India, Economic Survey 2017).

In India, universalisation of primary education has progressed significantly, in terms of the enrolment and completion rates of both girls and boys in both primary and elementary schools. The net enrolment ratio (NER) for boys and girls in primary education in India was at 100%, the youth literacy rate being at 92% for females and 94% for males. The trends of gross enrolment ratio (GER) at the levels of primary, second and higher education in India over 2005–2006 to 2015–2016 have been presented in Fig. 4.1.

Gender parity index (GPI) is used to measure the degree of gender disparities at different levels of education like primary, secondary and tertiary. GPI value of less than unity indicates that the enrolment for girls is higher than that for boys and vice versa if GPI value is more than unity. A GPI value lying between 0.97 and 1.03 is generally considered to reflect gender parity (ESCAP 2017). The trends of GPI at different levels of education in India are shown in Table 4.1. It shows that GPI has improved over years for all the sectors. The pupil–teacher ratio at different levels of education is given in Table 4.2. In Table 4.3, we have presented percentage expenditure on each subsector of education in India (Revenue Account) in 2014–2015 which shows that 3.75% of GDP is spent on education section sector as a whole. The expenditure shares of GDP are 1.93, 1.09 and 0.47, respectively, for primary, secondary and higher education in India. In Table 4.4, we have presented percentage of schools having total enrolment of 60 or less in 2010, 2014, 2016 and 2018 in India. It is discouraging to note that percentage of schools with total enrolment of 60 or less is rising over years.

Pupil–teacher ratio (PTR) is defined as the ratio of the number of pupils or students attending school and the number of existing teachers in the institution. This ratio is an important indicator to judge the quality of the institution. Firstly, it may be used as a tool to measure workload of teacher as well as the efficiency of allocation of resources in institutions, particularly in govern-

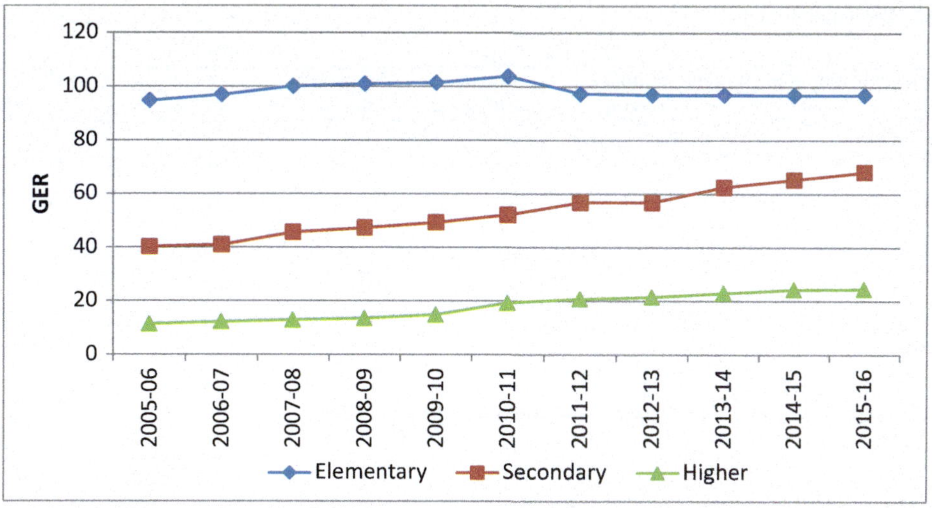

Fig. 4.1 GER in elementary, secondary and higher education in India. Source: Educational Statistics - At a Glance (Various Years), MHRD, GOI

Table 4.1 Gender parity index in education in India

Level/ year	Primary (I–V)	Upper primary (VI–VIII)	Elementary (I–VIII)	Secondary (IX–X)	Senior secondary (XI–XII)	(IX–XII)	Higher education
2005–2006	0.94	0.88	0.92	0.80	0.80	0.80	0.69
2006–2007	0.94	0.90	0.93	0.81	0.83	0.82	0.69
2007–2008	0.98	0.91	0.96	0.85	0.84	0.85	0.70
2008–2009	0.99	0.93	0.97	0.86	0.85	0.85	0.72
2009–2010	1.00	0.94	0.98	0.88	0.87	0.88	0.74
2010–2011	1.01	0.95	0.99	0.88	0.86	0.87	0.86
2011–2012	1.01	0.99	1.00	0.93	0.92	0.93	0.88
2012–2013	1.02	1.05	1.03	0.96	0.94	0.99	0.89
2013–2014	1.03	1.08	1.04	1.00	0.98	1.00	0.92
2014–2015	1.03	1.09	1.05	1.01	0.99	1.01	0.92
2015–2016	1.03	1.10	1.05	1.02	1.01	1.02	0.92

Source: Educational Statistics - At a Glance (Various Years), MHRD, GOI

Table 4.2 Pupil–teacher ratio

Highest level taught in the school/ years	Primary school	Upper primary school	Secondary school	Senior secondary school	Higher education
2007–2008	47	35	33	37	20
2008–2009	45	34	32	38	21
2009–2010	41	33	30	39	24
2010–2011	43	33	30	34	26
2011–2012	41	34	32	33	24
2012–2013	28	25	NA	NA	23
2013–2014	25	17	26	41	25
2014–2015	24	17	27	38	24
2015–2016	23	17	27	37	24

Source: Educational Statistics - At a Glance (Various Years), MHRD, GOI

Table 4.3 Percentage expenditure on each subsector of education in India (revenue account), 2014–2015

	% to total education expenditure	% to GDP
Elementary education	51.7	1.93
Secondary education	29.1	1.09
Higher (general) education	12.6	0.47
Higher (technical) education	4.7	0.18
Others	1.9	0.07
Total	100	3.75

Source: Analysis of Budgetary Expenditure on Education, 2014–2015, MHRD and DBIE, RBI

Table 4.4 Percentage of schools with total enrolment of 60 or less in 2010, 2014, 2016 and 2018 in India

	2010	2014	2016	2018
Primary schools (Std I–IV/V)	27.3	36.0	39.8	43.3
Upper rimary schools (Std I–VII/VIII)	2.7	7.2	8.9	10.7

School Report Card (Various Years), U-DISE & Educational Statistics - At a Glance (Various Years), MHRD, GOI

ment educational institutions. It also indicates the amount of individual attention a single student is expected to receive. The PTR should be 30:1 at primary level and 35:1 at upper primary level. As per the report of Unified District Information System for Education (UDISE), PTR at national level of India for primary schools was 23:1 in 2015–2016. Though some states of India are complying with PTR norms, states like Jharkhand, Bihar, Uttar Pradesh, and Madhya Pradesh possess higher percentages of schools with PTR greater than 30.

4.2.1 Status of Education in India

Education in India is provided by public schools (controlled and funded by three levels: central, state and local) and private schools. In the constitution of India, free and compulsory education has been given as a fundamental right for children between the ages of 6 and 14 years. The approximate ratio of public schools to private schools in India is 7:5. As per ASER 2018 report, 30.9% of children (age between 6 and 14) are enrolled in private school. The corresponding number widely varies with states; in Haryana 55.3% children enrolled in private school, in Kerala it is 46.9%, and in West Bengal it is only 7.9%.

As per the recent report by the National Commission for Protection of Child Rights (NCPCR), a huge number of Indian girls are found not attending presently any educational institutions, because of many reasons like doing household duties. The ASER of 2018 shows that 4.1% of girls (age 11–14) and 13.5% of girls (age 15–16) are not enrolled in school. This problem is very high in some states in India, like Gujarat, Chhattisgarh, Madhya Pradesh, Rajasthan and Uttar Pradesh.

According to the ASER of 2018, the percentage of children in Standard V who were able to read at Standard II level text is 50.3% on average. The corresponding number for the proportion of children who can at least do numerical division is 27.8%. It is obvious that the quality of education is not satisfactory. All these children are in the same grade and in the same age group, but their ability to read or do arithmetic varies widely. This variation has been referred to as one of the 'most critical constraints in the structure of the Indian education system today'. Table 4.5 also presents these indicators state wise. It may be noted that only 7.2% of students in Std V can do division in Meghalaya.

In school education, PTR is 38.9 in India in 2008, but it is 57.8 in 2018 with too much variation being present at state level. Most of the states have higher pupil–teacher ratio compared to standard level.

Table 4.6 also presents state-wise classroom–teacher ratio (CTR), and percentage of schools

Table 4.5 State-wise status of education in India

State	Private school	Not in school		Std V: Learning levels	
	% of children (age 6–14) enrolled in private schools	% of girls (age 11–14) not enrolled in school	% of girls (age 15–16) not enrolled in school	% of children who can read Std II level text	% of children who can do division
Andhra Pradesh	35.2	2.9	9.7	59.7	39.3
Arunachal Pradesh	35.2	2.9	8.6	37.1	27.3
Assam	24.8	2.6	9.6	40.1	17.8
Bihar	16.9	4.2	9.8	41.3	29.9
Chhattisgarh	20.0	5.6	21.2	59.5	26.9
Gujarat	12.4	3.6	24.9	53.7	20.1
Haryana	55.3	2.3	6.8	69.1	50.9
Himachal Pradesh	40.7	0.5	2.0	76.9	56.6
Jammu & Kashmir	40.1	2.4	12.5	41.9	25.0
Jharkhand	19.0	3.4	11.2	34.4	19.0
Karnataka	29.1	1.2	7.8	46.0	20.5
Kerala	46.9	0.5	0.6	77.2	43.7
Madhya Pradesh	26.1	7.7	26.8	41.6	19.8
Maharashtra	37.6	1.6	5.1	66.4	30.2
Manipur	70.4	1.6	5.4	67.5	50.5
Meghalaya	58.6	2.0	9.2	50.1	7.2
Mizoram	27.2	0.2	3.7	64.3	40.2
Nagaland	48.6	2.6	6.4	48.0	25.8
Odisha	10.5	2.1	12.3	58.4	25.4
Punjab	52.2	1.6	6.2	71.6	53.0
Rajasthan	35.8	7.4	20.1	49.1	23.3
Sikkim	30.7	0.9	5.1	41.7	12.5
Tamil Nadu	32.1	0.2	1.4	40.7	25.4
Telangana	41.8	0.9	6.2	43.7	27.1
Tripura	13.9	0.4	1.2	45.0	19.2
Uttar Pradesh	49.7	7.4	22.2	52.0	29.6
Uttarakhand	42.7	2.2	6.6	64.3	37.5
West Bengal	7.9	1.3	4.8	50.7	29.7
All India	*30.9*	*4.1*	*13.5*	*50.3*	*27.8*

Source: ASER Report, 2018

Table 4.6 Trends over time performance of schools with respect to selected right to education indicators by the state in years 2010 and 2018

	PTR and CTR				School facilities			
	% of schools complying with				% of schools that have			
	Pupil–teacher ratio		Classroom–teacher ratio		Office/store/office cum store		Playground	
State	2010	2018	2010	2018	2010	2018	2010	2018
Andhra Pradesh	60.0	61.5	53.9	70.0	57.0	58.6	61.2	61.9
Arunachal Pradesh	78.0	74.2	79.8	65.3	77.7	86.6	58.9	55.4
Assam	33.6	47.4	67.7	68.5	57.5	63.1	61.5	61.0
Bihar	8.8	19.7	48.2	59.6	69.0	83.7	48.3	51.1
Chhattisgarh	39.6	56.6	64.2	71.8	79.0	82.8	45.0	68.8
Gujarat	62.7	83.5	84.2	86.3	80.2	78.0	75.5	82.6
Haryana	40.3	76.3	75.1	79.2	85.8	88.3	79.7	83.4
Himachal Pradesh	60.6	56.8	76.7	78.9	75.9	79.8	75.6	81.7
Jammu & Kashmir		92.6		43.8		84.5		54.7
Jharkhand	11.2	28.3	81.2	83.1	84.9	86.6	37.9	39.1
Karnataka	69.4	79.4	82.8	82.6	72.1	81.8	66.0	78.9
Kerala	89.2	94.6	80.3	84.7	88.4	93.2	76.3	67.7
Madhya Pradesh	19.4	49.0	81.4	71.3	69.5	70.7	61.1	69.2
Maharashtra	58.9	77.6	87.6	83.7	34.3	38.7	84.7	86.8
Manipur	74.3	92.9	62.5	23.6	67.5	72.9	71.8	59.7
Meghalaya	54.3	45.2	84.2	78.1	34.6	47.5	45.8	54.4
Mizoram	89.1	75.4	57.6	69.1	78.5	84.1	39.0	65.5
Nagaland	91.9	97.6	78.6	56.1	83.8	82.4	64.2	52.1
Odisha	22.5	55.8	74.0	72.9	74.7	84.9	44.4	31.9
Punjab	34.9	76.1	76.9	72.8	78.5	80.1	69.3	72.0
Rajasthan	46.4	67.2	82.0	79.3	91.2	94.3	51.7	70.3
Sikkim	93.4	99.0	61.3	81.2	92.7	73.7	79.7	87.9
Tamil Nadu	47.0	59.9	75.2	81.2	54.8	50.9	68.7	72.5
Telangana	64.2	66.4	52.6	68.5	75.1	80.7	83.9	77.0
Tripura	68.5	82.2	60.0	70.3	89.6	84.2	89.5	83.5
Uttar Pradesh	16.1	33.3	81.6	71.3	88.6	86.3	60.8	71.3
Uttarakhand	13.7	31.3	87.4	81.8	87.7	90.7	67.0	68.6
West Bengal	26.2	64.3	64.8	54.8	79.0	81.5	42.1	52.8
All India	*38.9*	*57.8*	*76.2*	*72.9*	*74.1*	*77.2*	*62.0*	*66.5*

Source: ASER Report, 2018

with office/store and playground for the years 2010–2011 and 2017–2018.

Table 4.7 presents different physical infrastructure availability in schools, state wise in India, for the years 2010–2011 and 2017–2018. In India, 64.4% of schools have boundary wall. In Gujarat, 96% of schools have boundary wall, but in Meghalaya, it is only 12%. It is found that 91% of schools have kitchen sheds for cooking mid-day meal, which has improved compared to 2010. A lack of access to water and toilets in schools can seriously hamper with children's education and health. Though there is state-wise significant variation, 74.8% of schools have drinking water availability in India. Due to a lack of knowledge on good hygiene practices, many children fail to attend to schools regularly due to bouts of diseases like diarrhoea. In 2018, 74.2% of schools have usable toilet which has increased significantly from 2010. Girls are often found dropping out of schools when they reach puberty because of the embarrassment and the lack of private places in schools to maintain their dignity. One important indicator is the existence of facility girls' toilet in

Table 4.7 Facilities in schools for selected indicators by states in 2010 and 2018

	School facilities									
	% of schools with									
	Boundary wall		Kitchen shed for cooking mid-day meal		Drinking water available		Toilet available and useable		Girls' toilet available and useable	
State	2010	2018	2010	2018	2010	2018	2010	2018	2010	2018
Andhra Pradesh	47.2	55.1	64.2	72.9	64.8	58.1	38.6	86.4	25.4	81.1
Arunachal Pradesh	24.5	51.4	64.0	57.4	53.2	44.7	25.3	50.0	12.2	28.2
Assam	19.1	59.5	80.2	92.2	60.9	68.0	33.1	21.2	13.7	15.9
Bihar	48.1	55.7	64.0	91.6	78.7	89.7	33.6	75.6	18.1	63.0
Chhattisgarh	48.8	71.8	86.1	97.0	77.6	82.5	29.6	85.7	20.0	75.7
Gujarat	84.4	96.3	88.3	90.4	79.4	88.0	64.8	91.3	49.9	87.4
Haryana	82.7	90.8	51.0	88.2	74.6	82.0	67.9	90.8	52.8	84.4
Himachal Pradesh	37.9	63.6	82.5	99.3	83.2	89.4	56.0	94.2	38.7	86.3
Jammu and Kashmir		38.7		86.3		54.6		73.0		48.2
Jharkhand	27.0	34.8	73.5	88.7	73.8	82.6	26.8	74.9	20.9	72.5
Karnataka	59.3	84.7	92.9	93.0	75.8	76.8	38.4	70.8	31.8	66.4
Kerala	81.8	80.2	98.1	99.2	85.7	52.9	58.2	89.4	43.9	83.4
Madhya Pradesh	37.3	44.7	89.9	85.7	78.5	71.0	50.3	68.3	28.9	56.5
Maharashtra	57.5	74.0	78.2	94.9	69.0	70.9	53.0	70.1	43.2	63.9
Manipur	11.3	14.4	58.4	61.6	5.1	6.5	40.2	44.9	8.4	15.4
Meghalaya	14.2	12.7	60.6	84.5	23.9	15.5	24.5	44.8	14.8	29.9
Mizoram	37.7	35.5	96.2	96.1	48.5	57.4	55.6	44.6	30.8	34.9
Nagaland	42.8	36.4	81.7	83.0	37.0	27.3	53.9	61.8	30.6	47.0
Odisha	40.8	50.5	74.4	89.9	70.3	82.9	44.4	75.7	34.7	69.3
Punjab	82.8	92.6	94.7	99.1	83.1	82.7	61.2	89.5	49.4	83.9
Rajasthan	70.1	84.6	83.8	92.8	68.0	72.8	65.4	84.9	50.3	80.9
Sikkim	14.5	35.9	95.7	95.3	76.8	74.5	59.4	82.4	37.5	75.7
Tamil Nadu	60.7	75.6	96.7	96.2	80.5	80.2	44.6	90.2	35.1	86.2
Telangana	61.2	71.4	71.0	86.4	64.8	57.2	38.6	77.0	25.4	71.9
Tripura	19.4	31.6	88.2	98.3	40.0	45.6	43.0	53.0	30.3	32.7
Uttar Pradesh	44.4	72.4	89.3	95.4	82.2	85.1	47.4	72.7	33.9	67.2
Uttarakhand	66.8	58.3	96.3	98.0	68.3	75.6	53.4	85.8	24.0	67.2
West Bengal	34.5	55.1	86.3	94.0	67.2	81.3	52.1	81.1	23.7	67.7
All India	*51.0*	*64.4*	*82.1*	*91.0*	*72.7*	*74.8*	*47.2*	*74.2*	*32.9*	*66.4*

Source: School Report Card (Various Year), DISE

schools. In 2018, 66.4% of schools had girls' toilet in India with the presence of state-wise variation.

Table 4.8 presents trends over time performance of schools, for different states in India. In 2018, 29.4% of schools have a total enrolment less than 60. This indicator is very high for some states like Himachal Pradesh, Mizoram, Manipur, Meghalaya and Uttarakhand. In 62.4% of schools, Std II children were observed sitting with one or more classes and, in 53.9% of schools, Std IV children were observed sitting with one or more classes in India. Library books are available in 74.2% of schools in India with huge state-wise variation.

Table 4.9 presents state-wise computer availability in schools for the years 2010 and 2018. In 2018, 21.3% of schools had computer for children in India, with 75.4% in Kerala, 66.9% in

Table 4.8 Percentage of schools with respect to enrolment and library books

	Percentage of schools with									
	Total enrolment of 60 or less		Std II children observed sitting with one or more classes		Std IV children observed sitting with one or more classes		Library books available		Library books being used by the children on day of visit	
State	2010	2018	2010	2018	2010	2018	2010	2018	2010	2018
Andhra Pradesh	31.5	38.6	63.3	61.9	55.3	57.6	92.0	91.0	77.6	54.8
Arunachal Pradesh	33.9	49.0	31.7	37.3	26.9	27.2	13.0	24.1	6.3	4.4
Assam	40.9	41.0	43.4	52.2	40.8	46.5	20.8	73.1	10.5	38.8
Bihar	0.2	1.3	57.1	66.5	48.3	55.3	52.9	59.1	28.2	27.5
Chhattisgarh	16.1	40.2	64.9	71.3	51.1	53.3	72.9	89.7	36.5	23.8
Gujarat	4.6	12.8	36.5	50.9	33.0	45.4	83.8	85.3	48.5	40.5
Haryana	6.5	17.6	32.5	41.5	29.7	37.8	64.6	84.0	31.6	39.1
Himachal Pradesh	48.6	83.1	58.3	81.1	52.4	74.3	80.3	97.3	41.3	24.3
Jammu and Kashmir		52.4		61.3		52.1		58.9		26.6
Jharkhand	7.7	18.9	66.1	78.1	60.7	69.3	61.6	87.1	28.4	50.5
Karnataka	17.8	26.3	75.6	83.6	37.0	44.1	92.4	83.0	64.8	36.1
Kerala	19.9	24.1	7.1	17.0	5.4	21.0	83.1	90.0	62.4	30.5
Madhya Pradesh	10.4	33.8	66.9	82.6	57.4	75.0	56.3	84.0	29.1	43.8
Maharashtra	16.7	26.4	40.3	49.8	36.3	44.5	86.1	88.4	66.5	36.9
Manipur	35.3	63.6	37.7	43.9	32.1	38.4	9.2	9.0	5.9	3.2
Meghalaya	71.0	69.0	64.0	77.3	60.4	75.6	22.0	10.6	15.6	2.8
Mizoram	39.8	84.1	28.0	2.2	25.8	1.7	6.4	17.6	1.7	2.6
Nagaland	45.8	61.3	19.0	11.4	17.9	11.7	13.3	12.9	9.2	6.9
Odisha	21.4	31.3	72.8	78.9	62.1	69.8	65.3	80.3	46.8	54.0
Punjab	17.2	38.2	52.2	58.5	37.5	53.8	96.0	88.1	66.0	44.9
Rajasthan	13.0	17.7	66.2	72.9	52.9	60.3	63.7	81.8	23.3	34.1
Sikkim	23.2	53.3	9.0	23.8	9.2	20.4	44.1	52.3	26.5	31.8
Tamil Nadu	24.4	39.6	79.3	64.1	74.4	60.7	79.1	83.8	57.8	52.4
Telangana	17.2	34.8	57.1	60.5	48.3	49.0	92.0	77.7	77.6	55.7
Tripura	9.4	28.7	40.0	53.5	21.5	27.4	35.4	41.2	19.8	29.8
Uttar Pradesh	4.6	10.4	51.0	62.2	45.9	58.9	48.7	63.1	22.9	35.7
Uttarakhand	69.0	73.1	61.9	75.7	56.8	71.7	47.7	84.8	20.4	26.1
West Bengal	10.1	20.2	42.5	46.1	33.9	38.9	49.5	66.1	31.8	38.4
All India	*17.3*	*29.4*	*54.8*	*62.4*	*45.9*	*53.9*	*62.6*	*74.2*	*37.9*	*36.9*

Source: ASER Report, 2018

Gujarat, 64.6% in Maharashtra and 57.9% in Tamil Nadu, and some other states have too much lower value of this indicator. In 2018, 6.5% of schools have computer for children which is observed using them on day of visit. Table 4.9 also presents state-wise percentage of schools 'mid-day meal served in school on day of visit' for the year 2010 and 2018.

Table 4.10 presents state-wise performance grading index in school education. The performance grading index (PGI) (MHRD, 2018) is a measure of education quality to provide insights of the status of school education in states and UTs of India. The purpose of the index is to help states and UTs to find the gaps in education and, accordingly, prioritize areas for intervention for

Table 4.9 Percentage of schools with respect to computer availability and mid-day meals

	Percentage of schools with					
	Computer available for children		Computer available and children observed usingthem on day of visit		Mid-day meal served in school on day of visit	
State	2010	2018	2010	2018	2010	2018
Andhra Pradesh	9.3	22.6	6.2	6.6	99.7	96.0
Arunachal Pradesh	14.3	7.7	8.0	1.3	47.1	36.2
Assam	1.8	6.5	0.2	1.6	67.3	64.0
Bihar	6.9	3.4	4.0	0.6	57.2	84.5
Chhattisgarh	4.1	2.4	1.7	0.4	94.6	91.7
Gujarat	52.2	66.9	27.9	24.0	96.2	94.1
Haryana	17.4	18.3	6.9	5.1	93.7	85.3
Himachal Pradesh	6.7	6.6	3.2	2.1	98.0	93.1
Jammu and Kashmir		17.2		4.6		77.3
Jharkhand	7.0	6.6	4.1	1.1	92.6	79.0
Karnataka	29.4	41.8	13.4	9.9	96.0	97.5
Kerala	82.8	75.4	66.7	22.4	100.0	96.1
Madhya Pradesh	7.5	3.8	1.7	0.7	94.7	82.9
Maharashtra	33.3	64.6	19.8	19.0	90.7	94.7
Manipur	8.5	9.0	2.5	3.2	47.8	46.4
Meghalaya	2.8	2.1	0.9	0.7	51.9	47.9
Mizoram	7.7	9.9	5.9	0.4	94.0	89.2
Nagaland	14.8	13.2	3.7	2.4	31.9	27.4
Odisha	7.1	18.7	4.4	6.1	88.8	98.8
Punjab	10.7	21.5	5.2	3.8	97.9	93.4
Rajasthan	15.7	38.6	5.3	11.6	94.8	95.1
Sikkim	39.1	33.6	24.6	9.4	98.6	78.5
Tamil Nadu	47.0	57.9	29.4	29.3	99.4	98.7
Telangana	9.3	10.6	6.2	3.1	98.4	95.8
Tripura	8.5	4.4	5.3	0.9	74.7	96.5
Uttar Pradesh	1.4	3.3	0.3	0.7	71.3	93.3
Uttarakhand	6.7	9.8	1.5	0.7	95.0	88.1
West Bengal	1.3	6.7	0.5	1.2	63.4	81.6
All India	*15.8*	*21.3*	*8.6*	*6.5*	*84.6*	*87.1*

Source: ASER, 2018

ensuring robust school education system. At the same time, it may act as a good source of information for best practices. From the view point of access to school education, Tamil Nadu (79) and Kerala (78) are at better position compared to other states. From the view point of infrastructure and facilities, Punjab (139) and Goa (138) are at better position; from the equity aspects, Tamil Nadu (218) and Kerala (217) are at upper position compared to other states. From the view point of governance process, Gujrat (279) and Kerala (254) are at upper position compared to other states in India.

With the socioeconomic advancement achieved during the last decades, many countries in Asia have called for highly skilled human resources who can further boost development of the economy in the backdrop of globalization and the knowledge of economy. In higher education, college density is defined as the number of colleges per lakh (0.1 million) of eligible population within age 18–23 years. The lack of institutional capacity with regard to population density also lessens the access opportunities for students. Such circumstances make the disadvantaged groups more vulnerable resulting in

Table 4.10 Performance grading index

	Access (highest 80)	Infrastructure and facilities (highest 150)	Equity (highest 230)	Governance process (highest 360)
Andhra Pradesh	70	99	194	211
Arunachal Pradesh	46	63	197	134
Assam	64	72	208	211
Bihar	56	86	203	140
Chhattisgarh	66	113	206	213
Goa	75	138	209	163
Gujarat	71	99	207	279
Haryana	74	116	211	252
Himachal Pradesh	76	96	209	215
Jammu & Kashmir	53	88	203	168
Jharkhand	53	95	204	144
Karnataka	69	100	212	165
Kerala	78	123	217	254
Madhya Pradesh	58	99	209	207
Maharashtra	76	113	212	155
Manipur	60	68	193	149
Meghalaya	50	57	186	165
Mizoram	57	102	184	218
Nagaland	43	63	195	130
Odisha	69	94	214	223
Punjab	74	139	200	214
Rajasthan	56	84	210	234
Sikkim	58	90	198	215
Tamil Nadu	79	121	218	224
Telangana	66	96	205	167
Tripura	67	70	207	173
Uttar Pradesh	62	73	202	134
Uttarakhand	72	102	194	188
West Bengal	53	58	195	189
A & N Islands	46	87	205	158
Chandigarh	75	133	213	260
Dadra & Nagar Haveli	73	110	221	202
Daman & Diu	73	99	208	163
Delhi	72	114	214	223
Lakshadweep	67	98	206	133
Puducherry	77	114	206	166

Source: Performance Grading Index (PGI) for States and UTs 2017–2018, MHRD

dropouts and discontinuation of higher education. Institutional distribution per lakh population should ensure greater opportunities for access. Disparities can also be observed between states of India in terms of the availability of colleges per lakh population. Table 4.11 presents number of colleges per lakh population for the year 2010–2011 and 2017–2018. In 2017–2018, numbers of colleges per lakh population in India are 28. The three states with the higher densities of colleges are Karnataka (51), Telangana (51) and Andhra Pradesh (48). On the other hand, Bihar (7), Jharkhand (8), Tripura (12) and West Bengal (12) are at the bottom in terms of college density. In Bihar and Jharkhand, college per lakh population is very low, that is why average enrol-

Table 4.11 Progress of higher education in India

		Number of colleges per lakh population		GER		GPI		PTR	
		2010–2011	2017–2018	2010–2011	2017–2018	2010–2011	2017–2018	2011–2012	2017–2018
1	A&N Islands	12	14	11.4	21.8	1.4	1.1	12.1	18.1
2	Andhra Pradesh	48	48	28.4	30.9	0.8	0.8	17.7	16.2
3	Arunachal Pradesh	11	19	26.9	29.7	0.6	0.9	31.0	26.2
4	Assam	13	14	13.4	18.2	1.0	1.0	24.0	28.0
5	Bihar	5	7	10.5	13.0	0.8	0.8	49.9	62.8
6	Chandigarh	18	13	41.4	56.4	1.0	1.4	18.4	22.8
7	Chhattisgarh	20	24	13.6	18.4	0.7	1.0	21.8	28.5
8	Dadra & Nagar	9	13	3.6	9.1	1.1	1.7	25.6	27.4
9	Daman & Diu	7	15	3.5	5.2	2.1	2.2	11.1	13.6
10	Delhi	8	8	32.5	46.3	0.9	1.1	61.0	89.7
11	Goa	25	32	33.2	28.0	1.2	1.3	13.0	15.7
12	Gujarat	27	30	21.3	20.1	0.8	0.8	25.2	26.3
13	Haryana	33	30	24.1	28.7	0.8	1.1	20.9	38.1
14	Himachal Pradesh	38	45	26.0	37.9	1.0	1.2	22.3	24.1
15	J&K	14	23	16.8	27.7	1.0	1.1	29.7	26.1
16	Jharkhand	5	8	8.1	18.0	0.9	1.0	46.2	53.1
17	Karnataka	44	51	25.5	27.8	0.9	1.1	14.0	15.0
18	Kerala	29	44	21.9	36.2	1.3	1.3	15.1	17.8
19	Lakshadweep	0	0	0.0	7.6		3.8	13.2	12.3
20	Madhya Pradesh	23	24	13.6	21.2	0.8	0.9	31.9	31.3
21	Maharashtra	35	33	27.6	31.1	0.8	0.9	22.3	26.3
22	Manipur	23	26	35.9	31.8	0.9	1.0	18.9	23.3
23	Meghalaya	16	18	17.5	24.7	1.3	1.0	13.6	19.8
24	Mizoram	21	23	21.6	22.9	1.0	0.9	13.7	14.1
25	Nagaland	20	27	21.5	17.8	0.7	1.0	21.3	16.7
26	Odisha	23	23	16.1	22.0	0.8	0.9	18.0	25.1
27	Puducherry	54	47	31.2	45.4	0.9	1.1	7.9	12.2
28	Punjab	29	33	19.4	30.3	0.6	1.2	18.2	16.7
29	Rajasthan	29	33	18.2	21.7	0.7	0.9	27.6	29.0

(continued)

Table 4.11 (continued)

		Number of colleges per lakh population		GER		GPI		PTR	
		2010–2011	2017–2018	2010–2011	2017–2018	2010–2011	2017–2018	2011–2012	2017–2018
30	Sikkim	14	22	24.2	37.4	0.9	1.2	14.7	16.4
31	Tamil Nadu	27	35	32.9	48.6	0.8	1.0	18.4	17.5
32	Telangana		51		35.7		0.9	20.8	17.2
33	Tripura	8	12	13.6	21.2	0.7	0.8	22.6	28.5
34	Uttar Pradesh	17	28	16.3	25.9	1.1	1.1	27.9	25.4
35	Uttarakhand	28	37	27.8	36.3	1.1	1.0	35.1	43.1
36	West Bengal	8	12	12.4	18.7	0.8	0.9	32.7	34.8
	India	23	28	19.4	25.8	0.9	1.0	22.9	26.0

Source: AISHE Report (Various Years), MHRD

ment per college is too much high compared to other states in India.

Gross enrolment ratio (GER) in higher education is the ratio of number of students enrolled in higher education institutions within the country and eligible population (within age 18–23 years) in a year. Though India aims to attain GER of 30% by 2020, it is still far behind countries like China having GER of 43.39% and US having GER of 85.8%. On the basis of geographical location, the disparities in gross enrolment ratio (GER) across states are also apparent. Table 4.11 presents state-wise GER in higher education in India for the years 2010–2011 and 2017–2018. In 2017–2018, GER in higher education is 25.8. Differences in GER among states highlight the regional disparities in access to higher education. Thirteen states had a GER less than the national average, and 16 states have registered GER higher than national average (25.8%). Tamil Nadu has the highest GER in India at 48.6%. States like Tamil Nadu (48.6%), Himachal Pradesh (37.9%), Kerala (36.2%), Andhra Pradesh (30.4%), Haryana (28.7%) and Punjab (30.3%) had GER higher than national average. However, states like Bihar (13), Nagaland (17.8), Jharkhand (18), Assam (18.2), Chhattisgarh (18.4) and West Bengal (18.7) had GER ratio far less than the national average. Bihar had the lowest GER with just 13% of its eligible population (in age group of 18–23 years) pursuing higher education.

The gender parity index (GPI) is a socio-economic index usually designed to measure the relative access to education of females compared to males. Here we present state-wise GPI in higher education in India for the years 2010–2011 and 2017–2018. From the report published by All India Survey of Higher Education (AISHE), we see in Table 4.11 that GPI in India was 0.97 in 2017–2018. Especially in the state like Goa, Kerala, Himachal Pradesh and Haryana, more women are enrolled than men. In the states like Andhra Pradesh, Gujarat, Mizoram, Odisha, Tripura and West Bengal, GPI is less than the expected level. In these states, more males are enrolled than females.

Table 4.11 also presents state-wise pupil-teacher ratio (PTR) in higher education in India for the years 2010–2011 and 2017–2018. In 2017–2018, PTR in higher education was 26 in India.

4.3 Financing Education

The global community has been advocating for countries, especially developing countries, to invest more in education to increase access and participation and also to improve the quality of education. Government investment on education should care for addressing equitable distribution of resources across education subsectors, and it should support programmes for marginalized and disadvantaged groups. An increased focus on innovative financing which supports good-quality public education without imposing burden is needed. Funding of education will need to be drawn from multiple sources.

In order to achieve all the targets of Goal 4, significant financial gaps are found in case of early childhood development and tertiary and higher education. There is a gap of USD 429 billion to ensure access to quality early childhood care and preprimary education. Further, India will require an additional USD 301 billion for ensuring quality vocational, technical and tertiary education. For enhancing the standards of Indian higher to make it comparable with world standards, additional finance is required (Technology and Action for Rural Advancement 2015).

In India, though education falls under concurrent list, the states bear major primary responsibility for education expenditure. The share of education expenditure in gross state domestic product (GSDP) is the most commonly used indicator for measuring the priority given to education by states of India. A higher percentage of GSDP invested on education denotes a higher level of attention in education. During 2004–2005, the expenditure on education by all states was 2.5% of the total GSDPs of the states. During 2014–2015, this was increased to 3%. For the time period under study (2004–2005 to 2014–2015), states like Arunachal Pradesh, Assam, Bihar, Himachal Pradesh, Kerala, Manipur, Meghalaya, Mizoram, Nagaland, Rajasthan, Sikkim and Tripura were spending relatively higher proportion of the GSDP on education among all the states of the country.

During 2004–2005, average higher education expenditure by all states was 0.28% of GSDP, which was increased to 0.32% of GSDP in 2014–2015. During 2014–2015, states like Assam, Bihar, Manipur, Mizoram, Odisha and Jammu & Kashmir were spending relatively higher proportion of the GSDP on higher education among all the states of India.

The data in Table 4.12 also present that in most of the states, proportionate expenditure on education in GSDP has increased in the study period. However, this has not happened in higher education. In 11 states (mostly in eastern and north-eastern states) out of 28 states, the share of higher education to SDP has increased.

Table 4.12 Education expenditure percentage to GSDP

Region	State	2004–2005	2014–2015 (BE)
Central	Chhattisgarh	2.3	3.8
	Madhya Pradesh	2.2	2.6
Eastern	Bihar	3.3	5.9
	Jharkhand	2.1	2.9
	Odisha	2.5	3.3
	West Bengal	2.3	2.6
North Eastern	Arunachal Pradesh	5.4	8.2
	Assam	4.5	7.2
	Manipur	6.1	5.5
	Meghalaya	4	4.2
	Mizoram	7.1	8.9
	Nagaland	3.8	5.4
	Sikkim	8.8	5.6
	Tripura	5.3	5
Northern	Haryana	1.7	2.5
	Himachal Pradesh	4.2	4.9
	Jammu and Kashmir	2.5	4.5
	Punjab	2.1	2.2
	Uttarakhand	4.2	3.6
	Uttar Pradesh	2.6	3.3
Southern	Andhra Pradesh	3.1	2.7
	Karnataka	2.6	2.9
	Kerala	3	3.3
	Tamil Nadu	2	2.2
Western	Goa	2.4	2.1
	Gujarat	1.9	2.1
	Maharashtra	2.4	2.3
	Rajasthan	3	3.4
All States		*2.5*	*3*

Source: Analysis of Budgetary Expenditure on Education, MHRD; DBIE, RBIBE = Budget Estimate

4.4 SDG4 and Indicators

In Table 4.13, we have presented the targets of SDG4 for ensuring inclusive and equitable quality education and promoting lifelong learning opportunities for all. The selected indicators have been taken from the Baseline Report (GOI 2019b).

4.5 State-Wise Performance and Ranking by SDG

We have used principal component analysis (PCA) to derive the index of education development for different states in India. The important factors have been extracted using PCA. Therefore, the importance of the factors in measuring education quality is not the same. Using the proportion of these percentages as weights on the factor score coefficients, a non-standardized index (NSI) has been developed using the formula:

$$\text{Education Quality Index}(\text{INEQ}) = w_1 * f_1 + w_2 * f_2 + \cdots + w_k * f_k$$

where w_i is the weight of the factor i and f_i stands for ith factor.

To determine the education quality index, we have considered the following 37 education quality variables as given in Table 4.14. In Table 4.14, we have presented the maximum, minimum and average values of all the 37 variables taken for consideration. As the range suggests, there is huge variation among the states in the development of education quality. The scree plot derived from the factor analysis (Fig. 4.2) reveals that there are eight factors with eigen value greater

Table 4.13 Goal 4. Ensure inclusive and equitable quality education and promote lifelong learning opportunities for all TARGETS

Targets	Indicators used in the present study
4.1 By 2030, it is to be ensured that all girls and boys complete free, equitable and quality primary and secondary education leading to relevant and effective learning outcomes.	1. Gross enrolment ratio (GER) in higher secondary education, *2015–2016 (in percentage)* 2. Percentage of students in grades III, V, VIII and X achieving at least a minimum proficiency level in terms of nationally defined learning outcomes to be attained by pupils at the end of each of the above grades during 2017–2018
4.2 By 2030, it is to be ensured that all girls and boys have access to quality early childhood development, care and preprimary education so that they are ready for primary education	–
4.3 By 2030, it is to be ensured equal access for all women and men to affordable and quality technical, vocational and tertiary education, including university	1. Gross enrolment in higher education, i.e. graduation and above including certificates, diploma etc. during 2015–2016 2. Gender parity index
4.4 By 2030, substantially increase the number of youth and adults who have relevant skills, including technical and vocational skills, for employment, decent jobs and entrepreneurship	Percentage enrolment in technical and vocational education
4.5 By 2030, eliminate gender disparities in education and ensure equal access to all levels of education and vocational training for the vulnerable, including persons with disabilities, indigenous peoples and children in vulnerable situations	Enrolment rate of children with disabilities during 2015–2016
4.6 By 2030, it is to be ensured that all youth and a substantial proportion of adults, both men and women, achieve literacy and numeracy	Percentage literacy rate of youth during 2011
4.7 By 2030, *ensure that all learners acquire the knowledge and skills needed to promote sustainable* development, including, among others, through education for sustainable development and sustainable lifestyles, human rights, gender equality, promotion of a culture of peace and non-violence, global citizenship and appreciation of cultural diversity and of culture's contribution to sustainable development	National indicator not yet evolved
4.a Build and upgrade education facilities that are child, disability and gender sensitive and provide safe, non-violent, inclusive and effective learning environments for all	Proportion *(in percentage)* of schools with access to (1) electricity, (2) computers for pedagogical purposes, (3) adapted infrastructure and materials for students with disabilities/disabled friendly ramp and toilets, (4) basic drinking water, (5) single-sex basic sanitation facilities and (6) basic hand washing facilities *(2015–2016)*
4.b By 2020, substantially expand globally the number of scholarships available to developing countries, in particular least developed countries, small island developing states and African countries, for enrolment in higher education, including vocational training and information and communications technology, technical, engineering and scientific programmes, in developed countries and other developing countries	National indicator not yet evolved
4.c By 2030, substantially increase the supply of qualified teachers, including through international cooperation for teacher training in developing countries, especially least developed countries and small island developing states	Proportion *(in percentage)* of trained teachers, by education level, *2015–2016*

Source: GOI (2019a, b)

Table 4.14 Statistical summary of the variables taken

		Average	Max	Min
4.1 GER	Secondary	68.60	93.23	46.34
	Higher secondary	60.46	98.16	21.54
4.1Percentage of students in grades III, V, VIII and X achieving at least a minimum proficiency level in terms of nationally defined learning outcomes to be attained by pupils at the end of each of above grades during 2017–2018	Class-3 Language	90.29	97.00	74.80
	Class-5 Language	85.55	95.40	69.40
	Class-8 Language	82.82	92.20	64.30
	Class-3 Math	88.69	96.20	74.00
	Class-5 Math	79.81	92.10	64.20
	Class-8 Math	60.93	84.00	43.30
4.3 Gross enrolment in higher education, i.e. graduation and above including certificates, diploma etc. during 2015–2016	Male	25.50	48.44	4.11
	Female	25.88	70.44	9.22
	Total	25.58	57.59	5.74
4.4 Percentage enrolment in technical and vocational education	Technical education	13.69	44.10	2.70
	Vocational education	3.13	10.60	0.40
4.5 (i) Enrolment rate of children with disabilities during 2015–2016	Primary	1.35	3.33	0.53
	Upper primary	1.36	3.70	0.48
	Elementary	1.32	3.28	0.12
	Secondary	0.76	2.66	0.10
	Higher Secondary	0.33	0.95	0.04
4.5 Gender parity index	Primary	1.01	1.13	0.90
	Secondary	1.03	1.24	0.83
	Higher Secondary	1.08	1.98	0.78
	Tertiary Education	1.07	2.49	0.70
4.6 Percentage literacy rate of youth during 2011	Male	92.17	99.04	79.56
	Female	86.48	99.03	63.69
	Total	89.50	99.04	72.29
4.a Proportion of children with access to:	Electricity	72.80	100.00	19.45
	Computers for pedagogical purposes	41.61	100.00	9.37
	Adapted infrastructure and materials	80.14	100.00	50.00
	Basic drinking water	95.79	100.00	63.81
	Single-sex basic sanitation facilities	98.03	100.00	83.94
4.c Proportion of trained teacher by education level	Primary	77.80	100.00	28.68
	Upper primary	83.60	100.00	21.14
	Elementary	70.27	100.00	9.85
	Primary	18.17	39.00	5.00
	Upper primary	12.81	31.00	5.00
	Secondary	23.14	66.00	7.00
	Higher secondary	31.19	97.00	11.00

Source: GOI (2019a, 2019b) and other relevant reports

than 1, and they together explain about 82.25% variation of the variables as revealed from factor analysis.

After obtaining the factor scores, we have used the following formula to find the index value of education quality (INEQ) for each state.

$$\text{INEQ} = \left(\text{Index} - (-866559)\right) / \left(0.9003696 - (-866559)\right)$$

The average value of index was calculated to 0.472 with standard deviation of 0.2405. Table 4.15 presents the derived values of index and the ranks of the states in the education quality.

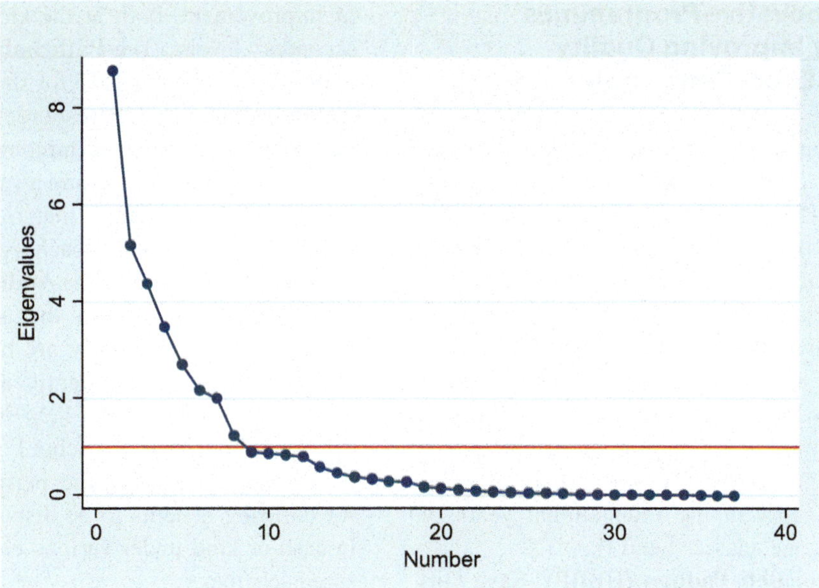

Fig. 4.2 Scree plot of eigenvalues after extraction of factors (Source: Own estimation)

Table 4.15 State-wise rank of quality of education in India

State	PCA score	Rank
Andhra Pradesh	0.470	18
Arunachal Pradesh	0.079	34
Assam	0.264	29
Bihar	0.000	36
Chhattisgarh	0.449	20
Goa	0.730	4
Gujarat	0.441	22
Haryana	0.442	21
Himachal Pradesh	0.774	3
Jammu & Kashmir	0.243	30
Jharkhand	0.284	27
Karnataka	0.705	6
Kerala	1.000	1
Madhya Pradesh	0.230	31
Maharashtra	0.533	14
Manipur	0.681	9
Meghalaya	0.169	33
Mizoram	0.668	10
Nagaland	0.278	28
Odisha	0.395	24
Punjab	0.429	23
Rajasthan	0.217	32
Sikkim	0.286	26
Tamil Nadu	0.683	8
Telangana	0.553	13
Tripura	0.523	15
Uttar Pradesh	0.031	35
Uttarakhand	0.588	12
West Bengal	0.465	19
Andaman & Nicobar Islands	0.481	17
Chandigarh	0.942	2
Dadra & Nagar Haveli	0.495	16
Daman & Diu	0.380	25
Nct Of Delhi	0.662	11
Lakshadweep	0.719	5
Puducherry	0.700	7

Source: Own estimation

4.6 Innovative Programmes for Improving Quality of Education in India

Many innovative programmes have been taken up in India for improving the quality of education (Pandey 2018). The following are some of the government programmes that have been taken up in recent years.

Padhe Bharat Badhe Bharat (PBBB) (Early Reading and Writing with Comprehension and Early Mathematics Programme): *PBBB* is a nationwide sub-programme of the *Sarva Shiksha Abhiyan* (SSA) initiated in 2014 to ensure quality at the foundational years of schooling, i.e. classes I and II.

Beti Bachao, Beti Padhao (BBBP) (Save Girl Child, Educate Girl Child): *BBBP* is one of the most leading and ambitious flagship programmes launched by the Government of India in 2015 to address the issue of decline in child sex ratio (CSR). The major objective of BBBP has been to prevent gender-biased sex-selective elimination and ensure survival and protection of the girl child and education of the girl child.

The Rashtriya Avishkar Abhiyan (RAA): It was launched in 2015 for strengthening learning of mathematics and science in upper primary classes for encouraging children of the age group 6–18 years.

Information and Communication Technologies (ICT): The Government of India has taken several initiatives for introducing good governance of school system through ICT-based initiatives in 2015 which includes the following: (1) ***Saransh*** (launched in 2015) is an initiative of Central Board of Secondary Education (CBSE) for allowing schools for online self-review to identify areas of improvement of students, teachers and curriculum. (2) ***Shaala Siddhi (self-assessment of all schools and external evaluation of sample schools):*** An initiative taken by NUEPA in 2015 aims to enable all schools to self-evaluate their performance and monitor their strengths and areas of improvement, both at the elementary and secondary levels. (3) **e-Pathshala** is an online platform launched in 2015 for showcasing and disseminating digital resources containing textbooks and other learning resources like audio, video, periodicals and a variety of other print and non-print materials for its stakeholders-students, teachers, educators, researchers and parents. (4) **Aadhar Link**: All school-going children in the age group of 5–18 years in the country are being covered under *Aadhar* (12-digit identification number) since 2017 which would help in tracking of dropout children from school and also for monitoring their academic performance and for ensuring benefits to be disbursed to them in cash or kind under various centrally sponsored schemes.

4.6.1 Accreditation of Educational Institutes

All universities and higher education institutes in India have been mandated to get themselves accredited either by National Assessment and Accreditation Council (NAAC) or National Board of Accreditation (NBA). The primary objective of accreditation is to ensure that an institution meets certain standards of quality. The accreditation is based on outcomes rather than inputs. NAAC was set up by University Grants Commission (UGC) in 1994 to monitor quality of Higher Education Institutes (HEIs) and it is presently considered as a major Quality Assurance (QA) body in India. Out of the 7876 higher education institutes assessed in first cycle by NAAC in India, 1844 institutes (23%) achieved 'A' grade denoting good quality. The National Institutional Ranking Framework (NIRF) approved by the Ministry of Human Resource Development (MHRD) was launched on 29th September 2015 in India, the fourth edition of NIRF rankings of the HEIs in India being published in 2019. The better performing states in terms of accreditation are: Tamil Nadu, Karnataka, Kerala and West Bengal.

4.6.2 Prime Minister's Research Fellowship

In order to attract bright students to do research in India to solve the country's daunting challenges, a Special Prime Minister's Research Fellowship Scheme for 1000 graduates, with CGPA of 6 and above on a 10-point scale, from IITs, NITs and IISERs, has been instituted. These meritorious students will fit to receive Rs. 70,000–Rs. 80,000 per month as fellowship grant.

4.6.3 Imprint

In order to tap the great research talent in IIT and IISc and to address the science and technology challenges to make India self-reliant, MHRD has launched in 2015 a catalytic scheme called Impacting Research, Innovation and Technology (IMPRINT). Many innovative research areas are being addressed under this scheme.

4.6.4 GIAN

Global Initiative of Academic Networks (GIAN) has been formulated by the MHRD in 2015 to attract the reputed international faculty from all over the world to Indian universities to provide opportunity to Indian faculty to share knowledge in cutting edge areas, to provide opportunity to Indian students and to gather knowledge from reputed faculty abroad. A large number of international faculties have visited different institutions in India for conducting relevant courses.

4.6.5 SWAYAM

SWAYAM is a programme initiated by the Government of India (GOI) in 2017 to achieve the three cardinal principles of education policy, viz. access, equity and quality. SWAYAM seeks to give the opportunity to students to join the mainstream of the knowledge economy through digital revolution. A National MOOC portal study webs of Active-Learning for Young Aspiring Minds (SWAYAM) has been created that has more than 1000 courses from the best faculties of India in the domain of engineering science, mathematics, humanities and social sciences, economics, management, arts and recreation and languages. These are available free of cost or at very low cost to any interested learner. These courses are available anytime, anywhere on any device. Students wanting certifications shall be registered and shall be offered a certificate after successful completion of the course with a little fee. Status of SWAYAM in India is presented in Table 4.16.

4.6.6 Smart India Hackathon

In 2017, the AICTE and MHRD have embarked upon a national level Smart India Hackathon to provide students a platform to solve some of the pressing problems they face in their daily lives. The major purpose of the initiative is to harness creativity and expertise for innovative solution of India's daunting problems.

4.6.7 Skill India

The campaign of 'Skill India' started in 2015 which aims to train 400 million Indian people in different skills by 2022. India needs huge skilled manpower in the coming decades to use its demographic dividend to its best. Hence, skills are required in new technology domains in the immediate future Mohanty & Zaidi (2012). IoT (Insert on Things), AI (Artificial Intelligence), Robotics, Data Analysis and Club Computing are being imparted as part of the curriculum. Some are being imparted to school drop-outs under Pradhan Mantri Kaushal Vikas Yogana (PMKVY), launched in 2015. The AICTE has imitated several other schemes for improving the quality of technical education and providing enriching experience to students in colleges. It has also initiated awards for clean and green campuses and *Chhatra Vishwakarma* awards for innovative

Table 4.16 National coordinators in SWAYAM

National coordinators	Appointed for	Partnering institutes	Completed courses	Student enrolment	Exam registration	Successful certification
AICTE	Self-paced and international courses	–	86	120,141	16	–
CEC	Nontechnical undergraduate education	19	265	263,560	5221	1627
IGNOU	Out of the school students	3	36	47,529	483	–
IIMB	Management studies	3	36	74,627	1121	206
NCERT	School education	8	75	78,375	–	–
NIOS	School education	1	130	2,944,430	–	–
NITTTR, Chennai	Teacher training programme	–	22	66,179	1377	–
NPTEL	Technical and engineering (UG and PG)	26	1300	6,335,382	627,866	305,530
UGC	Nontechnical post-graduation education	5	220	164,997	4965	1853

Source: Based on SWAYAM, MHRD, https://swayam.gov.in/ (data as on 15-5-2019)

projects and best student start-up. The AICTE has mandated all colleges to adopt at least five villages each under *Unnat Bharat Abhiyan*.

4.7 Conclusion

We have employed the methodology of factor analysis to construct the index of education quality development of the states in India. The results suggest there is huge variation in the education sector development of the states in India with poor education development in some states like Bihar, UP, Arunachal Pradesh, Meghalaya, Rajasthan and Madhya Pradesh. The states which are in better position in terms of education quality index are Kerala, Chandigarh, Himachal Pradesh, Goa, Lakshadweep, Karnataka, Puducherry and Tamil Nadu.

There is a pressing need for strengthening public education system in the country. Certain violations of RTE Act at the state level have weakened the implementation of the Act. It is extremely important to strengthen community and local institutions for bridging the gaps and providing free and compulsory 12 years of school education. The Central Government needs to substantially increase the share of education in the total central budget. Pupil–teacher ratio (PTR) needs to be improved. In the past two decades, post the onset of the so-called economic liberalization, the role of private sector in education has grown rapidly in India. It should be noted that marginalised people or economically weaker section of the population does not get deprived in getting the quality education.

Digital initiative has tremendous potential to impact the education of students and capacity building of teachers and educators and provide solutions to mitigate challenges being faced in the Indian education system. An enhanced focus on innovative financing strategies that do not burden the poor but do support good-quality public education is needed (ESCAP 2015). The efficiency of public investment on education can be enhanced through numerous measures like child-oriented teaching methods focussing on skill formation, using languages that students understand and lowering absenteeism with efficient utilisation of class time by teachers and responsible school management (ESCAP 2019). Some other suggestions for the improvement of school education are universal functional and numeracy of the school students by the end of class 3, universal preschool education to support readiness for class 1 in school, inclusion of extensive practical training through apprenticeship for teachers, introduction of charter schools (public funded schools but managed by private entities) and introduction of vocational education (Muralidharan 2019).

References

ASER (2018) Annual status of education report . ASER Centre, New Delhi. http://www.asercentre.org/

CII (2018) Indian solutions for the world to achieve SDGs. Confederation of Indian Industry (CII)

Didham RJ, Paul OM (2015) The role of education in the sustainable development agenda: empowering a learning society for sustainability through quality education. In: IGES (2015) achieving the sustainable development goals: from agenda to action

DISE, Elementary State Report Cards, http://udise.in/

ESCAP (2015) Statistical yearbook for Asia and Pacific, SDG4, United Nations

ESCAP (2017) Asia and the Pacific SDG progress report 2017, United Nations

ESCAP (2019) MFFD policy brief no. 86, April

Goel D, Vijay P (2017) Technical and vocational education and training (TVET) system\in India for sustainable development. https://unevoc.unesco.org/up/India_Country_Paper.pdf. Accessed 19 June 2019

GOI (2017) Chapter 10: Social infrastructure, employment and human development. Economic survey 2017-18. Ministry of Finance

GOI (2018) Educational statistics at a glance. Government of India, MHRD, New Delhi

GOI (2019a) An overview of SDGS, MOSPI, CSO. http://www.mospi.gov.in/overview-sdgs

GOI (2019b) SDG national indicator framework baseline report, 2015-16, MOSPI, CSO. http://www.mospi.gov.in/sites/default/files/main_menu/SDG_Framework/SDG_Baseline_report_4.3.2019.pdf

Griggs D, Stafford-Smith M, Gaffney O, Rockström J, Öhman MC, Shyamsundar P et al (2013) Policy: sustainable development goals for people and planet. Nature 495(7441):305

Kopnina H, Meijers F (2014) Education for sustainable development (ESD) exploring theoretical and practical challenges. Int J Sustain Higher Educ 15(2):188–207

Mohanty NK, Zaidi SMIA (2012) Innovation models for skills enhancement: South Asia. Innovative Secondary Education for Skills Enhancement (ISESE). Results for Development Institute, NUEPA

Muralidharan K (2019) Reforming the Indian School Education System (2019). In: Banerjee A, Gopinath G, Rajan R, Sharma MS (eds) What the economy needs now, Juggernaut Books, New Delhi

NUEPA (2014) Education for all-towards quality with equity: India. Ministry of Human Resource Development, Government of India

NITI Aayog, GoI (2015) SDG

Pandey B (2018) Achieving SDG 4 in India: moving from quantity to quality education for all, RIS-DP # 232, August 2018

Sachs JD (2012) From millennium development goals to sustainable development goals. Lancet 379(9832):2206–2211

Technology and Action for Rural Advancement (2015). Achieving SDGs in India: a study of India's financial requirements and gaps. https://www.devalt.org/images/L3_ProjectPdfs/AchievingSDGsinIndia_DA_21Sept.pdf

United Nations (2019) SDG indicators: global indicator framework for the Sustainable Development Goals and targets of the 2030 Agenda for Sustainable Development. https://unstats.un.org/sdgs/indicators/indicators-list. Accessed 19 June 2019

Varghese NV, Pachauri A, Mandal S (2018) India Higher Education Report 2017, Sage Publications, New Delhi

Waage J, Yap C, Bell S, Levy C, Mace G, Pegram T et al (2015) Governing the UN Sustainable Development Goals: interactions, infrastructures, and institutions. Lancet Glob Health 3(5):e251–e252

Wada Na Todo Abhiyan (WNTA) (2017) Civil Society report on sustainable development goals: agenda 2030

5 Gender Equality and the Sustainable Development Goals: Focussing on the Adolescent Girls in India

Amrita Patel, Sumanta Banerjee, and Asha Hans

5.1 Introduction

The discourse on the Sustainable Development Goals (SDGs) has gained momentum presently and in India; with the publication of the National Indicator Framework and the baseline report in 2018, SDGs are high on the agenda for the national and state governments. In this backdrop, the status of SDG5 on gender equality is a vital one as it is well understood that SDG5 is not a standalone goal as all the goals are interconnected and gender is an integral part of all goals. The 2019 SDG Gender Index has brought forth the gender equity aspect in each of the goals which have substantiated the need for making gender a basic and essential part of any planning to achieve the Sustainable Development Goals.

Along with the principle that all SDGs have a gender angle, there is another imperative that cannot be undermined, and that is the cross sectionality of gender especially women. The social determinants that influence the pathway towards sustainable development and especially for gender equity include factors such as age, level of education, social group and occupation (Sharma 2015).

5.1.1 About the Paper

This paper draws on the concept of intersectionality of SDGs and the cross-cutting aspects of gender to focus on the adolescent girls as a vital demographic constituency and stakeholder in the pathway to achieve the SGDs in India. Specifically, indicators of health, education, skill, violence and gender equity are being examined in the Indian context to understand the situation of the adolescent girls in the context of SDG5. The centrality of Goal 5 is the underpinning theme of the paper which argues for giving a concerted focus on the adolescent girls in any sustainable development programme and policy.

The paper is based on secondary sources reviews which include policies, plans, programs and reports of Government and Private and International NGOs. Through desk review of existing literature, the paper drew substantially from the information; data on interventions with regard to adolescent girls including evaluative and formative research studies and surveys on demography, education and skills training; reproductive health; economic empowerment of girls; youth enterprises and finally employment. Published articles and reports on adolescents in India, specifically Government of India reports

A. Patel
Sansristi, Bhubaneswar, India

S. Banerjee
Ravenshaw University, Cuttack, India

A. Hans (✉)
School of Women's Studies, Utkal University, Bhubaneswar, India

S. Hazra, A. Bhukta (eds.), *Sustainable Development Goals*, Sustainable Development Goals Series,
https://doi.org/10.1007/978-3-030-42488-6_5

and documentations including the National Family Health Survey (NFHS), National Sample Survey, the Indian Human and Development Survey, Youth in India: Situation and Needs Assessment by IIPS and Population Council and the Annual Status of Education Report (ASER) were studied along with the National Crime Records Bureau (NCRB) Reports. National Sample Survey Organization, the India Human and Development Survey, Youth in India: Situation and Needs Assessment by IIPS and Population Council, and the Annual Status of Education Report (ASER) were studied along with freely available records such as National Crime Record Bureau (NCRB) Reports. Current policy and strategy documents pertaining to development of adolescents implemented by various departments of the Government as well as laws and international convention addressing the protection of adolescents were considered.

The paper is divided into four sections: this section sets the background of the SDGs in general in India and presents the status of the SDG5 in the Indian context as well as traces the initiatives on gender equality undertaken in the country. Section 5.2 dwells on the adolescent girls, in particular, and plans and policies meant for them as well as pinpoints the adolescent girl's angle in the SDG National Indicator Framework; Sect. 5.3 presents evidence on some key indicators on the status of the adolescent girls in the country relevant to SDGs 3, 4, 5 and 16; and Sect. 5.4 concludes with the highlights of the gap areas on the sustainable development trajectory for the adolescent girls and emphasises the necessity for focussing on adolescent girls as a fundamental stakeholder who will influence the success of the SDGs in India specifically on gender equality.

5.1.2 Understanding the SDGs

In September 2015, Governments of all countries agreed on a path towards sustainable development of the world through the adoption of the 2030 Agenda for Sustainable Development. This agenda contains 17 Sustainable Development Goals with 169 targets, which set out quantitative objectives across social, economic and environmental dimensions of sustainable development—all to be achieved by 2030. The SDGs build upon the eight Millennium Development Goals (MDGs) agreed upon earlier in 2000. SDG agenda responds to the newer challenges such as that of climate change and conflict and is more broad-based and more comprehensive than the MDGs. The SDG agenda moves away from segmented approaches to development and promotes the integration and intersectionality of the economy, environment and society.

In India, the national framework of the SDG indicators has been developed by the Ministry of Statistics and Programme Implementation (MoSPI). NITI Aayog[1] is mandated with the task of coordinating work on SDGs by adopting a synergistic approach, involving central ministries, state, civil society organizations, academia and business sectors. A comprehensive mapping of SDG targets has been undertaken, and in November 2018, Niti Aayog has brought out a baseline report for the country.[2] In this report, NITI Aayog has calculated the SDG India Index covering 13 goals out of 17 (leaving out 12, 13, 14 and 17). This index evaluates the status of all the states and UTs on a set of 62 priority indicators, measuring their achievement on the outcomes of the various interventions and schemes of the Government. The SDG India Index has the objective to provide a holistic view on the social, economic and environmental status of the country, the states and the union territories. The overall SDG India Index score is 57, and 15 out of 29 states rank above the India score with Himachal Pradesh, Kerala and Tamil Nadu in the top positions. The states of Assam, Bihar and Uttar Pradesh are at the bottom of the ranking. All the seven union territories are, however, above the India score indicating a better social and economic status.

[1]National Institution for Transforming India https://niti.gov.in/. Accessed on 1 June 2019.

[2]https://niti.gov.in/writereaddata/files/SDX_Index_India_21.12.2018.pdf. Accessed on 30 April 2019.

5.1.2.1 Status of Goal 5

Specifically, for Goal 5, which is to achieve gender equality and empower all women and girls, there are nine targets with 29 national indicators as specified by MoSPI.[3]

For the purpose of the baseline report, Niti Aayog has taken up only 4 (5.1, 5.2, 5.5 and 5.6) out of 9 targets with 6 out of 29 indicators. The index score for goal of gender equality (Goal 5) ranges between 24 and 50 for states and between 27 and 58 for UTs. All India score is 36. Kerala and Sikkim among the states, and Andaman and Nicobar Islands, and Chandigarh amongst the UTs are in the performer category (with index score greater than and equal to 50 and less than 65). All the rest of the states and UTs have a score less than 50 which are termed as aspirant. There are no states which is an achiever (score of 100) or front runner (score 65–99). Bihar, Manipur and Uttar Pradesh are at the bottom of the list of states on achievement of SDG5.

In the six indicators selected for the purpose of the index calculation of SDG5, the country is yet to achieve the target set for the indicators (Niti Aayog 2018, p. 89). One of the vital indicators is sex ratio at birth (SRB).[4] While the target for India is 954, the actual SRB is 898. The lowest SRB is in Haryana (832). Similarly, the experience of spousal violence of women has a nil target, whereas 33% of ever married women in the country in the age of 15–49 years do experience spousal violence. The state of Manipur has reported the highest spousal violence of 54.70%. The target of usage of family planning methods is 100%, but only about half (53.5%) of women in the age group of 15–49 years use modern methods of family planning with only 24.10% usage in Bihar state. While it is aimed that 50% of women be in the state legislative assemblies, the ground reality is that a mere 8.7% women occupy the seats as members in the state assemblies. Thus, for the SDG5, the position of the country as a whole and the states is not an encouraging one.

5.1.2.2 Intersectionality and Cross-Cutting Aspects of Goal 5

It has to be understood that all the goals of SDGs are interwoven and the success of one goal is essential for the success of any other goal. The intersectionality and overlapping spaces between goals have to be factored strategically and effectively while developing an action agenda for attainment of the SDGs. By combining efforts across sectors and different goals, the final outcomes will provide more benefits and accomplishments across various sectors. The cross dependency of all the SDGs has been explained by categorising the SGDs 1, 3, 4, 5, 10 and 16 under the domain of well-being, while infrastructure domain comprises of SDGs 2, 6, 7, 8, 9, 11 and 12. Natural environment domain relates to the SDGs 13, 14 and 15, and SDG 17 (global partnership) is the all pervading background (Waage et al. 2015).[5]

There is undisputedly a gender angle in all the SDGs. If women and girls are denied rights and opportunities, progress will be hampered, and the SDG 2030 Agenda as a whole will be at risk. The needs and aspirations of half of the human population have to be considered, and a systematic mainstreaming of gender perspective in the targets and indicators of the Sustainable Development Goals is, therefore, crucial. There are interrelations of Goal 5 with other Goals, and the success of Goal 5 and its sustainability depend on a holistic approach that all the other goals are worked on with the same vigour from a gender perspective. For example, Goal 3 (good health and well-being) is dependent on the indicator of maternal mortality (which ought to be reducing). Expanding access to quality healthcare and ensuring universal access to sexual and reproductive health and rights are key for reducing mater-

[3]http://mospi.nic.in/announcements/sustainable-development-goalssdgs-national-indicator-framework. Accessed on 1 April 2019.

[4]Sex ratio at birth for children born in the last 5 years (females per 1000 males) for the country is 919 as per the NFHS 4 with the urban scenario being worse than the rural one (899 and 927, respectively) (International Institute for Population Sciences (IIPS) and ICF 2017).

[5]http://blogs.worldbank.org/transport/transport-or-gender-holistic-approach-critical-achieving-sdgs. Accessed on 5 June 2019.

nal mortality rates for women and girls. Goal 4 (quality education) has the indicator of eliminating gender inequality in education. There is evidence to recognise that an additional year of post-primary education for girls has important multiplier effects, expanding their employment outcomes, decreasing the chance of early marriage and improving their health and well-being.[6] While Goals 3 and 4 are in the domain of well-being, Goal 8 (providing inclusive and sustainable economic growth, employment and decent work for all) cannot be achieved unless it is accompanied by development of infrastructure, programs and policies that address women's equal participation in the workforce.

A significant step undertaken to assess the state of gender equality in the SDGs has been the development of the 2019 SDG Gender Index which evaluates 129 countries (covering 95% of the world's girls and women), 14 of the 17 SDGs, and 51 targets linked to issues inherent in the SDGs. India ranked 95th with a score of 56.2. Denmark has been ranked at the first place and Chad at the bottom most position. The index measures global gender equality looking at aspects such as poverty, health, education, literacy, political representation and equality at the workplace. India's highest goal scores are on SDG3 of health (79.9), SDG2 of hunger and nutrition (76.2) and SDG7 of energy (71.8). India's lowest goal scores are on SDG17 of partnerships (18.3); SDG9 of industry, infrastructure and innovation (38.1); and SDG13 of climate (43.4). None of the 129 countries are fully transforming laws, policies or public budget decisions on the scale needed to reach gender equality by 2030[7] (Equal Measures 2030 2019). This report reinforces the principle that gender equality is a cross-cutting aspect across all the SDGs, and so it is indeed vital that interdependency of goals specifically of gender equality is woven into the global and national agenda.

[6]http://www.unwomen.org/-/media/headquarters/attachments/sections/multimedia/2018/infographic-why-gender-equality-matters-to-achieving-all-17-sdgs-en.pdf?la=en&vs=3758. Accessed on 6 June 2019.

[7]https://data.em2030.org/wp-content/uploads/2019/05/EM2030_2019_Global_Report_ENG.pdf. Accessed on 10 June 2019.

5.1.3 Indian Context of Gender Equality

It is well recognised that gender equality is not only a fundamental human right but also a necessary foundation for a peaceful, prosperous and sustainable world. Advancing and encouraging gender equality is critical to all areas of a healthy society, from reducing poverty to promoting the health, education, protection and well-being of girls and boys as well as accelerating sustainable development both in economic and social front. The subject of gender equality is deeply enshrined in the Indian constitution. Article 14 of the Indian constitution, equality before law, states, 'The State shall not deny to any person equality before the law or the equal protection of the laws within the territory of India'. Article 15 prohibits discrimination on grounds of religion, race, caste, sex or place of birth. Article 15(1) stipulates that the state shall not discriminate against any citizen of India on the ground of sex. Additionally, the state is empowered to make any special provision for women. In other words, this provision enables the state to make affirmative discrimination in favour of women as per Article 15(3). In 1974, the Towards Equality Report by the Committee on the Status of Women in India was the first to talk about the discriminatory socioeconomic situation as well as analyse development from a gender perspective.

Over the years, the country has promulgated various legislations which contain several rights and safeguards for women on wages, labour welfare, domestic violence, medical termination of pregnancy, sex etc. The 73rd and 74th Amendments (1993) to the Constitution of India has provided a decision-making space for women by the reservation of seats for women in the urban and rural local bodies. Presently, many states such as Odisha and Bihar have half of such seats reserved in favour of women in the rural and urban local bodies.

The National Policy for the empowerment of women adopted in the year 2001 provided a holistic framework to work towards the empowerment of women, by all the stakeholders. In 2016, the draft national policy for women has the vision to create an effective framework to enable the process of developing policies, programmes and practices which will ensure equal rights and opportunities for women in the family, community, workplace and governance.[8]

As far as institutional mechanisms are concerned, there are multiple national- and state-level bodies such as the National and State Commission for Women, Ministries and Departments on women to advance these policies, and the implementation of these policies is decentralized to state- and district-level authorities and organizations.

Flagship schemes like the Mahatma Gandhi National Rural Employment Guarantee Scheme (MGNREGS)[9] have mandatory provisions for women,[10] while many states have developed specific programmes for the empowerment of women such as the Mission Shakti in Odisha,[11] Mahila Samakhya[12] and Kudumbashree in Kerala.[13] The planning process has generally been sensitive to the needs of the women, and in the 11th 5-year plan (2007–2012), women were recognized as equal citizens as well as the agents of economic and social growth. The vision for the 12th 5-Year Plan was to ensure improving the position and condition of women by addressing structural and institutional barriers as well as strengthening gender mainstreaming. In the last 5 years, some initiatives like the one-stop centres[14] and Sukanya Samridhi[15] are well-intentioned and have the potential of changing the ground situation.

Gender-based violence has always been top on the agenda for achieving women's rights, and in this regard, the Verma Commission set up after the Nirbhaya gang rape case in 2012 was a landmark moment. The Commission suggested reforms in management of cases related to crime against women along with electoral and education reforms. As a consequence, the Criminal Amendment Act, 2013, was adopted which amended the following laws: the Indian Penal Code, 1806; Code of Criminal Procedure, 1973; the Indian Evidence Act, 1872; and Protection of Children from Sexual Offences Act, 2012. New sections were inserted on consent, punishment and new forms of gender-based violence.[16]

In the international arena, Convention on the Elimination of all Forms of Discrimination Against Women (CEDAW) 1979, International Conference on Population and Development (ICPD) 1994 and the Beijing Platform for Action (BPFA) 1995 have provided the necessary framework towards achieving gender equality. India being a signatory to these international conventions is committed towards working for the empowerment of girls and women. CEDAW

[8]https://wcd.nic.in/sites/default/files/women%20empowerment%20poliy_Final_17May.pdf. Accessed on 10 June 2019.

[9]Initially named as NREGS, started since 2006.

[10]The women participation rate in Mahatma Gandhi National Rural Employment Guarantee Scheme (MGNREGS) has been 53% in FY 2017–2018 and 53% in 2018–2019 (as on 07.12.2018) (http://pib.nic.in/newsite/PrintRelease.aspx?relid=186368. Accessed on 7 June 2019).

[11]Started since 2001, the Mission Shakti programmes has created 6 lakh women self help groups with about 70 women members. The objective is economic and financial independence with https://www.facebook.com/missionshakti.odisha/. Accessed on 7 June 2019.

[12]Launched in 1988, Mahila Samakhya was present in 130 districts and 679 blocks/mandals in the country in 11 states by 2014. The programme has led to the emergence of leadership from the disadvantaged and marginalised women. (mhrd.gov.in/sites/upload_files/mhrd/files/upload_document/Report-MSP.pdf. Accessed on 7 June 2019).

[13]Kudumbashree is the poverty eradication and women empowerment programme implemented by the State Poverty Eradication Mission (SPEM) of the Government of Kerala (http://www.kudumbashree.org/pages/171. Accessed on 7 June 2019).

[14]A Govt of India Scheme, started since 2015 which is intended to support women and girls affected by violence, in private and public spaces, within the family, community and at the workplace, irrespective of age, class, caste, education status, marital status, race and culture will be facilitated with support and redressal. Operated by the Ministry/Department of Women and Child Development, this hospital/medical facility-based centre which offers a range of services including medical, legal, psychological and counselling support under one roof.

[15]A bank/post-office deposit in favour of the girl child which has the highest interest rate among small saving schemes.

[16]https://www.prsindia.org/uploads/media/Justice%20verma%20committee/js%20verma%20committe%20report.pdf. Accessed on 9 June 2019.

treaty which was adopted in 1979 had its purpose, outlined in Article 1, to focus on the forms of discrimination that women face and to help eliminate discrimination that either intends to or has the effect of limiting women from participating equally in public life. India ratified this treaty in 1993. In 1994, when the International Conference on Population and Development (ICPD) took place, it changed the world's population and development-related priorities and emphasised social inclusion, human rights and the importance of addressing the needs and developing the capacities of the young. The significance of the successful transition of the young to adulthood was emphatically stated and reiterated in the ICPD +15 Global Youth Forum in Bali in 2012, which focused on sexual and reproductive rights and inclusion of the most marginalised (Jejeebhoy Shireen et al. 2013). The Beijing Platform for Action of 1995 is an agenda for women's empowerment, and it suggests to take all necessary measures to eliminate all forms of discrimination against women and the girl child and remove all obstacles to gender equality and the advancement and empowerment of women (Article 24) and prevent and eliminate all forms of violence against women and girls (Article 29).[17]

In the last seven decades, post independence, myriads of work have been undertaken towards achieving the goal of gender equality. Whether by adopting laws to designing schemes and programmes or by ensuring women's political participation through reservation and implementing the international commitments in the domestic domain, the end result has been a progressive effort towards achieving gender parity. Significant achievements have been made towards removing gender inequality and the gender gap in literacy, education, skill building and access to financial resources amongst many other subjects, but there are scores of indicators of human development which are yet to be achieved by the women and girls of the country.

5.2 Focussing on the Adolescent Girls

The cross-cutting aspect of gender is an integral part of understanding gender equality. Just focussing on women alone is insufficient to measure progress.[18] The heterogeneity aspects of women across race, ethnicity, age, culture, religion and language play a role on their course towards achieving equity. Gender equity acquires meaning when they interact and intersect with other social relations. More than gender identities, the other cross-cutting identities of ethnicity, education, kinship relations and marital status are crucial in women accessing their rights (Rao 2005). In the SGD landscape, it is critical to move beyond averages and identify groups who are the most vulnerable yet key stakeholders towards achieving the SDGs (UN Women 2018). It is observed that the average aggregate figures on women's well-being often make invisible significant variations across regions, ethnic, racial and income groups. Thus the paper, having the core focus on the adolescent girls, aims to bring forth the distinct features and evidence of their status so as to gain a deep insight on the journey in the context of the SDGs in India.

5.2.1 Defining Adolescents

Adolescents are most often subsumed with youth or with children or with young people. At the global level, the World Health Organisation (WHO) defines 'adolescents' as individuals in the 10–19 years age group and 'youth' as the 15–24 years age group, while 'young people' covers the age range 10–24 years.[19] Adolescents up to the age of 18 are children and are protected under the Convention on the Rights of the Child.

While the Census and Ministry of Health and Family Welfare, Government of India, has

[17] http://www.indiaenvironmentportal.org.in/files/file/Beijing%20Declaration%20and%20Platform%20for%20Action.pdf. Accessed on 7 June 2019.

[18] https://gender.cgiar.org/un-women-2018-sdg-monitoring-report/. Accessed on 7 June 2019.

[19] http://www.searo.who.int/entity/child_adolescent/topics/adolescent_health/en/. Accessed on 10 June 2019.

adopted the WHO definition of adolescent, youth and young people, various agencies of the government use different definitions and age groups which have been fixed differently under different programmes keeping in view the objectives of that policy/programme.

National Youth Policy of 2014 defines youth age group as 15–29 years, while earlier in the National Youth Policy 2003, 'youth' was defined as a person of age between 13 and 35 years. Under Integrated Child Development Scheme (ICDS), adolescent girls are considered to be between 11 and 18 years, while Scheme for Adolescent Girls (SAG) addresses the adolescent girls in the age group of 11–14 years. As per the Child Labour (prohibition and regulation) Amendment Act of 2016, 'adolescent' means a person who has completed 14th year but has not completed 18th year, and 'child' means a person who has not completed 14th year.[20]

Adolescence is defined as a phase of life characterized by rapid physical growth and development; physical, social and psychological changes and maturity; sexual maturity; experimentation; development of adult mental processes; and a move from the earlier childhood socioeconomic dependence towards relative independence. The nature and quality of young people's future lives, as well as a country's future social and economic development, depend largely on how well adolescents navigate this transition from childhood to adulthood. By addressing their needs, one would not only be contributing to the socioeconomic development of the country but also to other societal concerns like social harmony, gender justice, population stabilization and improving the quality of life of the country (Population Council and UNICEF 2013). Today, every fifth person in India is an adolescent (10–19 years) and every third a young person (10–24 years). Investing in this cohort is the best way to leverage the nation's competitive advantage—its demographic dividend.

[20]https://labour.gov.in/sites/default/files/THE%20CHILD%20LABOUR%20%28PROHIBITION%20AND%20REGULATION%29%20AMENDMENT%20ACT%2C%202016.pdf. Accessed on 12 June 2019.

5.2.2 Planning and Policy

During the 10th 5-Year Planning (2002–2007) process, the Planning Commission had set up a working group for the Welfare and Development of Adolescents to provide inputs. Later, the Government of India launched the National Adolescent Health Strategy in 2014 to provide health, information and services to meet the diverse needs of adolescents in India and achieve improved health outcomes. The objective was to reach the adolescents in schools and communities, with the help of peer educators, counsellors, parents and the community. Key drivers of the strategy included communication for information and behaviour change programmes and adolescent-friendly health clinics. In the same year, the National Youth Policy of 2014 was announced with the aim 'to empower the youth of the country to achieve their full potential, and through them enable India to find its rightful place in the community of nations'. The Policy has five key objectives encompassing creation of productive workforce, instilling social values, participating in governance and creating equitable opportunities. There are 11 priority areas that define youth age group as 15–29 years.[21]

The spotlight on skilling in the country generated the need for a policy framework which was done in 2015 as the National Policy for Skill Development and Entrepreneurship of 2015. This policy acknowledges the need for an effective roadmap for promotion of entrepreneurship as the key to a successful skills strategy. The vision of the policy is 'to create an ecosystem of empowerment by skilling on a large scale at speed with high standards and to promote a culture of innovation based entrepreneurship which can generate wealth and employment so as to ensure sustainable livelihoods for all citizens in the country'. The policy is youth centric, and it suggests that state governments would be encouraged to set up centres at panchayat level for mobilising and imparting skills pertaining to local employment/livelihood opportunities to school dropouts, adolescent girls and rural youth.[22]

[21]http://www.rgniyd.gov.in/sites/default/files/pdfs/scheme/nyp_2014.pdf. Accessed on 8 June 2019.

[22]file:///E:/SDG_1/policy%20booklet-%20Final.pdf.

In any case, India is obligated to ensure national- and state-level initiatives under international law, especially the Convention on the Elimination of all Forms of Discrimination Against Women (CEDAW) 1979 and the Convention on the Rights of the Child (CRC) 1989. India is also obligated under regional standards created by organizations of which it is a member such as the South Asian Association for Regional Cooperation (SAARC) Convention Against Trafficking in Women and Children, 2002. These international and regional standards form a robust framework of complementary and mutually reinforceable protections and safeguards that uphold the rights of India's adolescents.

5.2.3 Gender in Adolescent Development

The gendered nature of adolescence and the full experience of adolescence for girls have been a particularly neglected area in policy. The language for this phase of life is gender neutral, as children differentiated as boys and girls become suddenly lumped together as 'adolescents', not to emerge as gender-differentiated beings again until they become men and women. As a result of the lack of importance placed on them, 'girls disappear as policy subjects after receiving their last childhood immunisation and do not reappear until they are pregnant and, in most cases, married' (Mensch et al. 1998).

To counter such a gender-insensitive policy knowledge and environment, the UN had identified five strategic priorities that highlight issues that would need to be taken into consideration in policies for adolescents, particularly the adolescent girls in the age group of 10–14 years: educate adolescent girls, improve adolescent girls' health, keep adolescent girls free from violence, promote adolescent girl leaders and count adolescent girls.[23]

Adolescent girls occupy an important and significant position in the overall framework of development which could be looked at through the 'capabilities approach' to human development. This places development as a process of expanding 'freedoms' or 'capabilities' that improve human lives by expanding the range of things that a person can effectively be and do, such as to be healthy and well nourished, to be knowledgeable and to participate in community life (Sen 1999). Development in this perspective is about facilitating the acquisition and use of such capabilities as well as removing obstacles to what a person can do in life (obstacles such as illiteracy, ill health, lack of access to resources or lack of civil and political freedom). This capability-based human development paradigm provides a more gender-sensitive agenda to public policy (Fukuda-Parr 2003). By gaining substantive freedoms of individuals, they (adolescent girls in this case) are active agents of change rather than passive recipients of dispensed benefits. There are evidences to establish that adolescent girls should not be thought of as passive recipients of help, unable to make change happen on their own. Social attitudes towards them can also be changed by providing appropriate exposures to communities, parents and society, but one of the very basic gaps has been that the designs of programmes for the girls have not yielded the desired results (Glennerster and Takavarasha 2010).[24]

5.2.4 Towards Gender Equity: Adolescent Girls in the SDGs

Taking into account the importance of the adolescent girls and their contribution towards national development, there has been a continuous advocacy globally that the indicators

Accessed on 8 June 2019.

[23] UN Joint Statement (ILO, UNESCO, UNFPA, UNICEF, UNIFEM, WHO) https://www.odi.org/sites/odi.org.uk/files/odi-assets/publications-opinion-files/8746.pdf. Accessed on 7 June 2019.

[24] https://www.povertyactionlab.org/sites/default/files/publications/Empowerment%20Literature%20Review%202011.04.05.pdf. Accessed on 4 June 2019.

of the SDGs be disaggregated at a minimum by sex by age in 5-year bands, in particular, for the age ranges of 10–14 and 15–19, as well as marital status.[25]

However, it is observed that in the Indian National Indicator Framework, there are only a few indicators which focus on the adolescents, youth and young people in general in a sex disaggregated manner. The SDGs on education, health, gender equity, water and sanitation; work and economic growth; and peace, justice and strong institutions have a total of 17 indicators which fall specifically within the age cohort of the adolescent and the youth. It is seen that out of these, only four indicators specify sex disaggregated data (in SDG 4,5 and 16); additionally, three indicators focus on the girls only (in SDG 3, 5 and 6), and the rest of the ten indicators do not mention about sex disaggregation. Thus, there is scope in only seven indicators in the National Indicator Framework for a spotlight on the adolescent/youth girls. The rest of the indicators being gender neutral do make it difficult to make the adolescent/youth girls visible (Table 5.1).

The General Comment of the Committee on Rights of the Child in April 2016, on the implementation of the rights of the child during adolescence, had highlighted that adolescent in general is minimally referenced in the SDG 2030 Agenda.[26] It is further said that there is also a significant lack of age and sex disaggregated data and indicators available to inform policy, identify gaps and support allocation of appropriate resources for adolescents. It is well established that sex disaggregated data is required for advocacy, policies and legislation and better data for SDG planning and implementation (Imrana 2017).[27]

5.3 Evidence of the Status of the Adolescent Girls in the Indian Context

For the SDG's 2030 agenda to be fulfilled, it is essential that a situational analysis is done on the status of the adolescent girls particularly on those relevant indicators which reflect on the outcomes of the development strategies, programmes and policies. This will help reflect on the gaps and provide vision on the road ahead.

In the country, subsequent to the South Asian Association for Regional Cooperation (SAARC) Decade of the Girl Child (1990–2000), for the first time in the post-independence period, issues related to 'girl child' became a subject matter of special inquiry in India. National schemes such as SABLA/SAG[28] and many state initiatives like Kanyashree[29] in West Bengal have given the necessary impetus to promote the nutritional and educational status of the adolescent girls. This section gives an overview of certain key indicators on the adolescent girls which are of immense significance in the context of the education, health and violence. Some of these indicators are already incorporated in the SDG's Indian National Indicator Framework, but many are not.

5.3.1 Demography and Sex Ratio

India has the largest adolescent population in the world[30] in the age group of 10–19 years, which constitutes about one fifth (20.9%) of the Indian population. The rural and urban population constitute 22% and 19% adolescents, respectively.[31] During the last four decades, the adolescent population has more than doubled from 116 mil-

[25]https://www.icrw.org/sites/default/files/images/Girl%20Declaration%20Joint%20Advocacy%20Group%20Core%20Indicators.pdf. Accessed on 3 June 2019.

[26]https://www.ohchr.org/Documents/HRBodies/CRC/GC_adolescents.doc. Accessed on 3 June 2019.

[27]https://events.development.asia/system/files/materials/2017/03/201703-importance-sex-disaggregated-data-tracking-gender-equality.pdf. Accessed on 10 June 2019.

[28]https://wcd.nic.in/schemes/scheme-adolescent-girls-sag. Accessed on 13 June 2019.

[29]https://www.wbkanyashree.gov.in/kp_4.0/kp_objectives.php. Accessed on 13 June 2019.

[30]UNICEF India. Adolescent Equity Profile. New Delhi, 2016.

[31]Census of India. http;//www.censusindiagov.in/2011census/population. Updated on 18-11-2014.

Table 5.1 List of goals and indicators from the National Indicator Framework which have focus on the adolescent, youth and young people

Sl num	Goal	Indicator	National indicator	Focus age group	Focus on sex disaggregated data	Focus on adolescent/youth girls
1	Goal 3	3.7.2	Percentage of women aged 15–19 years who were already mothers or pregnant	15–19 years	No	√
2	Goal 4	4.1.4	Percentage of students in grades 3, 5, 8 and 10 achieving at least a minimum proficiency defined learning outcomes to be attained by pupils at the end of each of the above grades	Students in grades 3, 5, 8 and 10	No	No
3		4.1.7	Out-of-school ratio in primary, upper primary, elementary, secondary and higher secondary	Primary, upper primary, elementary, secondary and higher secondary	No	No
4		4.3.1	Participation rate of youth and adults in formal and non-formal education and training in the previous 12 months	Youth and adults	No	No
5		4.3.2	Proportion of male-female enrolled in higher education and technical and vocational education	Students in higher education and technical and vocational training	√	√
6		4.3.3	Gross enrolment ratio for tertiary education	Students in tertiary education	No	No
7		4.4.1	Proportion of computer literate adults	Above 18 years	No	No
8		4.5.2	Gender parity indices for primary/secondary/higher secondary/tertiary education	Students in primary/secondary/higher secondary/tertiary education	√	√
9		4.6.1	Literacy rate of youth in the age group of 15–24 years	15–24 years	No	No
10	Goal 5	5.3.1	Proportion of cases reported under the Prohibition of Child Marriage Act (early marriage of children below 18 years of age) to total crime against children	Below 18 years for girls and below 21 years for boys	√	√
11		5.3.2	Proportion of women aged 20–24 years who were married or in a union before age 18	20–24 years	No	√
12		5.6.3	Proportion of population aged 15–24 years with comprehensive correct knowledge of HIV/AIDS	15–24 years	No	No
13	Goal 6	6.2.3	Proportion of schools with separate toilet facility for girls	Students in grades 1–10	No	√
14	Goal 8	8.5.6	Share of unemployed persons in population aged 15–24 (percentage)	15–24 years	No	No
15		8.6.1	Unemployment rate (15–24 years)	15–24 years	No	No
16		8.6.2	Proportion of youth (15–24 years) not in education, employment or training (NEET)	15–24 years	No	No
17	Goal 16	16.2.1	Number of victims of human trafficking per 100,000 population, by sex, age and form of exploitation	All ages including adolescent and youth	√	√

Source: National Indicator Framework

lion in 1971 to 253 million in 2011. Today, every fifth person in India is an adolescent (10–19 years) and every third a young person (10–24 years). In order to leverage the demographic dividend advantage,[32] investment in this age cohort is necessary. A demographic dividend does not result automatically from a large population of young people relative to the population of working-age adults without the needed population, social and economic policies.[33]

The adolescent sex ratio was 882 in 2001, and it increased to 898 (Census 2011). There has also been a small rise in the sex ratio of the youth population from 895 (2001 Census) to 908 (2011 Census). However, in general, it is highly skewed in favour of males. The deficit of females among the younger population does not augur well for equitable development and their empowerment.

5.3.2 Literacy, Education and Skilling

The literacy rates of the adolescent girls have been rising steadily, and more importantly, the gender gap is reducing, and in the 2011 Census, it is 3.5 points only (Table 5.2).

Table 5.2 Literacy rates of the adolescents

Census year	Male	Female	Gender gap
2011	91.7	88.2	3.5
2001	85.5	75.1	10.4

90% of the youth (males) and 82% of the female youth in the country are literates as per the Census 2011 data. Similarly, enrolment for the age group 15–16 years has also improved for both boys and girls, rising from 83.4% in 2014 to 84.7% in 2016[34] (Pratham 2017). Gender gap in educational attainment is decreasing, and the gender parity index[35] reveals there is increasing trend of female participation in all levels of school education, but the prevalence of gender-based discrimination is manifested in many forms, and one is that parents prefer to educate the sons in private schools which are perceived to be of higher quality (MHRD 2018).[36] On the other hand, the annual dropout rate of the girls is about 17% at the secondary level which again indicates the biases that confront the girls in schools. The National Family Health Survey (NFHS) 4 also shows that even today 10.4% of the girls in the age group of 15–24 years have not done any schooling.

As far as skilling is concerned, more than 12 million youth between 15 and 29 years of age are expected to enter India's labour force every year for the next two decades. It is estimated that by 2022, 109 million or so skilled workers will be needed in the 24 key sectors of the economy and the recently launched National Skill Development Mission aims to train approximately 400 million people across the country by 2022.[37] According to the National Skills Development Council Skill gap study[38] which analysed 24 sectors, overall projected employment by 2022 is 459.46 million. A World Bank project estimates that by 2023, about 8.8 million youth will have received some market-relevant training that will in turn open up

[32] Demographic dividend refers to the accelerated economic growth that begins with changes in the age structure of a country's population as it transitions from high to low birth and death rates.

[33] James N. Gribble and Jason Bremner: 'Achieving a Demographic Divided. Population Reference Bureau, Volume 67, Number 2 December 2012. (http://www.prb.org/pdf12/achieving-demographic-dividend.pdf).

[34] https://img.asercentre.org/docs/Publications/ASER%20Reports/ASER%202016/aser_2016.pdf. Accessed on 11 June 2019.

[35] The Gender Parity Index based on Gross Enrolment Rate (GPI- based on GER), is the ratio of Gross Enrolment Rate of female students enrolled at Primary, Secondary and Tertiary levels of education to the corresponding ratio of male students at that level. Thus GPI-based on GER is free from the effects of the population structure of the appropriate age group which provides picture of gender equality in education.

[36] https://mhrd.gov.in/sites/upload_files/mhrd/files/statistics-new/ESAG-2018.pdf. Accessed on 14 June 2019.

[37] http://www.worldbank.org/en/news/feature/2017/06/23/skilling-india. Accessed on 25 September 2017.

[38] REPORT OF THE SUB-GROUP OF CHIEF MINISTERS ON SKILL DEVELOPMENT, SEPTEMBER, 2015, NitiAyog file:///F:/ILO/Materails/Final%20report%20%20of%20Sub-Group%20Report%20on%20Skill%20Development.pdf. Accessed on 26 September 2017.

better job opportunities for them in a changing job market. Thus, a big opportunity lies ahead not only for the youth but also for the country in thriving on a skilled human resource, but sadly, there is no specific vision for the girls in the skilling roadmap.

5.3.3 Employment and Access to Credit

The adolescent girls and the young women in the country today have very restricted access, freedom and opportunity for gainful employment. The NFHS 4 has revealed that in India, only 31% of currently married women age 15–49 are employed, compared with 98% of currently married men in the same age cohort. An age-wise breakup reveals that for women employment increases with age, from 15% in the age 15–19 years to 38% amongst women in the 35–44 years age group, before falling to 35% in the oldest age group of 45–49 years (IIPS and ICF 2017). Delving deeper into their access to money and credit, it is observed that this age group is in the bottom of the ladder as older women particularly those above 40 years have the highest access. A very marginal percentage of girls in the age group of 15–19 years or 20–24 years have taken a loan from microcredit (IIPS and ICF 2017, p. 537).

The lack of opportunities or the unsupportive working conditions have led to a situation that 68.3% of women graduates (generally in the age of youth and young people) do not have paid jobs in urban India (AICTE 2019).

5.3.4 Health

The health condition of the adolescent girls in the country is pretty worrisome as only 53.9% of the girls in the age group of 15–19 years have a normal Body Mass Index[39] (BMI). Similarly, 54.1% of the girls in the same age group suffer from any type of anaemia.[40] Such undernutrition makes adolescent girls vulnerable to disease and early death and contributes to low birth weight of babies born to adolescent mothers, while also increasing maternal risk of haemorrhage and sepsis during childbirth. The prevalence of anaemia among girls is alarmingly high. In a study, significant association of anaemia was observed in low socioeconomic status, religion and infrequent or no consumption of animal protein. The micronutrient deficiencies were also found to be high in adolescent girls.[41]

Poor menstrual hygiene is one of the major reasons for the high prevalence of reproductive tract infections (RTIs)/sexually transmitted infections (STI). RTIs in the country contribute significantly to female morbidity. Most of the adolescent girls in villages use rags and old clothes during menstruation, increasing susceptibility to RTI. However, there are various other issues like awareness, availability and quality of napkins, regular supply, privacy and water supply, disposal of napkins, reproductive health education and family support which hamper the promotion of menstrual hygiene amongst the girls.[42]

5.3.5 Violence: Early Marriage, Early Pregnancy and Physical and Sexual Violence

Early marriage or child marriage is an endemic problem in the country with alarming proportions in certain states. On average, in the country, there has been a 20.6 point decline from the period of NFHS 3 (2005) to NFHS 4 (2015) in the percentage of women in the age group of 20–24 years who were married before the age of 18 years. For the men in the age group of 25-29 years who were married before the legal age of 21 years, it was 20.3% in NFHS which was a decline of 12 points. It is then observed that while adolescent girls do have higher incidences of early marriage, the gender is reducing. The prevalence in rural

[39] The body mass index (BMI) is expressed as the ratio of weight in kilograms to the square of height in metres (kg/m^2). Normal BMI means a range of 18.5–24.9.

[40] Less than 12.0 g/dL.

[41] Choudhary et al. (2006).

[42] Garg et al. (2012).

Table 5.3 Indicators of early marriage and early pregnancy

	Indicators	NFHS 4			NFHS 3	Decline (from NFHS 3 to NFHS 4)	Gender gap	
		Urban	Rural	Total	Total		NFHS 4	NFHS 3
1	Women age 20–24 years married before age 18 years (%)	17.5	31.5	26.8	47.4	20.6	6.5	15.1
2	Men age 25–29 years married before age 21 years (%)	14.1	24.4	20.3	32.3	12.0		
3	Women age 15–19 years who were already mothers or pregnant at the time of the survey %	5.0	9.2	7.9	16.0	8.1		

Source: NFHS 3 and NFHS 4

areas is higher than urban. Early pregnancy (of women in the age 15–19 years) is on the decline in the country which is good indicator of the impact of education, but the worrisome part is that still about 8% do have early pregnancy (Table 5.3). These indicators of early marriage and early pregnancy directly provide evidence that for the adolescent girls, while there has been significant improvement, still much more has to be achieved as the impact of these practices on the life of the girls is well established.

The NCRB report of 2016 presents a crime rate[43] of 0.1 under the PCMA 2006 with only 326 cases registered in the country (NCRB 2016).

The impact of violence on girls not only is long lasting and has a direct detrimental effect on their developmental milestones but also creates hindrances for their effective participation in development (UNICEF 2014). One in every four girls (24.8%) in the age group of 20–24 years has experienced physical violence since the age of 15 years, says the NFHS 4 survey data. For the younger girls of 15–19 years, the percentage is 17% of having experienced physical violence.

In the police records, only 50% of the missing girls below the age of 18 years[44] are traced which not only raises fundamental questions about their safety but also reinforces the subject of neglect and abandonment by the family and community due to discrimination. In the annals of the crime records, for the year 2016, 22.2% of the victims (girls) of sexual assault under the Protection of Children from Sexual Offences Act 2012 and Indian Penal Code (IPC) 376 were in the age group of 16–18 years. The highest age cohort of rape victims is 18–30 years[45] (Table 5.4). Such a scenario, even with very strict norms of investigation and penalty, reveals the social and economic vulnerabilities that young girls are exposed to.

Table 5.4 Women and girls victim of rape

		Percentage share of age group of victims
Victims of Sections 4 and 6 of POCSO Act r/w Section 376 IPC		
1	Below 6 years	1.3
2	6 years and above—below 12 years	4.1
3	12 years and above—below 16 years	15.6
4	16 years and above—below 18 years	22.2
	Total A (1 + 2 + 3 + 4)	43.2
Victims of Section 376 IPC		
5	18 years and above—below 30 years	42.1
6	30 years and above—below 45 years	13.3
7	45 years and above—below 60 years	1.3
8	60 years and above	0.1
	Total B (5 + 6 + 7 + 8)	56.8
	TOTAL A + B	100.0

Source: NCRB (2016)

[43]Crime Rate is calculated as Crime Incidence per one lakh of population.

[44]NCRB (2016, p. 526).

[45]NCRB (2016, p. 145).

5.4 Conclusion

The UN Sustainable Development Agenda framework provides an opportunity to renew and integrate efforts in order to meet, to a significant degree, national and global aspirations in a defined timeframe. The implementation strategy of SDGs has to be based on the priorities, as well as on the local challenges, capacities and resources available.

Financial resources are a key driver to various other capital and human resources. In a report done in 2015, it has been indicated that there is a first level estimation of financial shortfall of INR 533 lakh crores over the 15 years period till 2030 to achieve SDGs in India. Per year, on average, this works out to INR 36 lakh crores (TARA 2015). Though fiscal analysis was not within the scope of this paper, it is a vital aspect of any plan, programme or policy and, thereby, also of the SDG agenda.

Specifically, the adolescent girls who are the focus of this paper, positioned as key game changer in achieving the goals of sustainable development, have the potential which cannot be compromised by the failure to recognise or invest in the measures needed to enable them to enjoy their rights and participate and contribute effectively and substantially in the development process. The Lancet Commission[46] on adolescent health and well-being advocates that only through the improvement of the lives of the adolescents can the world in general progress. They argue that 'adolescents should be actively involved in working towards the SDGs at all levels—as agents for change at the school, community, societal, national, and international levels. Adolescents and young people are our best chance to achieve radical change for a prosperous, healthy, and sustainable world. The adolescents of today will be the policy makers of 2030'.

Additionally, it is high time that in India, new indicators be developed in order to analyse the status of the adolescent girls along with the gender disparities with respect to access to education, health, economic opportunities and impact of violence. This has a direct bearing on the monitoring framework of the SDGs. The targets and indicators have to be gender sensitive to focus on the adolescent girls across all the relevant SDGs (3, 4, 5, 8, 16 to name a few) so a holistic, forward-looking, sustainable development plan can be developed as well as outcomes measured. The significance of the gender equality aspect as a cross-cutting theme in all SDGs and focussing on the adolescent girls should be at the core of all planning and development agenda in India, or else the demographic significance of this age group will lose an opportunity to critically influence the development paradigm in the country.

[46] https://www.thelancet.com/action/showPdf?pi i=S0140-6736%2817%2931318-1. Accessed on 10 June 2019.

References

AICTE (2019) India Skill Report 2019, New Delhi

Choudhary A, Moses PD, Mony P, Mathai M (2006) Prevalence of anaemia among adolescent girls in the urban slums of Vellore, south India. Trop Doct 36(3):167–169. https://doi.org/10.1258/004947506777978253

Equal Measure 2030 (2019) Harnessing the power of data for gender equality, Introducing the 2019 EM 2030 SDG Gender Index, UK

Fukuda-Parr S (2003) The human development paradigm: operationalizing Sen's ideas on capabilities. Fem Econ. https://doi.org/10.1080/1354570022000077980

Garg R, Goyal S, Gupta S (2012) India moves towards menstrual hygiene: subsidized sanitary napkins for rural adolescent girls - issues and challenges. Matern Child Health J. Kluwer Academic/Plenum Press, New York. https://doi.org/10.1007/s10995-011-0798-5

Gribble J, Bremner J (2012) The challenge of attaining the demographic dividend, 6p. Retrieved from http://www.prb.org/pdf12/demographic-dividend.pdf

Glennerster R, Takavarasha K (2010) Empowering young women: What do we know? Cambridge, MA: Abdul Lateef Jamal Poverty Action Lab, MIT.

Inernational Institute for Population Sciences (IIPS) and ICF (2017) National Family Health Survey (NFHS-4), 2015-16: India. IIPS, Mumbai

Imrana J (2017) Tracking Progress on Gender Equality By 2030 the Importance of Sex-Disaggregated Data in Gender Equality & Social Protection Sdg 5: Achieve Gender Equality (pp. 14–15).

Jejeebhoy Shireen J, Francis Zavier AJ, Santhya KG (2013) Meeting the commitments of the ICPD Programme of Action to young people. Reprod

Health Matters 21(41):18–30. https://doi.org/10.1016/S0968-8080(13)41685-3
Mensch BS, Bruce J, Greene ME (1998) In: Judith AM (ed) The uncharted passage: girls' adolescence in the developing world. The Population Council, Inc, New York, p 115
MHRD (2018) Educational Statistics at a glance, Government of India, Ministry of Human Resource Development, Department of school education and Literacy, Statistics Division, New Delhi 2018. The web link is mhrd.gov.in/sites/upload_files/mhrd/files/statistics-new/ESAG 2018.pdf
NITI Aayog (2018) SDG INDIA INDEX: 2018 Baseline report. Retrieved from www.niti.gov.in; https://niti.gov.in/writereaddata/files/SDX_Index_India_21.12.2018.pdf
NCRB (2016) Crime in India. National Crime Records Bureau, India
Population Council, UNICEF (2013) Adolescents in India: a desk review of existing evidence and behaviours, programmes and policies, New Delhi
Pratham (2017) Annual Status of Education Report 2016
Rao N (2005) Questioning women's solidarity: the case of land rights, Santal Parganas, Jharkhand, India. J Devel Stud 41(3):353–375. https://doi.org/10.1080/0022038042000313282
Sen A (1999) Development as freedom. Oxford University Press, Oxford
Sharma B (2015) Level of women empowerment and it's determinates in selected South Asian countries, vol 20(4), Ver. V, pp 94–105. http://iosrjournals.org/iosr-jhss/papers/Vol20-issue4/Version-5/R0204594105.pdf
TARA (2015) Achieving the sustainable development goals in India: a study on financial requirement and gaps. UNDP & Ministry of Environment, Forest & Climate Change, Government of India
United Nations Children's Fund (2014) A statistical snapshot of violence against adolescent girls, p 40. Retrieved from https://www.unicef.org/publications/files/A_Statistical_Snapshot_of_Violence_Against_Adolescent_Girls.pdf
UN Women (2018) Turning promises into action: gender equality in the 2030 agenda for sustainable development, pp 1–346. Retrieved from www.unwomen.org/sdg-report%0ALas
Waage J, Yap C, Bell S, Levy C, Mace G, Pegram T, Unterhalter E, Dasandi N, Hudson D, Kock R, Mayhew SH, Marx C, Poole N (2015) Governing sustainable development goals: interactions, infrastructures, and institutions. In: Waage J, Yap C (eds) Thinking beyond sectors for sustainable development. Ubiquity Press, London, pp 79–88. https://doi.org/10.5334/bao.i

Amrita Patel was a Researcher, Trainer and Teacher on Women's Studies and has worked on issues of Gender Budgeting, land rights of women, women empowerment, climate change and declining child sex ratio.

Asha Hans is a former Prof of Political Science and Director School of Women's Studies Utkal University, Odisha. She is the founding President of Sansristi, a gender research organization.

6 Clean Water and Sanitation: India's Present and Future Prospects

Nilanjan Ghosh, Soumya Bhowmick, and Roshan Saha

6.1 Introduction

6.1.1 Sustainable Development Goals and India

Policymakers and academicians across the world have attempted to mainstream *the 2030 Agenda for Sustainable Development*, envisioning a pathway to reconcile the impossible trinity of equity, efficiency and sustainability (Ghosh 2017). The Sustainable Development Goals (SDGs) are aimed at optimizing the different, but interrelated, conjunctions of human–nature, human–human and nature–nature interactions. The importance of 'quality of life' parameters, such as access to healthcare, education, employment opportunities, food, drinking water and sanitation, in the overall development of a nation has been often put to the forefront by economists such as Andrew Oswald (1997) and Amartya Sen (2000). The SDGs reiterate the same, but are an overarching goal in the sense that inter-temporal considerations are also embedded in their objectives, and the financial gaps are huge in meeting these objectives especially for the low-income countries.

Overexploitation and indiscriminate utilization of resources have already led to several regions overshoot 'day zero' in terms of availability of natural resources in one form or other. The policy space between the planetary ceiling and the social institutions is shaped like a doughnut (two concentric circles). It is this region within the planetary and social boundaries where human activity ought to take place. Developmental activities which do not manifest in societal improvement, or positive externalities, can never be self-sustaining (Ghosh 2017). Neither must any developmental activity exceed the planetary boundaries nor should they fall shy of the social foundations (Raworth 2012). These concepts further highlight the importance of inculcating the SDGs in the national development agendas.

The SDGs highlight areas of, both, critical human deprivations and critical natural thresholds—hunger, poverty, diseases, illiteracy, poor sanitation, lack of drinking water, biodiversity loss, marine, soil and air pollution, and climate change. Some goals essentially encompass the human and natural capital stock of the planet, while the other goals aimed at ensuring clean energy, sustained economic growth, infrastructure development, reduction of inequalities, fostering innovation and community and urban planning capture elements of physical and social capital (Ghosh et al. 2019).

The UN Sustainable Development Solutions Network's SDG Index Report 2018 places India at a rank of 112 out of 156 countries (score of 59.1%) in terms of SDGs performance, lagging

N. Ghosh (✉) · S. Bhowmick · R. Saha
Observer Research Foundation, Kolkata, India
e-mail: nilanjanghosh@orfonline.org

S. Hazra, A. Bhukta (eds.), *Sustainable Development Goals*, Sustainable Development Goals Series,
https://doi.org/10.1007/978-3-030-42488-6_6

behind the East and South Asia's average regional score of 64.1%. India's large population and geography makes the implementation of human capital inducing objectives such as SDG 6, that is, access to 'clean water and sanitation' a rather arduous task. Despite such impediments, the status of SDG 6 is moderately increasing but is, however, insufficient to reach the 2030 targets according to the report. Recognizing the linkages of SDG 6 with the other goals, a report submitted to the Ministry of Environment, Forests and Climate Change, Government of India in 2015, underscores that the estimated financial gap in India for implementing SDG 6 stands at US$ 123 billion (Technology and Action for Rural Advancement 2015). It is important to note here that efficient implementation of such goals would require more accuracy in the federal governance mechanisms both at the State and Union levels.

6.1.2 Changing Paradigms

The history of development of the present civilization involves the history of the various types of human interventions in the hydrological cycle. This was made possible by human ability to build bigger and bigger engineering structures to modify the flows of streams and rivers. Human control over the aquifers was established through stronger and stronger pumping technologies to take water out from deeper and deeper levels of aquifers. Dams were effectively used for controlling floods and generating hydro-electricity at a very large scale. This offered a reasonable protection against seasonal water shortages and even spatial inequities in water availability. The irrigation canals made it possible for humans to grow food in newer and newer areas as much as it enhanced the growing seasons for crops.

On the other hand, as demand for water for meeting the basic human needs started being satisfied, forces of development started showing its signs. Perhaps, the gravest effect of the escalating urbanization was felt in the agricultural water use, which encountered manifold increase, over the last two centuries, in order to meet needs of the burgeoning urban population. Traditionally, water has been looked at as a resource occurring in 'abundance' in nature, and hence, increasing demand was never seen as posing any potent threat. Hence, the impression that became predominant emanated from the idea that water scarcity is spatial, and more water can be diverted to the water-scarce zones from the water-rich zones, through appropriate supply augmentation plans. In order for 'water to be distributed equitably', the traditional thought process provoked the idea of supply expansion plans through interventions in the natural hydrological flows (Rao 1975). Eventually, water resource planning was generally reliant on linear projections of future populations, per capita demand, agricultural production and levels of economic productivity (Gleick 2000a).

Towards the middle of the last century, serious concerns started to being expressed on the long-term wisdom of following such a strategy that is focused exclusively on the increasing intervention into the hydrological cycle. Despite its impressive short-term successes in providing larger supplies, it is increasingly being realized that addressing the new and emerging challenges is no more possible over the long term, unless some fundamental changes take place in the way humans have looked at water resources so far. The 'business as usual' process has started to be feared as counterproductive. There emerged the need for a fundamental change in terms of a new interdisciplinary paradigm that has been constantly gaining ground over the years. The new ways of managing water on the basis of a holistic knowledge base have increasingly been identified as Integrated Water Resource Management (IWRM).

The origin of such comprehensive efforts to address issues of water management finds its allusion in the 1977 Mar del Plata conference on Water. The Rio Summit in 1992 expanded the agenda to include ecological water needs, which have been adopted in the current context of SDG 6. The Dublin statement reinforced this view. Implicitly, these documents identified 'basic water requirements' and 'sustainable water requirements', where the former essentially refers to drinking water for survival, water for human hygiene, water for sanitation services and modest household needs for preparing food. Unless these basic requirements are fulfilled by the state, large-scale human misery and suffering will contribute to the risk of social and military conflict (Gleick 1996). The latter talks about the usage of water in the purview of environmental limits.

The realization of the need for holistic modes of water management has been reflected in some of the policy actions of the developed world, primarily with the dawning of the ecological concerns (Gleick 2000b). Continued investments in huge engineering interventions are being challenged by those who believe a higher priority should be assigned to projects that meet basic and unmet human needs for water (Gleick 1996). The United States, the country which started the global trend of building large dams, is following '... a new trend to take out or decommission dams that either no longer serve a useful purpose or have caused such egregious ecological impacts so as to warrant removal. Nearly 500 dams in the USA and elsewhere have already been removed and the movement towards river restoration is accelerating' (Gleick 2000a).

The Murray-Darling Basin Commission in Australia is seriously contemplating on extending financial encouragement to farmers for saving on their allocation of irrigation water and to allow the savings to remain instream (Bandyopadhyay and Perveen 2004). In another instance, Chile's National Water Code of 1981 established a system of water rights that are transferable and independent of land use and ownership. The most frequent transaction in Chile's water markets is the 'renting' of water between neighbouring farmers with different water requirements (Gazmuri 1992). Helming and Kuylenstierna (2001), while cautioning against the damages that can be caused by supply augmentation plans, emphasized that '... Demand side management is therefore slowly becoming a new paradigm for water governance'.

6.1.3 Entry Point of This Chapter

One needs to note here that the discussions in Sect. 6.1.2 relate with a holistic paradigm of water management that does not explicitly show up in SDG 6. To be more specific, SDG 6 is more related to human household use and may apparently seem 'anthropocentric' in its delineation. But what is often missed out is that a holistic water management paradigm that looks at water as a 'flow' and not as a stock of resource to be used for storing and consumption as per need is one of the pre-requisites for achieving this goal. The various ecosystem functions and eventually the ecosystem services of a free-flowing system entail various provisioning services as also regulating services like clean water provisioning through natural purification processes. While human interventions are needed to provide for clean water, the nature's ability to do so also needs to be acknowledged, as one addresses SDG 6.

While we are aware of this aspect, it is practically impossible in India to find a 'free-flowing' river, except for some smaller stretches. In this chapter, therefore, that aspect is assumed away. Rather, the critical entry point happens to be the performances of the states in terms of the various initiatives at the level of the citizens. In this chapter, we report on the performances of the Indian states by developing an index on SDG 6 on the basis of various parameters and using statistically determined weights on the basis of principal component analysis. In Sect. 6.2, we talk of the various SDG targets and the initiatives of the central government with respect to this goal in India. In Sects. 6.3 and 6.4, we explain the datasets and methodology and report on the results of the ranks of various Indian states with respect to their performances. Section 6.5 provides the concluding remarks.

6.2 Water and Sanitation for All

SDG 6 aims to ensure 'availability and sustainable management of water and sanitation for all'. It recognizes that social development and economic prosperity are built upon the foundations

Table 6.1 SDG 6 targets

SDG 6 sub-goals	Objectives
6.1	Universal and equitable access to safe and affordable drinking water
6.2	Adequate and equitable sanitation for all
6.3	Improvement of water quality through reduction of water pollution
6.4	Increase of water-use efficiency across sectors and reduce number of people suffering from water scarcity
6.5	Implementation of integrated water resource management at all levels
6.6	Protect and restore the health of water-related ecosystems
6.a	International cooperation and capacity building in developing countries through waste water treatment, desalination, recycling and reuse technologies, etc.
6.b	Participation of local communities for improvement of water and sanitation

Source: United Nations General Assembly Report (2015)

of sustainable management of freshwater resources. Water resources and sanitation are embedded in most forms of development targets, such as food security, health promotion and poverty reduction, agriculture and industrial growth, energy generation and ecosystem services (United Nations 2018). Table 6.1 illustrates the specific targets delineated under SDG 6.

Water and sanitation are strongly related to public health (Roy and Pramanick 2019). While HIV/AIDS, tuberculosis and malaria attract most attention of international public health community, diarrhoea, a water-borne disease occurring mostly in the poorer nations, alone kills more children in a year than the former three combined (Boshci-Pinto et al. 2008). According to the United Nations, one fifth of the child deaths in the world are in India due to severe diarrhoea. Hygiene, sanitation and water (HSW) interventions can reduce incidences of diarrhoea, ascariasis, cholera, scabies, trachoma, amoebiasis, etc. (Bartram and Cairncross 2010). The benefits of HSW are far greater than these disease-specific statistics. Malnourished children when recovering from diarrhoea are usually vulnerable to pneumonia. This diarrhoea-induced susceptibility is associated with 26% of all childhood pneumonia cases. Reducing incidences of diarrhoea will have secondary impacts in reducing other diseases as well (Schmidt et al. 2009). Reduction in risks to health, especially malnutrition, also results in better school performance by children and timely entry into the labour market (Acharya and Paunio 2008), directly influencing hunger reduction and lowering poverty levels in the long run. It is, thus, evident that hygiene, sanitation and water are foundations of a well-functioning public health system and good health, which accounts for one of the main problems in India's development trajectory. Access to basic sanitary needs will improve the overall health and well-being of people, especially children. Focussing on SDG 6 will improve the performance of other SDG indicators through the interlinkages.

In the context of SDG 6, however, India has progressed a lot from the early 1990s until now. The major problems in this aspect include issues of open defecation, garbage disposal in water resources and non-access to clean drinking water. Although these issues mostly come under the radar of the Ministry of Jal Shakti- formed by merging the erstwhile Ministry of Drinking Water and Sanitation (MDWS) and the Ministry of Water Resources, River Development and Ganga Rejuvenation (MoWR) in 2019- the benchmark schemes implemented by the Government of India to tackle the related issues in rural India include the *Swachh Bharat Abhiyan* (Clean India Campaign), the National Rural Drinking Water Programme, and *Namami Gange* (River Ganga Conservation). The major Union level initiatives in the past two decades on urban water and sanitation needs are outlined in Table 6.2.

6.3 Measuring Progress of SDG 6

Prior to the SDGs, the Millennium Development Goals (MDGs) for 2015 identified access to safe drinking water and basic sanitation as one of its

Table 6.2 Central government initiatives for urban water and sanitation

Initiatives	Objectives
1. Jawaharlal Nehru National Urban Renewal Mission (JNNURM) 2. Ministry of Urban Development—High Powered Expert Committee, 2008 3. 12th Five Year Plan Committee	Urban infrastructure, water supply, drainage
1. Ministry of Urban Development—Advisory Note on Improving Water Supply and Sanitation Services, 2012 2. National Water Policy 2012 3. MoUD and MoHUPA Centres of Excellence and National Resource Centre	Development of water supply and sanitation businesses, service improvement plan, capacity building, reducing leakages in water supply and re-use of water
Steering Committee on Drinking Water Supply and Sanitation, Planning Commission, GOI, 2002	Levy of water charges for maintenance and future improvement schemes, supply of healthy drinking water
1. Swachh Bharat Mission 2. National Urban Sanitation Policy 3. Service Level Benchmarking Initiative 4. National Sanitation Ratings of Cities 5. Central Public Health and Environmental Engineering Organization Manuals 6. Septage Management Advisory 7. Advisory on Water and Sanitation Services	Healthy sanitation practices, awareness and behavioural change in people with regard to urban water and sanitation, waste management and drainage, private sector participation

Source: compiled from various MoWR and MDWS reports

targets under the broader goal of 'Ensure Environmental Sustainability'.[1] While both rural and urban India met the MDG targets for improved drinking water, its performance in terms of sanitation was far from satisfactory. Even though households might have had access to an 'improved' source of water, it does not indicate adequate supply of water of acceptable quality.[2] Therefore, performances under the MDGs were not well founded. SDG 6 is a distinct improvement over the MDGs (Wankhade 2016). Not only does it consider infrastructural facilities that ensure clean and safe drinking water and sanitation for all, but it also looks into aspects of governance and efficient utilization of water resources. In other words, it has brought the whole cycle of water and sanitation in the governance discussion.

Green Indian States Trust's seminal study (2007) on freshwater quality outlines some key issues related to surface water and groundwater in India. It highlights the importance of natural capital assessments with regard to water, especially in a country like India where almost 70% of the geographical area is classified as arid and semi-arid. Therefore, data monitoring is crucial in measuring progress of SDG 6 in India. As SDG 6 is multi-dimensional in nature, it would require indexation of the various parameters to a single composite index to make any such analysis meaningful. In order to represent the broad category of targets, some indicators have been chosen to represent the key aspects of the goal-demand, supply, quality, management and governance.

This study entails developing two indices[3] for 'clean water' and 'sanitation' separately, across the 23 states[4] of the country and then discussing

[1]MDG 7 refers to 'Ensure Environmental Sustainability'. Target 7.c. states: 'Halve, by 2015, the proportion of the population without sustainable access to safe drinking water and basic sanitation'.

[2]According to the Joint Monitoring Programme (JMP), WHO-UNICEF, an 'improved' drinking water source is one that, by nature of its construction and when properly used, adequately protects the source from outside contamination, particularly faecal matter. 'Improved' drinking water sources include piped water into dwelling, public taps or standpipes, tube wells or boreholes, protected dug wells, protected springs and rainwater collection. 'Unimproved' sources include unprotected spring, unprotected dug well, cart with small tank/drum, tanker-truck, surface water and bottled water.

[3]All computations have been performed on Stata 12.0.

[4]The latest available state-wise data for each of the indicator variables have been chosen. The Union Territories and the North Eastern States (except Assam) have been excluded from the study due to unavailability of data.

the intersection of these two scores with the help of a matrix analysis. At the very first stage, the quantified/quantitative indicators reflecting each aspect are taken, and their respective weights in the context of 'clean water' or 'sanitation' are determined by principal component analysis[5] (PCA). For this purpose, it is first necessary to apply transformation functions to the raw data. Subsequently, weights are attached to the transformed values of each indicator. Ghosh et al. (2014) suggest using PCA for weight distribution since it stands out as one of the best practices globally, to enhance statistical robustness in assigning weights. Following this method, weights are attached to different indicators without exposing them to 'subjectivity' and 'sub-optimal representation'. Finally, statistically computed weights and transformed indicator values are aggregated using the additive function to obtain the index scores.

Table 6.3 illustrates the different indicators[6] chosen to measure the status of the 23 Indian states with respect to SDG 6. One of the indicators, Composite Water Index, of the NITI Aayog is essentially a measure of the performance of different Indian states in terms of management of water resources. It depicts the importance of the governance aspect of SDG 6 as several water-scarce states are the top performers. Water Body and Water Withdrawal as a Percent of Water Availability capture the supply and demand side of the water respectively. Subsequently, indicators reflecting adequate sanitation facilities represent the provision of infrastructure for waste management and disposal.

6.4 Results Across Indian States

The data for the 23 Indian states on the above-mentioned indicators have been used to estimate the following two separate indices for 'clean water' and 'sanitation' respectively. Figure 6.1 represents the performance of different states in providing access to clean water. Since it is a baseline study, it reflects a profile of the current status. Gujarat, Jammu and Kashmir, Madhya Pradesh, Andhra Pradesh, Odisha, Maharashtra, Karnataka and Telangana represent the top eight states, while Delhi, Uttarakhand, Haryana, Jharkhand, Uttar Pradesh, Bihar, Kerala and West Bengal represent the bottom eight. Most of the states in the bottom eight are well endowed with water resources as they lie along the Ganges basin, while the better performing states are relatively water scarce. Gujarat, Maharashtra, Madhya Pradesh, Andhra Pradesh, Karnataka and Telangana have suffered from severe droughts in recent years. This drives the focus of water issues in India from an endowment perspective towards better management and governance of water resources.

Similarly, Fig. 6.2 represents the index scores for provision of sanitation facilities. Goa, Delhi,

[5]Calculation of weights for each of the seven indicators under SDG 6 has been conducted by principal component analysis (PCA), to rank the states. This methodology has been chosen over a simple average technique to understand the dominant patterns in the data set in terms of weights that should be assigned to each parameter. Now, let us define the weight attached, by PCA, to an indicator '*m*' in 'clean water' or 'sanitation' index as ω_m where $\omega_m = [\max\{\text{component}^1, \text{comp}^2, \ldots\ldots, \text{comp}^m\}]^2 * $ explanatory power of [max{comp1, comp2,…,comp *m*}]. Each of the weights has been scaled up to sum up to 1 (100%) for both 'clean water' and 'sanitation', in order to avoid under representation.

[6]The data collected across 23 Indian states, owing to either the population size or geographical area in most cases, have inherent scale biases. The variables have been converted into per unit format (to remove the relevant scale bias). Further, all the data points have been normalized by the following formula to range from 0 to 1 (unit free). Now, for the negative indicators, the complement of 1 for their respective normalized values has been taken, so as to convert them into a positive indicator. This has been done so as to form a uniform, unit and direction free, composite index score for 'clean water' and 'sanitation' respectively.

$$Y_{kj} = \frac{y_{kj} - \min_j(y_{kj})}{\max_j(y_{kj}) - \min_j(y_{kj})}$$

where y_{kj} denotes the value of the component indicator k of 'clean water' or 'sanitation' for state j; Y_{kj} denotes normalized value of the component indicator k of 'clean water' or 'sanitation' for state j; $\min_j(y_{kj})$ denotes the minimum value of the row vector of y_{kj} values across the states; $\max_j(y_{kj})$ denotes the maximum value of the row vector of y_{kj} values across the states; with '*k*' denoting an indicator, it is defined by the closed set $k = [1, 3]$ for 'clean water' and $k = [1, 4]$ for 'sanitation' and '*j*' denoting a state, which is defined by the closed set $j = [1, 23]$

Table 6.3 Indicators of clean water and sanitation in India

SDG 6 components	Target category and indicator chosen	Data source
Clean water	6.a. Composite Water Index Score (2016–2017)[a]	NITI Aayog, June 2018—Composite Water Management Index: A Tool for Water Management[b]
	6.6. Water Body (2018)[c]	School of Oceanographic Studies, Jadavpur University (2018)
	6.4. (Complement of) Water Withdrawal as a Percent of Water Availability (2012)[d]	Q. No. 2131, Dated: 24/07/2014, Ministry of Water Resources, River Development and Ganga Rejuvenation, Lok Sabha & Lok Sabha Unstarred Question No. 4426, dated 03.05.2012
Sanitation	6.b. (Complement of) Slum Population (2011)[e]	Registrar General of India, Census of India, 2011
	6.2. Number of Households Having Access to Water for Toilets (2016)[f]	Swachhata Report, 2016 (MOSPI)
	6.2. Wards Having Access to Liquid Waste Disposal for Community and Public Toilets (2016)[g]	Swachhata Report, 2016 (MOSPI)
	6.2. Solid Waste Disposal—Total Waste Processed (2016)[h]	Swachhata Report, 2016 (MOSPI)

[a]Originally expressed as composite water index scores, the values have been normalized to range from 0 to 1. The missing value of Delhi is substituted by the average values of its neighbouring states that are Haryana and Uttar Pradesh. The missing value of West Bengal is substituted by the average of its neighbouring states—Odisha, Jharkhand, Bihar, Sikkim and Assam. The missing value of Jammu and Kashmir is substituted by the average values of its neighbouring states that are Himachal Pradesh and Punjab

[b]The Composite Water Management Index (CWMI) uses data collected at central and state level from 2016 to 2018. Their findings show that the water scarce states (Gujarat performs best) perform much better in terms of the CWMI than the relatively water abundant states (Meghalaya performs worst)

[c]In order to capture the total water availability in a state, we have used the total area under water bodies as a measure. These figures were originally in hectares which have been converted into square metres and then the per capita values have been calculated (as per Census 2011) to remove scale biases. This data set has then been normalized to range from 0 to 1

[d]Individual data for ground water availability and ground water withdrawal have been obtained from the mentioned sources. Water withdrawal as a percentage of water availability for each state has been calculated from the given data sets. These figures have then been normalized to range from 0 to 1. Andhra Pradesh values have been used as a proxy for Telangana. The normalized values have been subtracted from 1 to convert into its complement to make the score positive

[e]Originally expressed figures are in percentage of total urban population. The values have been normalized to range from 0 to 1. The value for Andhra Pradesh has been used for Telangana. The normalized values have been subtracted from 1 to convert into its complement to make the score positive

[f]Original figures were expressed in percentage. These values have been normalized to range from 0 to 1

[g]Original figures were expressed in percentage. Missing value for Delhi has been replaced by the average of Uttar Pradesh and Haryana, while missing value of Goa has been replaced by the average of Karnataka and Maharashtra. These values have then been normalized to range from 0 to 1

[h]Original figures are expressed as percentage of waste processed. We have normalized these data to range from 0 to 1

Kerala, Gujarat and Telangana are the best-performing states, while Chhattisgarh, Bihar, Odisha, Andhra Pradesh and Jharkhand represent the poorest five performers. Impact of inadequate sanitation facilities contributed to an increasing prevalence of malnourishment and stunting among children in Jharkhand, Bihar and Uttar Pradesh (Manisha 2018).

These indices, individually, suggest the existence of widespread inequalities across states in terms of providing access to clean water and sanitation. Very few states have been able to undertake holistic planning in order to achieve the basic objectives. It is interesting to observe that Delhi, which was the poorest performer in terms of providing clean and safe water, is one of the

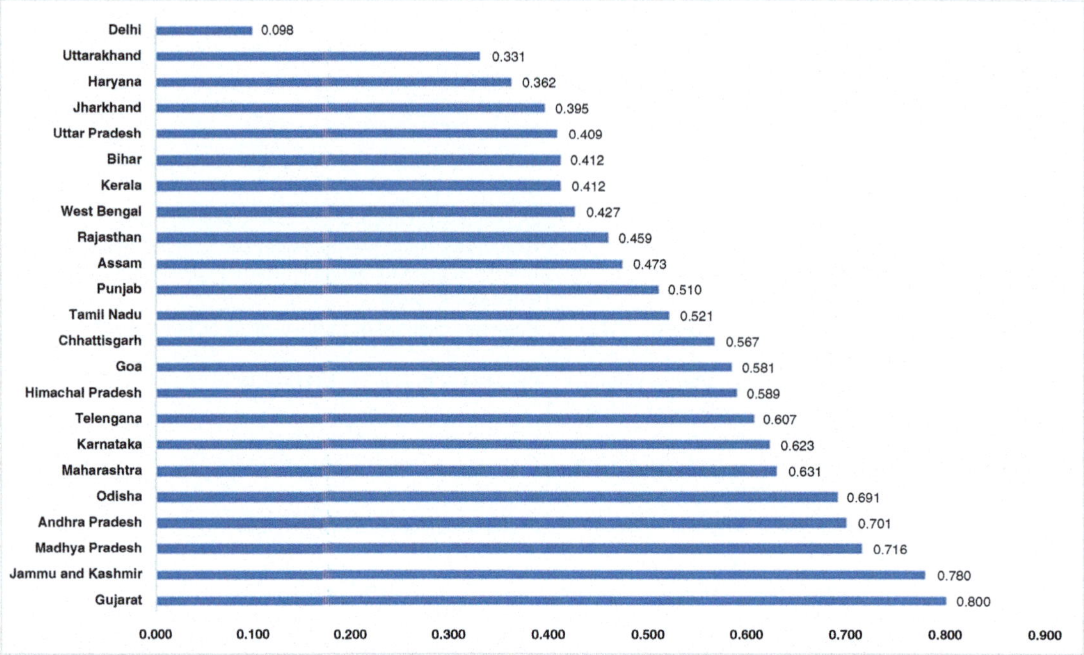

Fig. 6.1 Providing access to clean water. (Source: Computed Index Scores)

highest ranked states in terms of providing access to sanitation facilities. Gujarat, on the other hand, has been relatively outstanding in both parameters.

In terms of overall performance in providing 'clean water and sanitation', Fig. 6.3 represents the 'clean water–sanitation' matrix. The intersection between water and sanitation issues shows the placement of the Indian states in terms of SDG 6. The states have been classified in terms of four categories, with A and D representing the worst and best respectively[7] (A, B, C and D are arranged in ascending order of performance in terms of both 'clean water' and 'sanitation' indices).

Performance keeps improving as one moves from the top-left grid to the bottom-right grid of the matrix. The top-left corner grid is empty suggesting that almost every state has been successful in providing at least a basic minimum level of clean water and sanitation in synchronization with the MDGs. Gujarat fares as the best performer of the SDG 6. The states can be classified into four categories in terms of SDG 6 performance—*Under Performers (Grids AA, AB, BA, BB), Better Water Facilities (Grids AC, AD, BC, BD), Better Sanitation Facilities (Grids CA, CB, DA, DB)* and *Top Runners (Grids CC, CD, DC, DD)*. The states belonging to these categories are as follows:

- *Under Performers*: Jharkhand, Bihar, Rajasthan, West Bengal, Assam, Uttar Pradesh and Punjab
- *Better Water Facilities*: Chhattisgarh, Andhra Pradesh, Odisha, Maharashtra, Jammu and Kashmir and Madhya Pradesh
- *Better Sanitation Facilities*: Uttarakhand, Haryana, Tamil Nadu, Delhi and Kerala
- *Top Runners*: Himachal Pradesh, Telangana, Karnataka, Goa and Gujarat

[7]Categories have been identified based on the following:

A: $X_j < (\mu - \sigma)$;
B: $(\mu - \sigma) < X_j < (\mu)$;
C: $(\mu) < X_j < (\mu + \sigma)$;
D: $X_j > (\mu + \sigma)$;

where μ is the mean and σ is the standard deviation of X_j, and X_j refers to the state-wise computed index values for 'clean water' and 'sanitation' and '*j*' denoting a state, which is defined by the closed set $j = [1, 23]$.

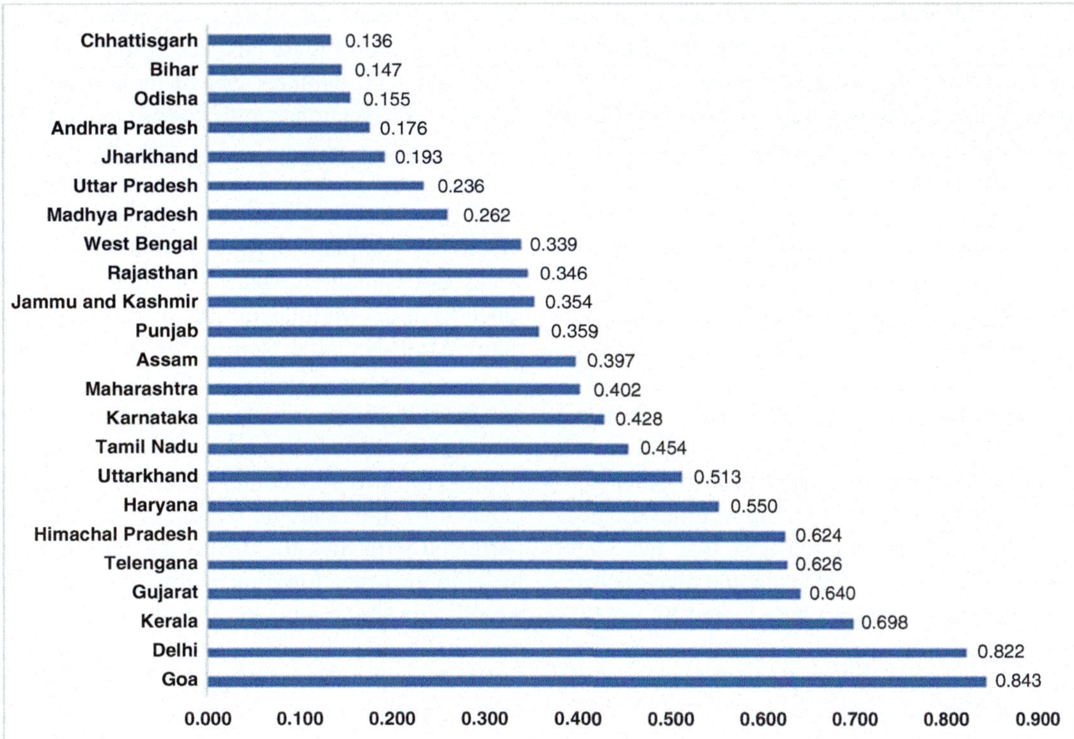

Fig. 6.2 Providing access to adequate sanitation facilities. (Source: Computed Index Scores)

SDG 6		CLEAN WATER			
		A	B	C	D
SANITATION	A		Jharkhand, Bihar	Chhattisgarh	Andhra Pradesh, Odisha
	B		Rajasthan, West Bengal, Punjab, Assam, Uttar Pradesh	Maharashtra	J&K, Madhya Pradesh
	C	Uttarakhand, Haryana	Tamil Nadu	Himachal Pradesh, Telengana, Karnataka	
	D	Delhi	Kerala	Goa	Gujarat

Fig. 6.3 Clean water and sanitation matrix (authors' computation)

6.5 Concluding Remarks

This chapter is a modest attempt to rank states in terms of their performances with respect to SDG 6 that essentially covers two dimensions: clean water and sanitation. Though often clubbed together as they are generally important cornerstones of local government's responsibilities, they may occur independently. This is prevalent from the classification that shows states with good performance in water supply, but not being good performers in sanitation, and vice versa. There are 11 states in such categories. On the other hand, one needs to note here that the entire idea of ranking states in terms of their SDG performance is yet another attempt to promote 'competitive federalism' in the same vein in which the 'ease of doing business' rankings are conducted. However, one needs to note here that these ranks are across the Indian states in relation to each other, and as such, India's performance in the context of SDG 6 has not been too encouraging in the global context. The newly formed ministry, Jal Shakti, can take a leaf from this exercise. At the same time, it is important that a holistic approach to water governance is undertaken keeping in mind the ecosystem functions and services that the water body renders. At a basin scale, the idea of 'free-flowing rivers' needs to be adopted by the new body to the extent possible. A secure hydrological future requires keeping water instream through demand management.

On the other hand, as argued by Ghosh et al. (2019), almost all the SDGs are embedded in one form of capital or the other, that is, human (SDGs 1–5: reflecting on poverty, hunger, health, education and gender equality), physical (SDGs 8 and 9: employment, growth, industry, innovation and infrastructure), natural (SDGs 14 and 15: life below water and land respectively) and social (SDGs 10 and 16: social institutional variables, etc.). In this context, SDG 6 contributes to the important factor of human capital, as stated earlier. All these are factors of creating enabling business environment Ghosh et al. (2019). As such, there is a two-way causality between business performance and SDGs. While the role of the private sector and multilaterals is being seen as important drivers for achieving SDGs globally, many private organizations are transcending the unidimensional goals of short-term profit maximization and focusing on sustainability parameters in an attempt to create a long-term business strategy.

The Business and Sustainable Development Commissions Report 2017 identifies immense business opportunities associated with the SDGs and estimates their aggregate global potential value in 2030 at US$ 12 trillion in current prices. More than half of these business solutions reside in developing economies such as India with large markets. International public–private, public–public, private–private partnerships leveraging on individual comparative advantages are the way forward to incentivise projects and successfully implement projects relating to clean water and sanitation. Apart from financial assistance from the Multilateral Development Banks, many Multi-National Companies in this sector such as AquaFed, Cargill and P&G are actively collaborating with state and non-state actors in developing nations for drinking water and sanitation facilities. For example, according to the US Council for International Business, Gap Inc. has partnered with Swasthi Health Resource Centre in building water filtration plants in rural India which already caters to approximately 5000 households and 29 schools in 30 villages across India. Another policy which holds much water in contemporary India is pricing water across all urban regions for efficient utilization and progressive distribution of funds for areas that are lagging behind in terms of water and sanitation facilities.

References

Acharya A, Paunio M (2008) Environmental health and child survival: epidemiology, economics, experiences. Washington, DC, The World Bank (Environment Department)

Bandyopadhyay J, Perveen S (2004) Interlinking of rivers in india: assessing the justifications. Econ Pol Wkly 39(50):5308–5316

Bartram J, Cairncross S (2010) Hygiene, sanitation and water: forgotten foundations of health. PLoS Med 7(11):e1000367

Boshci-Pinto C, Velebit L, Shibuya K (2008) Estimating child mortality due to diarrhoea in developing countries. Bull World Health Organ 86(9):710–717

Business & Sustainable Development Commission (2017) Better business better world, London

Gazmuri R (1992) Chilean water policy experience. Paper presented at the ninth annual irrigation and drainage seminar, Agriculture and Water Resources Department. The World Bank, Washington, DC

Ghosh N, Sinha N, Jhunjhunwala S (2014) Indexing Covariation in Base-Metals Prices. The Journal of Index Investing 5(3):87–93

Ghosh N (2017) Ecological economics: sustainability, markets, and global change. In: Mukhopadhyay P et al (eds) Global change, ecosystems, sustainability. Sage, New Delhi

Ghosh N, Bhowmick S, Saha R (2019) A 'social' index for Ease of Doing Business. The Hindu Business Line

Ghosh N, Bhowmick S, Saha R (2019) SDG Index and Ease of Doing Business: a Sub-National Study. Observer Research Foundation Occasional Paper 199

Gleick PH (1996) Basic water requirement for human activities: meeting basic needs. Water Int 21(2):83–92

Gleick PH (2000a) The changing water paradigm: a look at twenty-first century water resources development. Water Int 25(1):127–138

Gleick PH (2000b) The world's water 2000-2001: the biennial report on freshwater resources. Island Press, Washington, DC

Helming S, Kuylenstierna J (2001) Water – a key to sustainable development. Issue paper for the International Conference on Freshwater, Bonn, 3–7 December

Manisha M (2018) SDG 6 in India: challenges and innovation for sustainable sanitation. The Global Network for Economics of Learning, Innovation, and Competence Building System

NITI Aayog (2018) Composite Water Management Index: a tool for water management

Oswald AJ (1997) Happiness and economic performance. Econ J 107: 1815–1831

Rao KL (1975) India's Water Wealth. Orient Longman, New Delhi

Raworth K (2012) A safe and just space for humanity: can we live within the doughnut? Oxfam Discussion Papers

Roy A, Pramanick K (2019) Analysing progress of sustainable development goal 6 in India: past, present and future. J Environ Manage 232:1049–1065

Schmidt W-P, Cairncross S, Barreto ML, Clasen T, Genser B (2009) Recent diarrhoeal illness and risk of lower respiratory infections in children under the age of 5 years. Int J Epidemiol 38:766–772

Sen A (2000) Development as freedom. Anchor Books, New York

Technology and Action for Rural Advancement (2015) Achieving the sustainable development goals in india: a study of financial requirements and gaps. Development Alternatives Group, New Delhi

United Nations (2018) Sustainable development goal 6 synthesis report 2018 on water and sanitation, New York

United Nations Council for International Business (n.d.) Business + SDGs. Retrieved from Business for 2030: Forging a path for business in the UN 2030 Development Agenda. http://www.businessfor2030.org/goal-6-ensure-water-sanitation

United Nations General Assembly (2015) Transforming our world: the 2030 agenda for sustainable development

Wankhade K (2016) Operationalising SDG 6 in Urban India. Indian Institute of Human Settlements

7 Enabling Energy Sustainability

Suvra Majumder

7.1 Introduction

Access to equitable, affordable, reliable and environmentally sustainable modern energy services is at the helm of human development (Gaye 2007) and key to achievement of inclusive social development, poverty alleviation and economic empowerment of the deprived and marginalised populations. It is, therefore, 'energy resources and services' and has been accorded with uttermost priority under the 2030 agenda as a dedicated and stand-alone goal (SDG 7) (United Nations Department of Economic and Social Affairs (UNDESA) 2019). Furthermore, SDG-7 compliance is linked to the achievement of 125 of 169 targets of other 16 SDGs (International Bank for Reconstruction and Development 2017), aimed at fuelling human development, addressing the heightened challenge of social inequality, resource starvation, climate change and compliance of NDC commitment (UNFCC 2015).

This article aims at assessing strategic governance approach adopted in implementing of specific SDGs and its compliance, identifying key challenges and propose enabling framework for transition from incremental progress to transformational acceleration in adoption of 2030 agenda and its objective. In addition to the introduction, the article has six more sections. Sections 7.2, 7.3 and 7.4 assess the approach in which Goals 7.1, 7.2 and 7.3 are being institutionalised. Section 7.5 focuses on projected energy scenario and pathways. Section 7.6 focuses on the key strategies to be adopted for implementing of SDG agenda.

7.2 Access to Energy Services

7.2.1 Electricity Access

From nearly 240 million population, without access to electricity (IEA 2015) in 2014, to 18,734 unelectrified households, GoI has made an enormous stride in ensuring last mile electrification. 'Pradhan Mantri Sahaj Bijli Har Ghar Yojana—Saubhagya' has been instrumental in ensuring electricity access to 26.2 million unelectrified rural households (Pradhan Mantri Sahaj Bijli Har Ghar Yojana—'Saubhagya' 2019), in addition to 0.85 million marginalised urban households (BPL families) (Lok Sabha Question No 3721 2019) (Table 7.1).

The strategic shift, from the approach of village electrification (village deemed to be electrified with 10% households being electrified) to total household electrification under 'Saubhagya' by means of extending free electricity connec-

S. Majumder (✉)
Technical Expert, UNDP, India

S. Hazra, A. Bhukta (eds.), *Sustainable Development Goals*, Sustainable Development Goals Series,
https://doi.org/10.1007/978-3-030-42488-6_7

Table 7.1 Percentage of HH electrification

SDG baseline (October 2018)	Target by 2030	Achievement (March 2019)
94.57%	100%	99.99%

Source: Saubhagya; SDG India Index, NITI Aayog

tions to unelectrified rural households and urban BPL families (REC India 2019), colossal expansion of the distribution network, off-grid DRE (HLS) or micro-grid applications, has resulted in the transformational achievement. The massive leap in electrification has in turn reduced the consumption of highly subsidised PDS kerosene, use of diesel in diesel-micro-grid or stand-alone genset (Abhishek Jain, November 2018), thereby addressing the concern over energy security, oil import dependency and energy subsidy budget. **In line with the SDG target (100% electrified population), compliance to SDG 7.1.1. can already be deemed to be achieved.**

Paradoxically, the enhanced access is yet to address the pervasive energy poverty, with India significantly lagging behind global average in terms of energy consumption (Ministry of Finance 2019) and the disparity in the per capita electricity consumption across the states. Disparity in electricity usage also exists across urban and rural areas, including uneven spatial dispersion across different income groups (Ministry of Finance 2019). Specific to rural areas, electricity access is yet to result in social upliftment and livelihood sustainability (Rockefeller Foundation, Smart Power India, ISEP 2019) and thus far from realising the SDG objective of fulfilling the 'essential needs of the world's poor (Wikisource n.d.)', inequality and discriminations. Jeopardised with the persistent challenges of reliability, affordability and unintentional prioritisation of households over enterprise has forced rural enterprises to stack upon alternate source for electricity (Rockefeller Foundation, Smart Power India, ISEP 2019), impacting their growth possibilities and sustainability.

> India needs to increase its per capita energy consumption by four times to reach the HDI level of 0.8.

7.2.1.1 Reliability of Supply

The roll out of the flagship programme '24 × 7 Affordable and Quality Power for All' (PIB, GoI, MoP 2017) is yet to address the reeling challenges of unscheduled power cuts and poor-quality supply in many of the rural and peri-urban areas coupled with irrational consumer burden of high electricity tariff, impacting industrialisation and societal upliftment. Studies (Abhishek Jain, November 2018) (Rockefeller Foundation, Smart Power India, ISEP 2019) across rural areas of Bihar, Jharkhand, Madhya Pradesh, Uttar Pradesh, Odisha and West Bengal reflected the criticality of power supply. MoP also substantiates the aforesaid claim, declaring 13 states with supply below 20 h/day (National Power Portal 2019) (Fig. 7.1).

7.2.1.2 Affordability of Supply

Affordability of electricity services is a major bottleneck for the economically disadvantaged in post-adoption consistent usage. While subsidised connectivity (under Saubhagya) has ensured accessibility and adoptions, its consistent use has still been a constraint. With 4% of the economically disadvantaged household's expenditure on electricity considered as affordable (Abhishek Jain, November 2018), households subjected to unmetered fixed billing and non-periodic billing cycle face the challenge of inflated electricity expenditure (Rockefeller Foundation, Smart Power India, ISEP 2019) and envisage electricity services as unviable commodity (Fig. 7.2).

7.2.1.3 Equitability of Access

Electricity access to 0.5 million household in remotest areas under Saubhagya and previously under remote village electrification programme were provisioned through solar-powered home lighting system of 200–300 kWp (Palit, April 2018). These are in addition to households electrified through DRE grid (more than 14,000

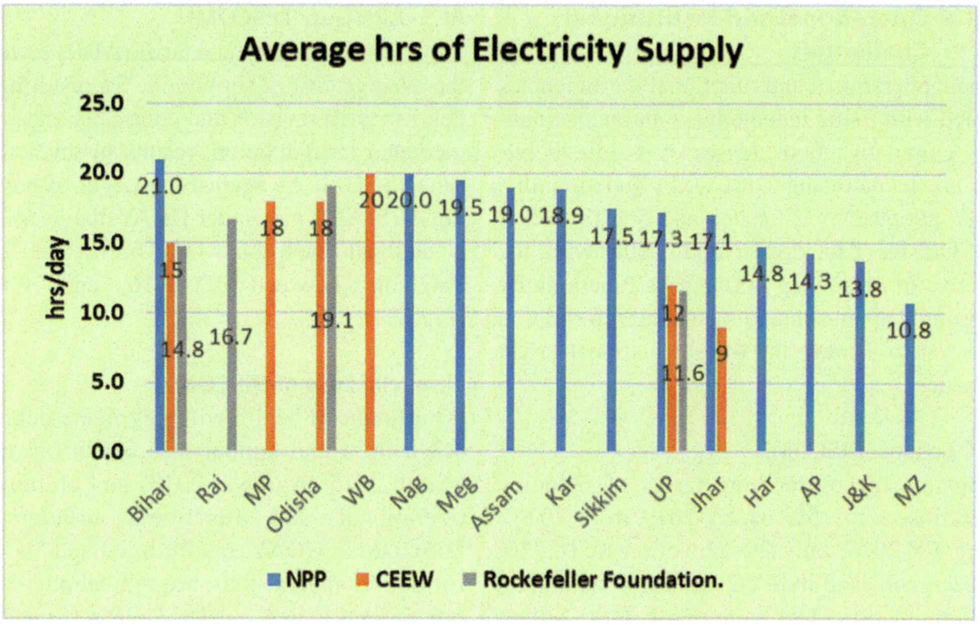

Fig. 7.1 Average hours of electricity supply. (Source: NPP, CEEW and Rockefeller Foundation)

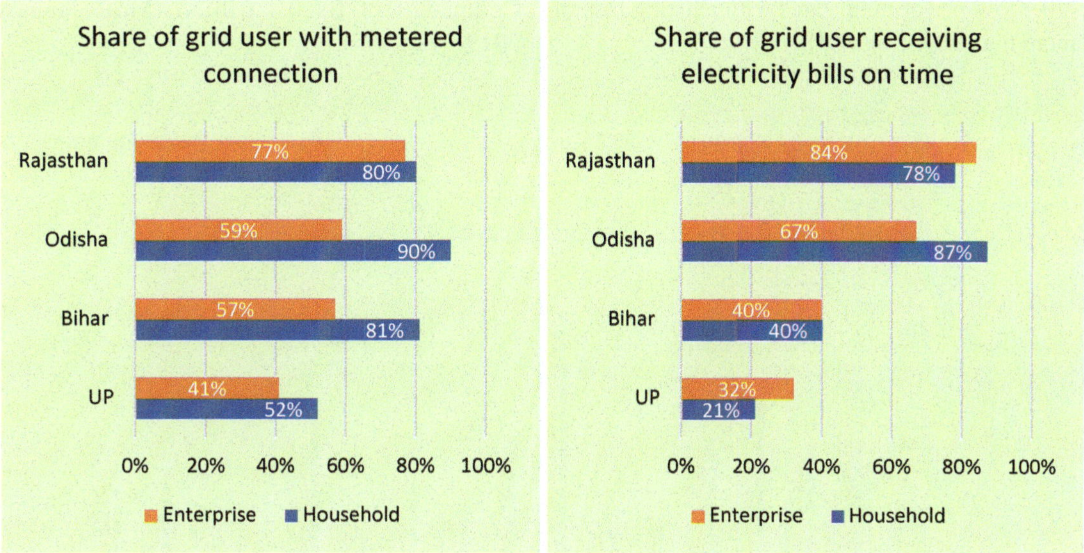

Fig. 7.2 State-wise share of unmetered connection and timely billing. (Source: Rockefeller Foundation, Smart Power India, ISEP. (2019). Rural Electrification in India)

micro-/mini-grid (Rajarshi 2018)) and over two million stand-alone system (Rockefeller Foundation, Smart Power India, ISEP 2019). Although the stand-alone system has ensured the access but has eclipsed the need to provide quality access, equitable supply and required services and critical access for rural micro-enterprises restrict rural economic empowerment (Rockefeller Foundation, Smart Power India, ISEP 2019).

7.2.1.4 Operational and Institutional Challenges

Chronic operational and functional inefficiencies coupled with rising indebtedness and poor financials (cumulative book losses of ₹ 150.49 billion) are jeopardising DISCOM's functionality. UDAY adopted by 27 states and 5 UTs except WB, Odisha, Chandigarh and Delhi with the objective of improving DSICOM's functionality has brought about some fundamental reformation but is yet to achieve the transformational target planned.

AT&C Losses (DISCOM)

As against the overarching target of reducing AT&C losses to 15% by FY-2019 from 20.8% during FY-2016, only 2% reduction to 18.72% has been achieved by FY-2018 at the nationally resulting in estimated savings of ₹ 90 billion (MoP, January 2019). The national AT&C loss trajectory is a mix of high-performing states achieving the loss reduction target of 15% and other low-performing states with existing loss of more than 20% (Fig. 7.3).

ACS–ARR Gap (DISCOM)

Supplying electricity at a tariff (ARR) lower than the average cost of supply (ACS), resulting from delay in tariff revision and politically and socially mediated tariff fixation, results in financial loss for DISCOM. As against the target of neutralising ACS-ARR gap under UDAY, the average gap at national level could only be reduced by Rs. 0.42/unit between FY-2016 and FY-2018 (Fig. 7.4).

Financial Loss of DISCOMs

Poor financial health with aggregate debt of Rs 4.3 trillion and annual loss of Rs 0.6 trillion March 2015 (Agarwal 2018) had cluttered the operational and functional efficiency of DISCOMs. UDAY institutionalised as transformative scheme has brought about sectoral reform and bailed out DISCOMs from its loss regime with book loss reduced from Rs 0.51 trillion in FY-2016 to Rs 0.15 trillion by FY-2018 and total outstanding to Generation Companies of Rs 0.4 trillion (MoP, January 2019) (Fig. 7.5).

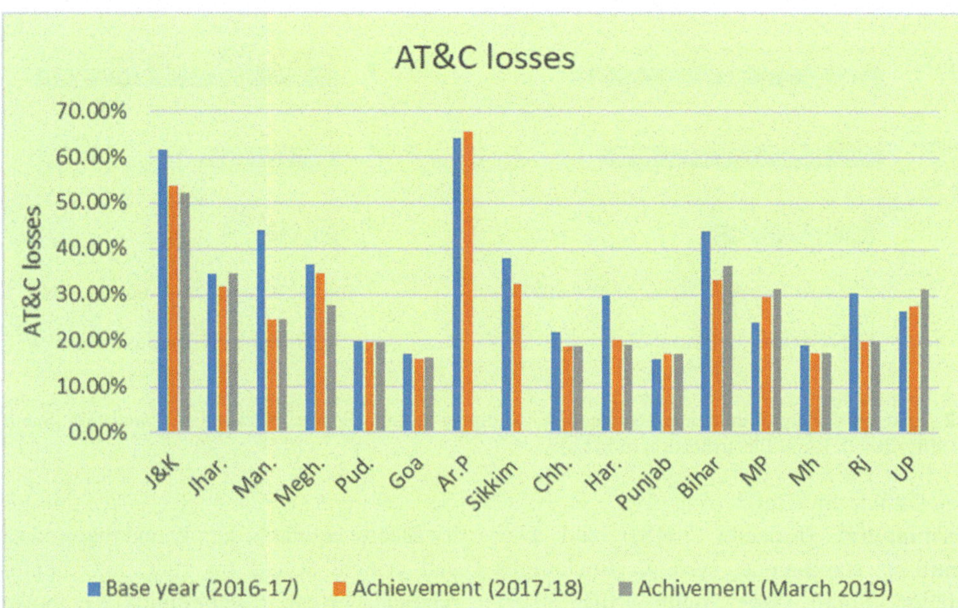

Fig. 7.3 AT&C loss trajectory. (Source: UDAY portal)

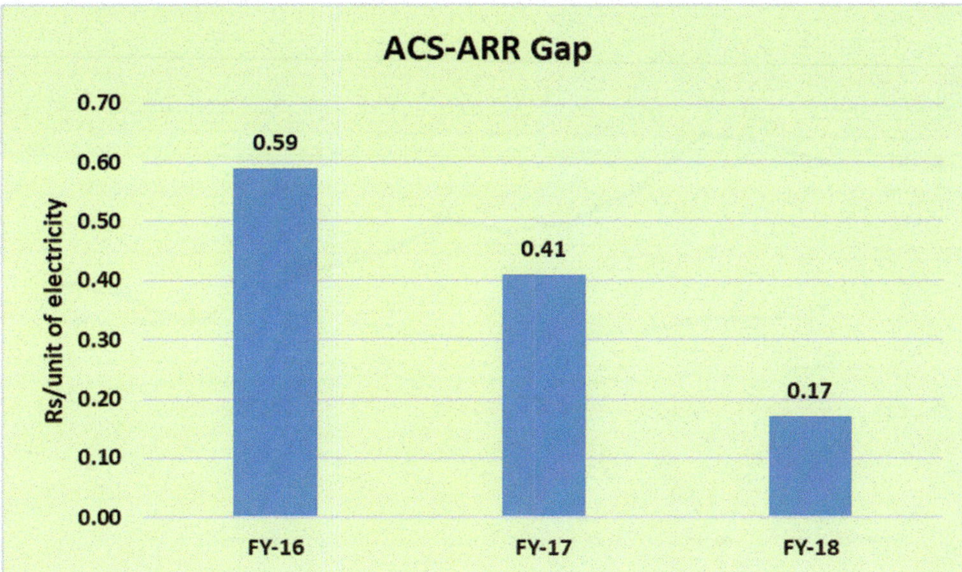

Fig. 7.4 ACS-ARR gap. (Source: UDAY newsletter)

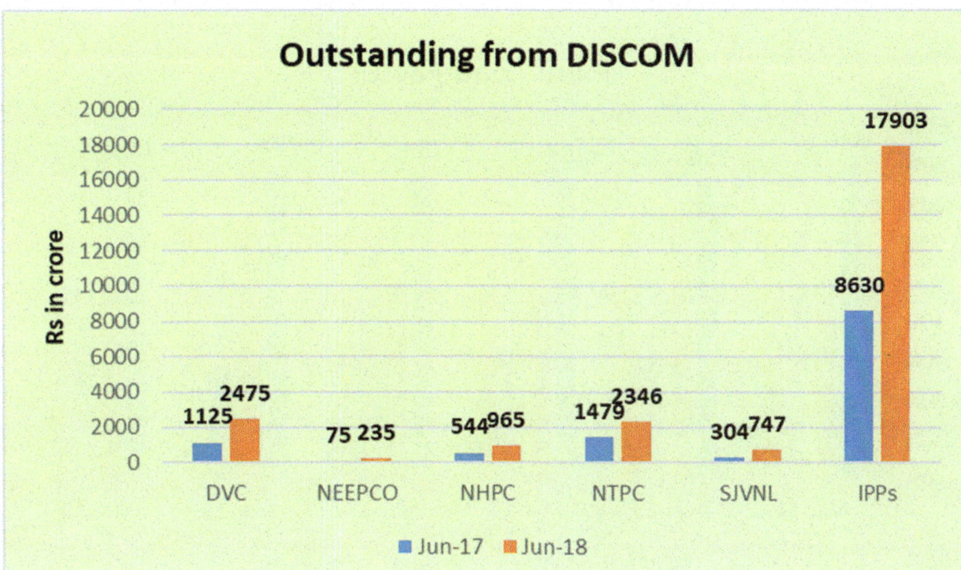

Fig. 7.5 Outstanding from DISCOM [20]

Energy Deficit and Deteriorating Operational Efficiency of Generating Stations

In spite of being a power surplus country and substantial reduction in demand supply gap from 3095 MU (3.6%) during March 2014, gap still persists in range of 884 MU (0.4%) during March 2019 (Fig. 7.6).

Stressed Generation Sector

The predominant coal-based generating units are stressed by multiple constraints like (a) coal sup-

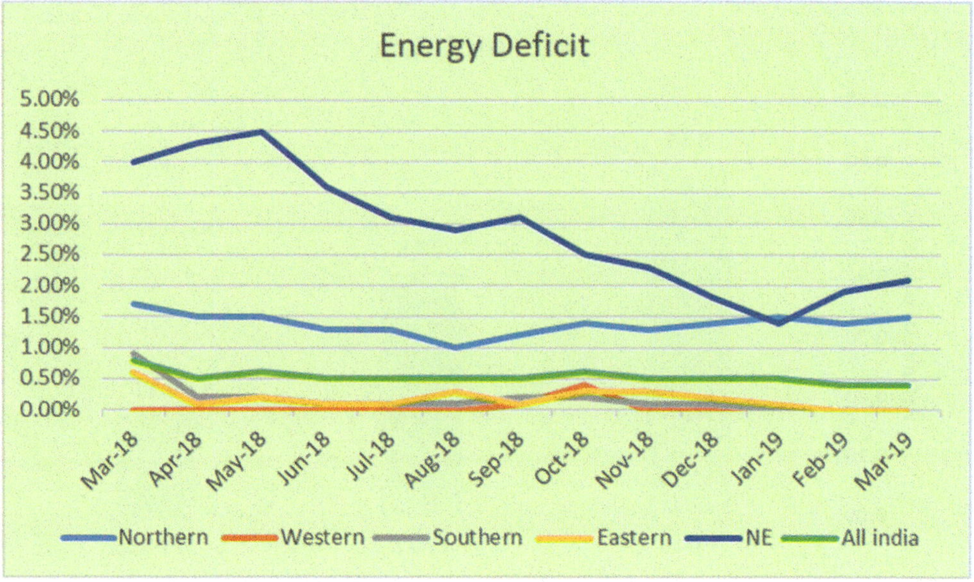

Fig. 7.6 Energy deficit. (Source: CEA)

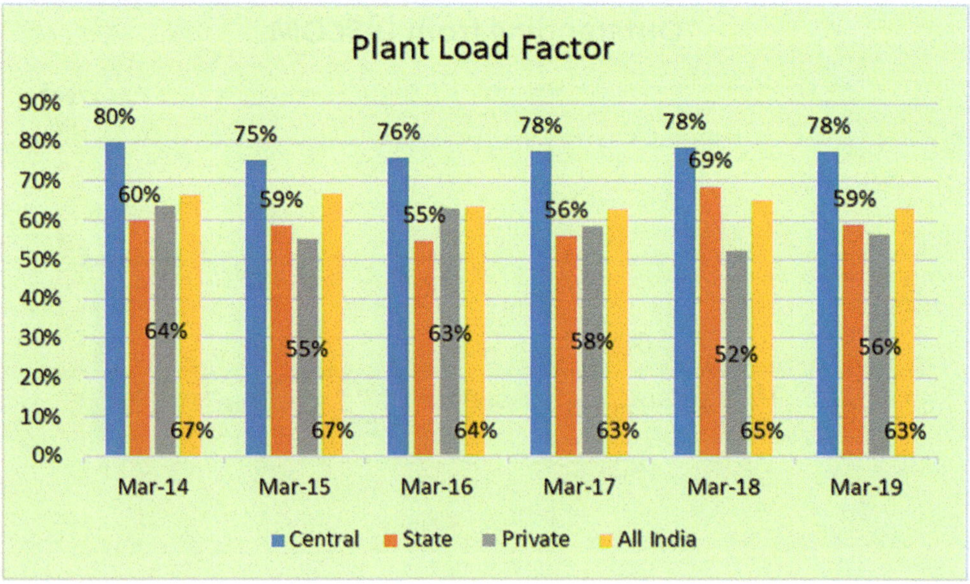

Fig. 7.7 Plant load factor. (Source: CEA)

ply and legal issues with auctioned coal mines, (b) delay/inability of DISCOM in payment, (c) dishonoured PPA and low offtake, (d) regulatory and contractual disputes leading to poor operational efficiency and (e) increased evacuation of unscheduled variable renewable energy to the grid (Fig. 7.7).

7.2.2 Clean Fuels and Technologies

Dependence on traditional fuel like coal, charcoal, wood and animal dung and its incomplete combustion in inefficient and highly polluting stoves for cooking is the most prevalent practice amongst the lower income groups (UN India n.d.;

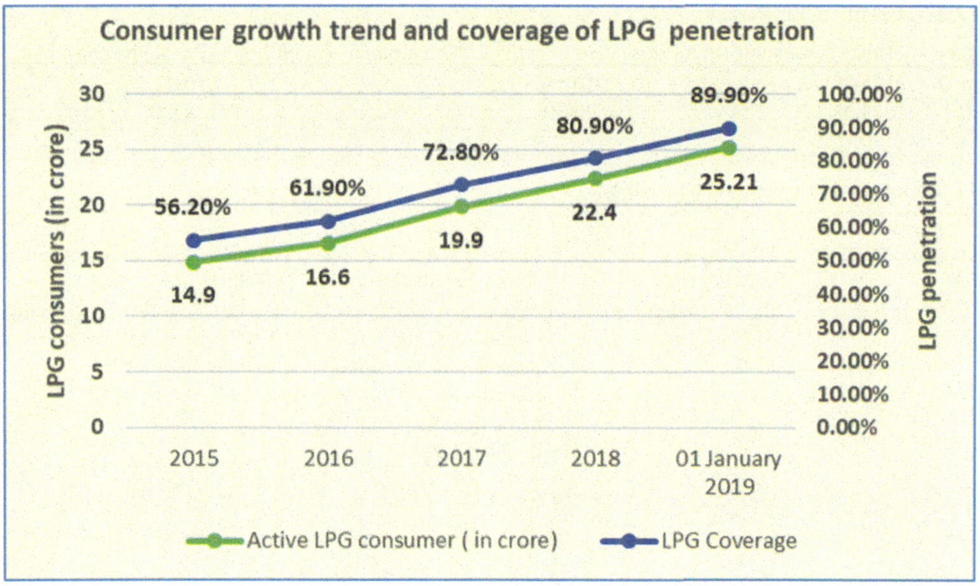

Fig. 7.8 LPG penetration and growth trajectory

World Health Organisation 2018). Such practice is a critical contributor to HAP and drivers of global environmental health risk. HAP in India alone has been the major cause of chronic cardiovascular and respiratory diseases and acute lung and respiratory infections in children resulting in over one million deaths per year (World Health Organisation 2018; Balakrishnan and Sambandam 2004) and has been regarded amongst one of the ten most important risk factors for mortality and morbidity (The Institute for Health Metrics and Evaluation (IHME) 2019). Use of wood specially entails drudgery of women and children spending about 5–8 h or more in a week and carrying head loads of 20–30 kg over a long distance, resulting in reduced available time for education and alternate economic activities (UN India n.d.). Access to clean cooking fuel has, therefore, been considered imperative for achieving the human development priorities. Through introduction and institutionalisation of several promotional schemes by the Union Government, the LPG coverage has increased to 89.9% household[1] by 2019 (Petroleum Planning & Analysis Cell 2019) as compared to 56.2% during 2015 (Fig. 7.8).

The schemes such as (1) PAHAL, (2) PMUY (59.41 million BPL consumers covered till January 2019) followed by Ujjwala Plus Scheme, (3) 'Give It Up' or 'Giveback' and (4) SAHAJ have acted as a transformative drivers in mainstreaming adoption of LPG across the country. Strengthening of the supply chain network through increased distribution network and introduction of extended service network like 'Rurban Vitrak', 'Gramin Vitrak' and 'Durgam Kshetriya Vitrak' (Petroleum Planning and Analysis Cell 2018) has further ensured the reliability of supply (Table 7.2).

Ironically, the aforesaid penetration does not resemble the true picture of accessibility in terms of (1) equitable distribution, (2) equitable adoption amongst all economic class and (3) sustained usage?

The penetration of LPG across the landlocked north-eastern states of Meghalaya (44.3%), Nagaland (60.6%) and Tripura (69.4%), eastern states of Bihar (68.2%), Jharkhand (66.6%), Odisha (67.9%), and western states of Chhattisgarh (72.1%), Madhya Pradesh (74.7%) and Gujarat (67.2%) is comparatively lower than the national average.

[1]As on 1 January 2019. Disclaimer: The content of this paper does not reflect the views or policy of UNDP

Table 7.2 Disparity in LPG usage

Category	No of households (in million)	No of active LPG consumers (in million)	Per capita cons. in kg/year
Urban	85.28	101.04	10.07
Rural	183.28	65.21	8.05

Table 7.3 Household expenditure on cooking

% of Population obtaining cooking fuel for free	Monthly exp. on cooking fuel	Monthly spending capacity across diff income group	Cost of LPG/14.2 kg cylinder (Delhi)
35% (37% rural and 25% urban)	Rs 358 (Rural—Rs 354 and Urban—Rs 372)	Less than Rs 2250/ month—Rs 334	Rs 419–Rs 432/14.2 kg cylinder in 2016 increased to Rs 495/14.2 kg cylinder in 2019
		Between Rs 2250–5000/ month—Rs 356	
		More than Rs 5000/ month—Rs 396	

Moreover, disparity also exists in terms of penetration and usage of LPG across the rural and urban areas in each state (Abhishek Jain, November 2018) (Lok Sabha 2016). Accelerated by subsidised upfront distribution under PMUY, the adoption rate amongst the economically disadvantaged class has been remarkable. However, ensuring its sustained usage has emerged as a challenge with only 80% of the beneficiaries opting for second refill in the first year itself (Lok Sabha 2018).

Mindset of prolonging LPG usage, possibly due to 'stacking' phenomenon, restricting instant transition from one fuel to another (Smith and Sagar 2014), affordability and availability of cheaper fuel are identified as the major drawbacks towards sustained LPG usage (Pandey 2018).

Affordability Increasing household expenditure and availability of free fuel led reluctance amongst the rural and BPL consumers forbid sustained LPG usage (CRISIL, June 2016) (IOCL, Product_Previous Price 2016; IOCL, Products 2019). Expenditure on LPG for a BPL family estimates to 9.5% of the monthly income in contrast to 6% of the monthly budget expenditure on cooking fuel as per NSSO (Kundan Pandey 2019) (Table 7.3).

7.3 Share of Renewable Energy in Energy Mix

7.3.1 Renewable Energy Share

Electricity production alone from conventional source accounts for 42% of the entire GHG emission (in 2014) (MoEFCC 2018). Reducing the emission footprint and sustainability of energy sector, GoI has been proactively pursuing the imperative of enhancing the share of renewable. As against the installed generation capacity of 360.78 GW (CEA, April 2019b), renewable installed capacity accounts for 22.34% (80.63 GW without large-scale hydro) (MNRE, April 2019) and 34.93% (126.03 GW with large-scale hydro) (PIB, March 2019) (Figs. 7.9 and 7.10).

Of the renewable capacity, addition wind and solar-based units account for the key share followed by biomass and small hydro. Buoyed by strong policy, regulatory, financial and fiscal initiatives and decline in the cost of solar PV and wind-based generation capacity has significantly increased during last 5 years (Table 7.4).

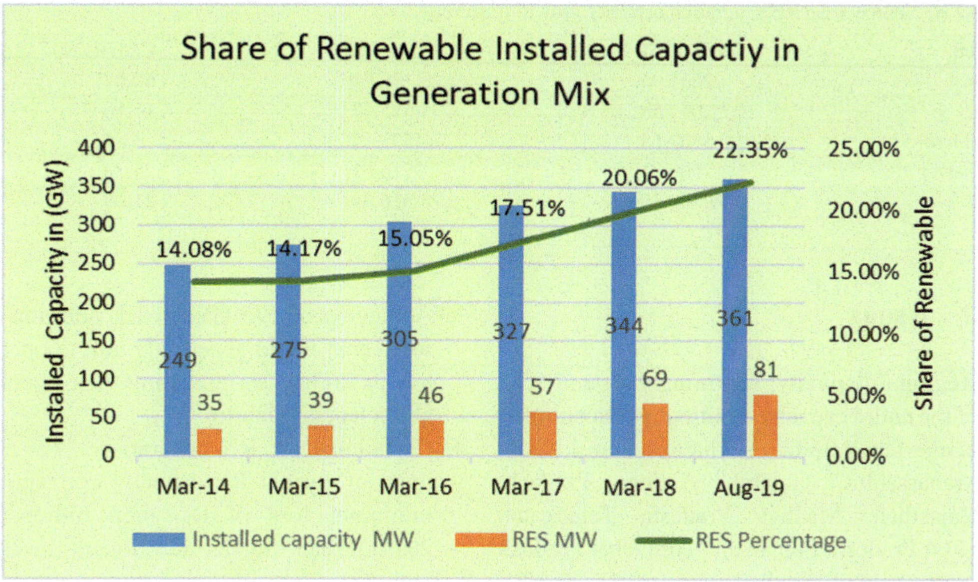

Fig. 7.9 Share of renewable. (Source: CEA)

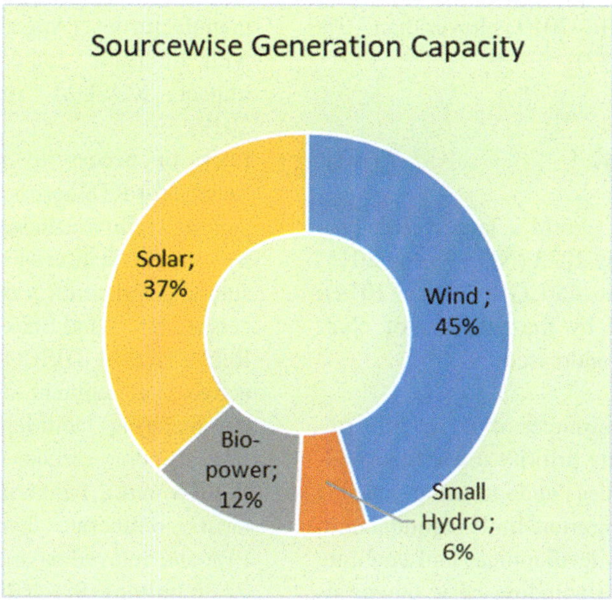

Fig. 7.10 Source-wise Renewable Energy Generation mix. (Source: MNRE)

Although the global indicator for Goal 7.2 is set towards 'Renewable energy share in the total final energy consumption', the monitoring indicator proposed by NITI Aayog in its indexing report pertains to share in generation mix.

Increase in renewable energy capacity has resulted in addressing energy security, reduce fuel import and reduction in environmental degradation in addition to the GHG avoidance (around 335.5 $MtCO_2$ between 2014 and 2018) (MoEFCC 2018).

Table 7.4 Source-wise RE generation capacity growth trajectory

Source	March 2014	March 2019	CAGR (%)
Solar	2631.93	28,180.71	60.67
Wind	21,042.58	35,625.97	11.10
Small hydro	3803.68	4593.15	3.84
Bio-power	7509.81	9916.61	5.72
Total	**29,462.55**	**78,316.44**	**21.60**

7.3.2 Issues

(a) **Regionaldisparity**: Alarmingly around 95% of the entire capacity addition is concentrated across few state like Karnataka, Tamil Nadu, Maharashtra, Gujarat, Andhra Pradesh, Rajasthan, Madhya Pradesh, Telangana, Uttar Pradesh, Punjab and Himachal Pradesh (with share of installed capacity of over 1%).

(b) **Lower renewable energy consumption**: With over 20% share energy mix, the average contribution of energy generated from renewable during FY2018–2019 is lower than 10% (Figs. 7.11 and 7.12).

7.3.3 Challenges

Operationalisation of world's largest RE programme of 175 GW by 2022 (NITI Aayog 2015) (presently escalated to 450 GW (Singh 2019)) could be jeopardised by the plethora of challenges addition if not addressed.

1. Limited financial resources of the federal government with strong priority towards providing basic amenities and meeting of the developmental aspiration limit capital intensive infrastructure development and enabling market ecosystem (including basic amenities like land and evacuation facilities) required for harnessing of renewable energy potential.
2. Improper forecasting and scheduling facilities coupled with the variability of RE generation and absence of adequate flexible peaking units (gas based, pump storage, etc.) and storage facilities impacts the stability of the grid.
3. Achieving of 175 GW of RE addition might force coal-based generating unit to operate as low as 26% PLF, impacting their operational and financial viability.
4. Curtailment of RE generation[2] (1% curtailment of current RE capacity corresponds to minimum loss of Rs. 6.58 billion (CEA 2019a)) can impact developers investment interest.
5. Changes in regulatory norms towards safeguard duties have driven cost and time although marginally boosting the domestic manufacturing (Aggarwal and Dutt 2018), with domestic content requirement might impact KUSUM implementation (Bajaj 2019).
6. Lack of access to preferential consumer finance for RTS across residential, C&I sector lacking creditworthiness due to perceived risk by financial institutions and variance in understanding and comfort of lenders with RE projects (CRISIL and Shakti n.d.).
7. Reluctance of DISCOMs for RTS adoption across C&I segment, delayed mobilisation of net metering facilities (Aggarwal and Dutt 2018), poor offtake of renewable power (CEEW n.d.), hankering for lowest possible tariffs, deliberate disregard for committed PPA and delayed payment (payment due of Rs 97.36 billion) (Singh 2019).
8. Disorganised and underdeveloped supply chain operators for O&M of RTS and DRE units.

[2]Curtailment relates to situation where power grid operators instruct to limit the power output of specific RE generators.

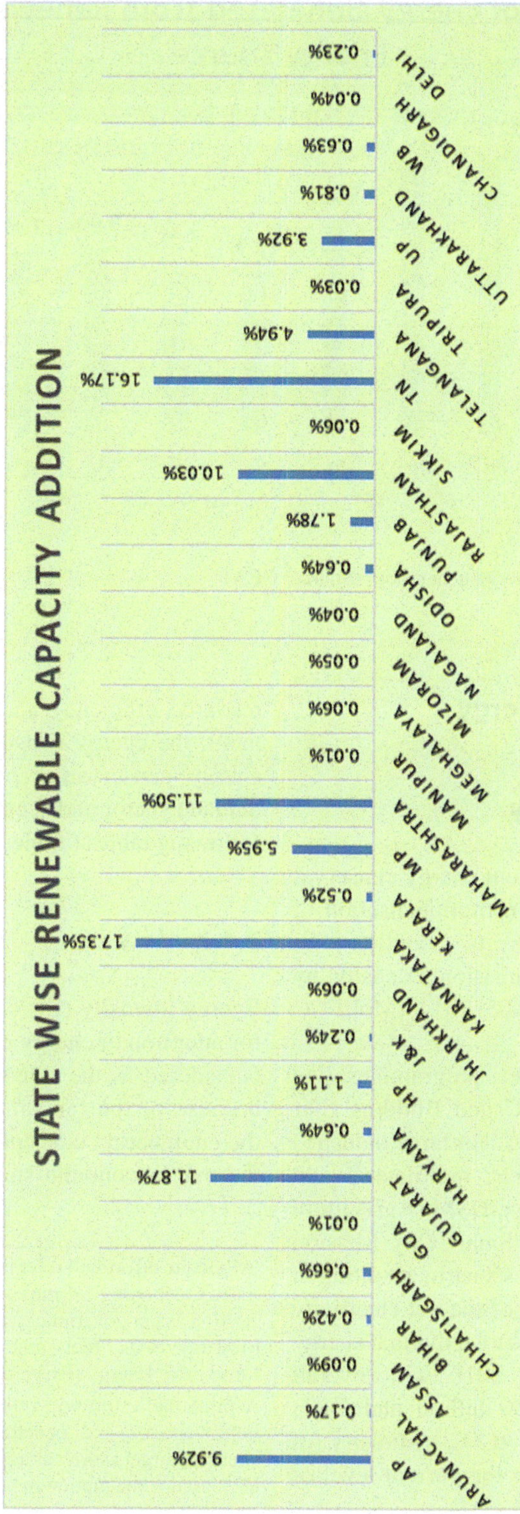

Fig. 7.11 State-wise renewable capacity addition. (Source: MNRE, August (2019))

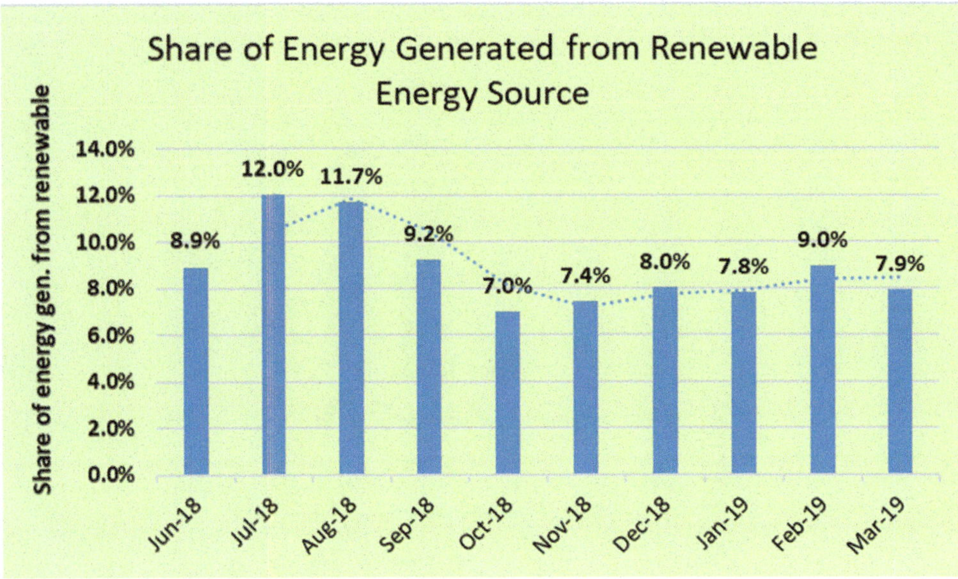

Fig. 7.12 Share of renewable energy generation. (Source: CEA)

7.4 Adoption of Energy Efficiency

7.4.1 Energy Intensity

Institutionalisation of stringent energy conservation measures fostered by institutionalisation of enabling environment for its adoption has resulted in reduction of emission intensity by 21% over the period of 2005–2014 (MoEFCC 2018) (Fig. 7.13).

Institutionalisation of key programmes like UJALA, SLN, S&L, ECBC and Building Star rating programme, CCT, PAT (for large industry) and SME programme (BEE, multilateral and bilateral programme, MDBs), DSM (Agriculture and Municipal DSM) and Smart City Mission has reformed the energy use scenario, resulting in electricity saving of 7.21% of the net electricity consumption and 2.7% of the total thermal energy consumption during 2017–2018 with overall monetary saving of Rs 530 billion and GHG avoidance of 108 million tCO_{2e} (Ministry of Finance 2019). Considering the current level of resource efficiency and technological advancement, the energy consumption (BEE 2019) and energy conservation potential across different demand sector and scenario[3] are presented in the following tables (Tables 7.5 and 7.6).

7.4.2 Issues

Energy intensity (EI) conceived as the indicator for adoption of energy efficiency cannot alone be considered as the sole indicator (IEA 2018a, b) because of the fundamental difference between the engineering concept of energy efficiency and the macroeconomic statistic of energy intensity

[3]Moderate effort will refer to Moderate/Business as Usual (BAU) technological improvements and technology penetration as per government/other agencies target as well as Moderate/BAU fuel mix shift from fossil fuel to Renewable Energy (RE)/electricity-based consumption.

Aspirational effort will refer to aggressive technological improvements and penetration over government/other agencies target as well as aggressive fuel mix shift towards RE-based consumption in sector.

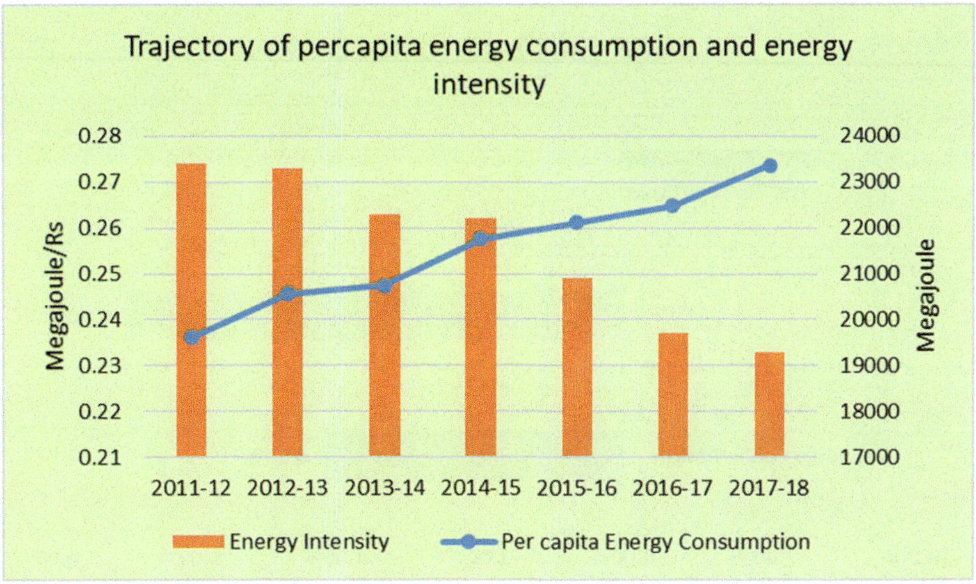

Fig. 7.13 Trajectory of per capita energy consumption and energy intensity

Table 7.5 Sectoral energy consumption

Mtoe (2016–2017)	Domestic	Commercial	Industries	Municipal	Transport	Agriculture	Others	Total
Primary energy excluding electricity	26.8	0	303.8	0	44.1	0.9	73.6	449.1
Electricity	22.3	8.5	36.7	2.6	1.5	16.8	3.3	91.7
Total energy	**49.1**	**8.5**	**340.4**	**2.6**	**45.6**	**17.7**	**76.9**	**540.8**

Table 7.6 Sector-wise and scenario-wise energy saving potential

	Energy consumption 2031 (least effort)	Moderate savings—2031		Aspirational savings—2031	
Sector	Mtoe	Mtoe	%	Mtoe	%
Agriculture	64.4	5.7	9	9.9	15
Transport	232.9	15.8	7	23.8	10
Domestic	98.6	12.1	12	15.1	15
Commercial	29.5	4.9	17	6.4	22
Municipal	8	0.9	12	1.5	19
Industries	443.4	47.5	11	72.3	16
Total (mtoe)	**876.8**	**86.9**	**10**	**129**	**15**

(Proskuryakova and Kovalev 2015). Changes in EI can also be attributed to movement in economic activity from energy intensive heavy industries to less energy intensive service sectors, variation in the trajectory of growth in energy consumption, GDP and others (Fig. 7.14).

7.4.3 Challenges in Adoption of EE Measures

(a) **Industrial and utility sector**: While PAT being institutionalised towards accelerating adoption of EE measures across large-scale

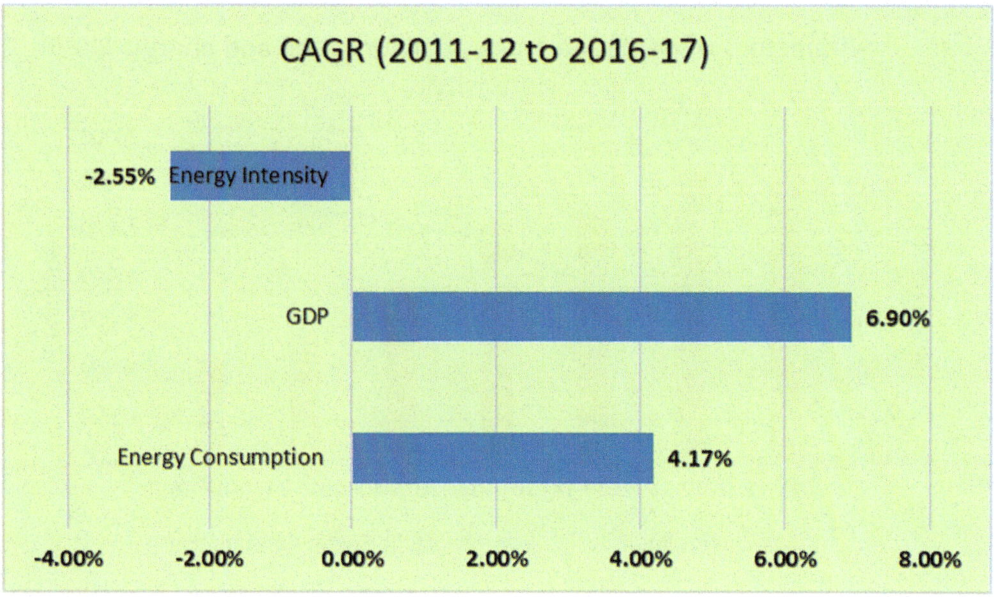

Fig. 7.14 Energy intensity. (Source: [45])

industrial units and utilities, lack of regulatory or policy instruments forbids mainstreaming of EE adoption across the energy guzzling and resource inefficient MSME sector (Industrial and MSME sector has energy saving potential of 47 MToe (BEE 2019)). EE adoptions in MSMEs are, in general, restricted by (1) lack of sensitisation and awareness regarding EE measures and benefits, (2) lack of performance benchmarking info (Biswas et al. 2018), (3) poor vendor and service provider network, (4) unwillingness to revamp existing industrial operation using age-old technologies and practices, (5) unskilled workforce in most cases, (6) poor financial health and lack of business vision and (7) poor access to finance.

(b) **Building sector (residential and commercial)**: Even with energy saving potential in the range of 20–40% (Vasudevan et al. n.d.), adoption of EE measures across building sector has been poor. Lack of regulatory enforcement mechanism (PAT included large commercial buildings only with energy consumption of 1000 MTOE) for EE adoptions, poor institutionalisation of ECBC, lack of energy performance benchmarking and poor adoption of star-rating programme are the major bottlenecks. Although 'UJALA' has mainstreamed adoption of energy efficient lights and fans across domestic sector, poor enforcement of S&L programme and continuation of manufacturing of inefficient products have restricted the market of EE products and services.

(c) **DSM sector**: Poor adoption of DSM measures aimed at reduction and phasing out of energy demand owes to (1) institutional disinterest, capacity and financial strength of the DISCOMs and (2) lack of enforcing regulatory mechanism on accelerating DSM implementation and cost recovery mechanism (Shakti 2013).

(d) **Access to finance**: (1) Lack of non-recourse finance as EE initiatives/products is not considered as stand-alone financing product by Fis, (2) small ticket EE proposals often result in high transaction cost, (3) unpopularised ESCO-based business model and (4) absence of dedicate product line (Table 7.7).

Table 7.7 Investment requirement for sectoral energy efficiency improvement (BEE 2019)

Sector	Investment for moderate savings—2031 (Rs in billion)	Investment for aspirational savings—2031 (Rs in billion)
Agriculture	914.67	1582.29
Transport	811.54	1057.01
Domestic	1202.33	1451.33
Commercial	148.22	241.59
Municipal	3071.37	4637.29
Industries	2260.39	3657.06
Total	**8408.52**	**12,626.56**

7.5 Projected Energy Scenario and Pathways

Electricity demand in India is likely to grow at a CAGR of 6.18% till 2021–2022 from 2016 to 2017 level and at a CAGR of 5.51% between 2021 and 2026 (CEA 2018a, b). In addition, the current per capita energy consumption of 24 Gigajoules in 2017 (Ministry of Finance 2019) needs to be quadrupled to achieve the HDI of 0.8 and enter the group of countries with high human development without invoking multidimensional challenges of energy security, energy equity, environmental sustainability and incremental concerns of climate change. The pathways for achieving energy sustainability are as follows:

(a) Balancing the growth in energy consumption by coupling with clean energy transition (increasing share of renewable in generation mix and reducing energy intensity) to address the concern of energy security and environmental degradation.
(b) Integrated solution towards decarbonisation of the energy consumption and transition in energy ladder from higher to lower specific energy consumption and emission intensity.
(c) Fostering the reliability and equality of energy access at all scale and reach.
(d) Coherent and innovative approach towards regulating energy cost and thereby addressing the heightened challenge of energy equity, energy poverty and deprivation of energy services.
(e) Mainstreaming and localisation of SDGs within centralised and decentralised planning.

7.6 Targeted Strategies Towards SDG Compliance

Considering the existing scenarios and challenges in translating the commitments of 2030 SDG Goal-7 agenda, key catalytic and accountable strategies are proposed below:

Area	Target strategies and recommendation accelerating SDG adoptions
Policy, Foal 7.a.1, 7.b.1	(a) Mainstreaming of SDG agenda under the ambit of decentralised planning and programme formulation
	(b) Integrating customised SDG target within the outcome budget of the federal and central government
	(c) Strong political will towards transit from silo type to converged approach in addressing the interlinked challenges of energy, addressed by different department through different scheme leading to duplication of effort and resources and ensuring ownerships of the targets
	(d) Periodic analysis of gaps and accelerators for adoption of SDG agenda and develop policy insights using customised tools (UNDP)
	(e) Despite of synergies between Goal 7 targets of SDG and quantifiable targets of NDCs, both have been approached separately, other than the recent initiatives of MoEFCC in linking SDG and NDC under the SAPCC (State Action Plan of Climate Change) umbrella. Successful implementation of both the agendas will require for more coordinated effort in terms of amalgamating and mainstreaming the SDG and NDC target into policy planning along with coherent joint responsibility sharing amongst the implementing agencies and integrated fiscal and regulatory reforms for both agendas [60]
	(f) With an estimated budgetary requirement of USD 854 billion between 2015 and 2030 and viability gap of USD 406 billion [61], strong policy advocacy is required in tapping of international financial assistance on a predictable basis (including enhanced allocation under GEF, funding allowance under GCF, bilateral and multilateral funding assistance and market-based mechanism)
	(g) Develop an integrated platform for shift from traditional closed-ended to open-ended governance model for SDG mainstreaming by embedding collectivised approach of federal and sub-national governments and development communities
Monitoring	(a) The existing indicator set out against Goal-7 should be further broadened and incremented to achieve the objective of SDGs and address the challenges of inequality and discrimination 7.1.1: Should include the factor of reliability, equitable access and affordability over and above the access factor Indicator 7.1.2: Should also factor in the sustained usage Indicator 7.2.1: % of RE energy consumption should be the indicator Indicator 7.3.1: Should normalise the factor of GDP and bank upon the actual sectoral savings as a percentage of saving potential, and mapping energy efficiency at physical, technological and other sectoral and demographic level without reference to economic and financial parameters [44]
	(b) Institutionalising SDG review mechanism at the decentralised governance level through aggregation of people-centred and gender-sensitive data and community feedback and establish of both top-down and bottom-up reporting approach. Strengthening of national statistical systems through capacity building, developing indicators in new areas of measurement (UNDP)

Area	Target strategies and recommendation accelerating SDG adoptions
Institutional	(a) Encouraging SDG localisation (sub-national context) to address the equality-focused SDG progress and strengthen governance institutions towards reducing of inequality and exclusions (IPR 2019)
	(b) Fostering localised awareness and stakeholder's sensitisation along with advocacy for SDG implementation (Niti Aayog and United Nations)
	(c) Facilitating technology transfer including manufacturing techniques and facilities along with capacity building of the domestic manufacturing and service sector towards reducing import reliance and managing cost dynamics
	(d) Strengthening of SDA and SNA for institutionalising EE and RE measures and policies
	(e) Strengthening of supply chain
	(f) Capacity building of workforce
Goal 7.1./ Indicator 7.1.1	(a) Improving operational and functional efficiency of DISCOMS: (1) Ensuring 100% metering (consumer, feeder and DT), (2) IT enablement of the distribution sector and mainstreaming smart metering, (3) reducing AT&C losses, (4) offset ACS-ARR gap and non-biased regularity of tariff revision and (5) bailing out of DISCOMs debts
	(b) Augmenting of operational and functional efficiency of generating unit (1) ensuring reliability of coal linkage and rationalisation of coal escalation index, (2) enforcing payment rationalisation and security for government sector and IPPs and (3) reduced generation cost through adoption of energy efficient and technology upgradation initiatives, rigorous fuel quality monitoring [50] and increasing PLF between 80 and 85% [52]
	(c) Augment and strengthen transmission and distribution infrastructure to address capacity addition and grid outreach; development of dedicated energy corridor
	(d) Mandatory institutionalisation of EE and DSM measures
	(e) Enabling policy proposition for differentiation in tariff based on electricity load and electricity consumption so that consumer under lower bracket of electricity usage can be subsidised by consumer under higher load and consumption bracket (IPR 2019)
Goal 7.1./ Indicator 7.1.1	(a) Behaviour embedded in the fabrics of societies plays a significant role in adoption of modern energy options [20]. Information towards health gain of LPG usage, health cost and impact in terms of morbidity and mortality form HAP should, therefore, be communicated towards consumer behavioural change through trustworthy communication channel to gain consumers' confidence leading to accelerated uptake
	(b) Innovative pricing options for LPG such as telescopic pricing, cross-subsidies, promotions of smaller gas cylinders, pre-paid coupons and targeted subsidies could ensure sustained usage [53]
	(c) Introduction of innovative consumer financing options including micro-financing opportunities for refilling of LPG cylinder [20]
	(d) Delay in subsidy realisation under DBT often discourages LPG usages amongst the marginalised consumer and, therefore, requires advocacy towards institutionalising of effective tools for delivering of timely subsidies [20]
	(e) Success of 'Pay-as-you-go' model for LPG in sub-Saharan countries could be a viable model for economically disadvantaged class in India and could be opted for [20]
	(f) Considering the limitation of LPG affordability amongst the marginalised focus should also be on mainstreaming of efficient biomass cook stoves [54]

Area	Target strategies and recommendation accelerating SDG adoptions		
Goal 7.2	Considering the projected generation capacity and proposed RE capacity addition proposition by 2026–2027 (CEA 2018a, b), the compliance to SDG target of 40% generation mix appears to be achievable. The revised RE capacity addition target of 450 GW as proposed by GoI by 2022 will further escalate the RE share in the total generation mix by 2030 and increase the share of RE-based energy in the total energy consumed		
		2021–2022	**2026–2027**
	Installed capacity (GW)	479.41	619.06
	RE capacity (GW)	175.00	275.00
	Hydro (GW)	51.30	63.30
	RE share (without hydro)		
	% RE (generation capacity)	36.5%	44.4%
	% RE (energy generated)	19.24%	23.31%
	RE share (with hydro)		
	RE capacity (with hydro)-GW	226.3	338.3
	% RE (generation capacity)	47.20%	54.65%
	Draft CEA report (Central Electricity Authority (n.d.)) projects of 831.5 GW of installed capacity with share of renewable of 64.9% (540.33 GW) by 2030 with CO_{2e} emission of 1154 million tonnes further substantiating the plan for RE capacity compliance		
	With substantial percentage of the RE capacity addition will be through private sector participation, a series of market enabling strategies has been proposed		
	(a) Institutionalising and mandating letters of credit for RE developers to ensure timely payments by DISCOMs and stronger standard PPA for wind and solar projects enforcing stringent penalties against its dishonouring		
	(b) Enforcement of stronger and reliable forecasting regime and regulation for all interstate grid connected solar and wind power plant developers for day-ahead and week-ahead periods towards ensuring grid stability along with imposition of deviation settlement mechanism. Long- and short-term forecasting has been a challenge considering weak scientific models and unavailability of weather forecast data		
	(c) Provisioning of adequate infrastructure (like land bank, evacuation facilities, dedicated green corridors) and mainstreaming of modalities for ensuring grid stability under high RE scenario through mainstreaming the concept of flexible operation of gas-based and hydro-based power plants, pump storage units, cost-effective storage technologies, flexing operation of coal-based generation unit [43, 58] and use of Automatic Generation Control Mechanism [37]		
	(d) Addressing technical hiccups of net metering, gross metering and curtailment to boost investor confidence for grid interactive DRE applications		
	(e) Mainstreaming of policy instrument like net metering regulation, waiver of interstate transmission charges [38], innovative policy instruments like relaxing of contract demand norms for RTS (e.g. for MSME sector in Gujarat) and adjusting fixed charges of tariff against solar capacity (e.g. for MSME sector in Madhya Pradesh)		
	(f) Institutionalising of financial and fiscal instruments like preferential feed-in-tariff, generation-based incentives, VGF (viability Gap Funding), M-SIPS (Modified Special Incentive Package Schemes), foreign direct investment, capital and interest subsides and fiscal instrument like accelerated tax benefits and back-loaded tariff [59]		

<table>
<tr><th>Area</th><th colspan="3">Target strategies and recommendation accelerating SDG adoptions</th></tr>
<tr><td rowspan="16">Goal 7.3</td><td colspan="3">The primary energy intensity has already declined from 0.0004 toe in 1990 to 0.0002 toe in 2017, with series of energy savings targets been proposed till 2026–2027 with a cumulative saving potential of 337.12 BU (CEA 2018a, b)</td></tr>
<tr><td>Programme</td><td>2021–2022</td><td>2026–2027</td></tr>
<tr><td>S&L (BU)</td><td>78.49</td><td>102.11</td></tr>
<tr><td>Building (BU)</td><td>6.89</td><td>8.36</td></tr>
<tr><td>Agriculture (BU)</td><td>2.93</td><td></td></tr>
<tr><td>Industries (PAT) (BU)</td><td>98.9</td><td>136.1</td></tr>
<tr><td>Others (BU)</td><td>76.8</td><td>90.5</td></tr>
<tr><td>Total (BU)</td><td>263.92</td><td>337.12</td></tr>
<tr><td colspan="3">With substantial potential to be realised towards complying to the SDG and NDC target, a series of strategies has been proposed:</td></tr>
<tr><td colspan="3">(a) Augmenting the resource use and specific energy consumption of power utilities and industrial facilities through adoption of energy efficient and technology upgradation measures</td></tr>
<tr><td colspan="3">(b) Energy performance benchmarking and industrial sensitisation</td></tr>
<tr><td colspan="3">(c) Widening and deepening of PAT and its successful operationalisation guided by strong M&V framework</td></tr>
<tr><td colspan="3">(d) Mainstreaming and extending outreach of UJALA, S&L (including enforcement of mandatory provision for restricting manufacture of inefficient products), star rating of building (for existing building) and ECBC (including IT enable building energy management system, etc.)</td></tr>
<tr><td colspan="3">(e) Institutionalisation of Ag-DSM (by facilitating adoption of star-rated EE agriculture pumps, promotion of solar pumps, ensuring metered connection, segregation of agricultural feeder, facilitate Ag-DSM financing in ESCO mode/DSM mode and promotion of water conservation techniques) and Mu-DSM initiatives (bundling of EE initiatives across street lighting and drinking water pumping sector and facilitating its implementation through either ESCO mode or internal finance)</td></tr>
<tr><td colspan="3">(f) Reducing energy usage in transportation sector through adoption of e-mobility, setting of energy consumption norms for vehicle, enhancing reliance on public transport and incorporation of sustainable transport principles into urban design</td></tr>
<tr><td colspan="3">(g) Promotion of energy saving bonds and insurance products for mobilisation of affordable finance and reduce implementer risk</td></tr>
<tr><td>Goal 7.a and 7.b</td><td colspan="3">(a) Strengthening research and institutionalisation of IoT application, blockchain, embedding artificial intelligence in industrial process management, net zero building, etc.</td></tr>
<tr><td>Mobilising finance</td><td colspan="3">(1) Fostering concessional finance by allowing long horizon pension funds; insurance companies and sovereign funds to invest in RE and EE projects; support capital raising by RE and EE investors through issuance of tax-free bond; reduction of sovereign guarantee fee and allowance of additional fund flow to financial institutions; opening of government/intermediary guarantees for private sector investors; allowance of funding from Infrastructure Debt Fund for non-PPP-RE or EE projects; (2) sensitisation of the financial institution on risk assessment of RE and EE projects; (3) ensuring regulatory certainty and continuity [40]; (4) promotion of alternate business model like RESCO and ESCO and (5) demand aggregations and clustered approach for financial syndication</td></tr>
</table>

References

Jain, A. T. (November 2018). Access to Clean Cooking Energy and Electricity Survey of States 2018. CEEW

Agarwal H (2018) Quality power generation in India and health of DISCOMs. Quality Power IEEMA

Aggarwal M, Dutt A (2018) State of Indian renewable energy sector: drivers, risks and opportunities. CEEW

Bajaj S (2019) National Solar Federation Says Domestic Content Requirement May Hurt KUSUM Program. September 17. https://mercomindia.com/national-solar-federation-kusum-program/

Balakrishnan K, Sambandam S (2004) Exposure assessment for respirable particulates associated with household fuel use in rural districts of Andhra Pradesh, India. J Expo Anal Environ Epidemiol 14(Suppl 1):S14–S25

BEE (2019) Unlocking national energy efficiency potential

Biswas T, Sharma S, Ganesan K (2018) Factors influencing the uptake of Energy Efficiency Initiatives by Indian MSMEs. CEEW

CEA (2018a) Committee on optimal energy mix in power generation on medium and long term basis

CEA (2018b) National electricity plan

CEA (2019a) Flexible operation of thermal power plant for integration of renewable generation

CEA (2019b) Monthly installed capacity. April

CEEW (n.d.) Addressing renewable energy curtailment

Central Electricity Authority (n.d.) Draft report on optimal generation capacity mix for 2029-30

CRISIL (2016) Assessment report: primary survey on household cooking fuel usage and willingness to convert to LPG. Petroleum Planning and Analysis Cell, Ministry of Petroleum & Natural Gas. June

CRISIL, Shakti (n.d.) Enabling low-cost financing for renewable energy in India

Gaye A (2007) Access to energy and human development. UNDP

IEA (2015) India energy outlook

IEA (2018a) Energy efficiency 2018

IEA (2018b) Status of power system transformation

International Bank for Reconstruction and Development (2017) State of electricity access report. The World Bank, Washington, DC

IOCL (2016) Product_Previous Price. Indane14.2 kg subsidised Previous Price: https://www.iocl.com/Product_PreviousPrice/Indane14kgsubsidPrevPrice.aspx

IOCL (2019) Products. April 28. Indane gas: https://www.iocl.com/Products/Indanegas.aspx

IPR (2019) Annual policy review

Pandey, K. J. (2019) Down to earth. Ujjwala scheme: are cleaner cooking fuels affordable and accessible? April 24. https://www.downtoearth.org.in/coverage/energy/india-steps-on-the-gas-58502

Lok Sabha (2016) Starred Question No. 1. LPG in Rural Areas. April 25. http://164.100.47.190/loksabhaquestions/annex/8/AS1.pdf

Lok Sabha (2018) Unstarred Question No. 2657. March 12. http://164.100.47.190/loksabhaquestions/annex/14/AU2657.pdf

Lok Sabha Question No 3721 (2019) January 13

Ministry of Finance, G. o. (2019) Economic Survey 2018-19, vol I. Department of Economic Affairs, Economic Division

MNRE (2019) Physical progress (achievements). April. MNRE: https://mnre.gov.in/physical-progress-achievements

MoEFCC (2018) India Second BUR to UNFCCC. MoEFCC

MoP (2019) UDAY. UDAY Newsletter. January

National Power Portal (2019) February. Status of Rural Power Supply: https://npp.gov.in/dashBoard/rd-map-dashboard

NITI Aayog (2015) Report of the Expert Group of 175 GW Renewable Electricity by 2022

NITI Aayog (2018) SDG India index. India

NITI Aayog and United Nations (n.d.) Localising SDGs

Palit D (2018) Universal energy access and Saubhagya scheme, connecting the unconnected. Akshay Urja, April, pp 16–19

Pandey K (2018) Down to earth. May 31. Down To Earth/News: https://www.downtoearth.org.in/news/energy/pmuy-successful-in-terms-of-number-but-is-it-the-same-with-intent%2D%2D60701

Petroleum Planning and Analysis Cell (2018) Ready Reckoner, oil industry information at a glance. Ministry of Petroleum & Natural Gas

Petroleum Planning and Analysis Cell (2019) LPG PROFILE. Ministry of Petroleum & Natural Gas, India

PIB (2019) March. http://pib.nic.in/newsite/PrintRelease.aspx?relid=189300

PIB, GoI, MoP (2017) Vision of the government is '24x7 power for all'—all the states on board to achieve target by March 2019. India, December 7

Pradhan Mantri Sahaj Bijli Har Ghar Yojana—'Saubhagya'. (2019) April 30. Saubhagya—dashboard: https://saubhagya.gov.in/

Proskuryakova L, Kovalev A (2015) Measuring energy efficiency: is energy intensity a good evidence base? App Energy 138:450–459. Elsevier

Sen, R. (2018) A detailed manual on lead acid battery operation & maintenance for solar PV plants. USAID

REC India (2019) April 30. Saubhagya: https://www.recindia.nic.in/saubhagya

Rockefeller Foundation, Smart Power India, ISEP (2019) Rural electrification in India

Shakti (2013) Utility CEO Forum on DSM

Singh M (2019) PM Modi's ambition for 450 GW renewable energy comes amidst slowdown. Down to Earth. September

Smith KR, Sagar A (2014) Making the clean available: escaping India's Chulha trap. Energy Policy 75:410–414. Elsevier

The Institute for Health Metrics and Evaluation (IHME) (2019) Health data for India. April 27. http://www.healthdata.org/india

UDAY (May 2019) UDAY dashboard. https://www.uday.gov.in/home.php
UN India (n.d.) Towards accessibility, availability, affordability and accountability, sustainable energy for all in India. UN India, India
UNFCC (2015) NDC Registry. India INDC to UNFCCC: https://www4.unfccc.int/sites/ndcstaging/PublishedDocuments/India%20First/INDIA%20INDC%20TO%20UNFCCC.pdf. Accessed 24 Apr 2019
United Nations Department of Economic and Social Affairs (UNDESA) (2019) Energy: sustainable development goals knowledge platform. April 24. Sustainable Development Goals Knowledge Platform: https://sustainabledevelopment.un.org/topics/energy
Vasudevan R, Cherail K, Bhatia R, Jayaram N (n.d.) Energy efficiency in India, history and overview
Wikisource (n.d.) Brundtland Report. Chapter 2. Towards Sustainable Development: https://en.wikisource.org/wiki/Brundtland_Report/Chapter_2._Towards_Sustainable_Development
World Health Organisation (2018) Opportunities for tranition to clean household energy. India

Structural Change in Employment and Unemployment in India

8

Arup Mitra and Jitender Singh

8.1 Employment

As per the Periodic Labour Force Survey, 2017–18, total employment (PS + SS) in the economy is estimated at 472 million, which is almost the same as estimated in the 68th NSSO Employment and Unemployment, 2011–12. This stagnation in job creation in the economy is more worrying when during the same period the production, in terms of GDP, in the economy has grown at a good rate. This clearly indicates that the additional production in some of the sectors is not able to create additional jobs. It is important to find out which these sectors are.

Agriculture continues with the trends of shedding jobs. There is a decline of about 16 million jobs in the sector between 2011–12 and 2017–18, compared to about 19 million between 2011–12 and 2009–10 and 14 million between 2009–10 and 2004–05. Its employment share in the economy has also declined to 44% in 2017–18. It is expected that in the process of development the additional labour force will shift out of agriculture. However, this magnitude of decline in employment in the sector requires careful interpretation (Table 8.1).

In the course of the development process, it is expected that the industry sector will absorb the additional labour force as the productivity is higher in the sector than the agriculture sector. However, the performance of the industry sector on employment front does not seems to be creating jobs to compensate for the decline in agriculture. The industry sector could add only about one million jobs between 2017–18 and 2011–12, which have been inadequate to absorb the shift in the labour force from the agriculture sector. Moreover, about four million jobs have declined in the manufacturing sector, which is a vibrant sector of the economy and expected to provide quality jobs. The construction sector has created an additional about four million jobs during this period.

The service sector has added about 14 million jobs between 2017–18 and 2011–12 compared to 16 million between 2011–12 and 2009–10. It contributes about one-third (31%) of the total employment in the economy (Table 8.2).

One of the reasons for the decline in the jobs in the manufacturing sector is the decline in the employment elasticity in the sector. As per estimates, the employment elasticity in the sector has declined to 0.10 between 2004–05 and 2011–12 (Table 8.3).

A. Mitra (✉)
South Asian University, New Delhi, India
e-mail: arupmitra@sau.int

J. Singh
Member of Indian Economic Service, Government of India, Delhi, India

S. Hazra, A. Bhukta (eds.), *Sustainable Development Goals*, Sustainable Development Goals Series,
https://doi.org/10.1007/978-3-030-42488-6_8

Table 8.1 Sectoral employment (person in Million PS + SS)

	Agriculture	Industry	Manufacturing	Construction	Services	All
1993–94	239.4	56.0	39.7	12.2	78.6	374.2
1999–00	239.9	64.7	43.7	17.5	92.9	397.1
2004–05	257.9	86.3	56.2	26.0	112.7	456.7
2009–10	243.6	98.5	50.4	44.0	116.3	457.9
2011–12	224.3	116.4	61.4	49.9	132.0	472.6
2017–18	208.2	117.5	57.1	55.2	146.4	472.1

Source: Computed from various rounds of NSSO

Table 8.2 Sectoral share in employment (in %)

Year	Agriculture	Industry	Manufacturing	Construction	Services	All
1993–94	64.0	15.0	10.6	3.3	21.0	100
1999–00	60.4	16.3	11.0	4.4	23.4	100
2004–05	56.5	18.9	12.3	5.7	24.7	100
2009–10	53.2	21.5	11.0	9.6	25.4	100
2011–12	47.5	24.6	13.0	10.5	27.9	100
2017–18	44.1	24.9	12.1	11.7	31.0	100

Source: Computed from various rounds of NSSO

Table 8.3 Manufacturing sector

Periods	GDP growth (%)	Employment growth (%)	Employment elasticity[a]
1993–94 to 1999–00	7.4	1.68	
1999–00 to 2004–05	6.0	5.72	0.80
2004–05 to 2009–10	10.1	−2.06	−0.27
2009–10 to 2011–12	8.1	10.91	1.74
2004–05 to 2011–12	9.5	1.32	0.10

Source: Computed from various rounds of NSSO
[a]Estimated using the CAGR Approach by Sangita Misra and Anoop K Suresh, "Estimating Employment Elasticity of Growth for the Indian Economy", W P S, RBI, (DEPR): 06/2014

8.2 Unemployment

As per the latest report on employment–unemployment situation (Periodic Labour Force Survey, 2017–18 of the National Statistical Office), the unemployment rate is estimated at 6.1% at the all India level as per the Usual Status criterion, which confirms that the unemployment rate is not only on the rise in the country, but also is at an all-time high. It has been highest since 1977–78 among the males in the rural and urban areas and highest since 1983 among the females in the rural as well as urban areas (Fig. 8.1). Further, between the sexes, the unemployment rate is higher among the urban females at 10.8%, than their male counterparts, estimated at 7.1%. In the rural areas, the male unemployment rate estimated at 5.8% exceeds the female rate at 3.8%. The rate being the highest among the urban females does not come as a surprise because many of the urban females are educated and can afford to wait for long till a job of desirable status becomes available as they may not be the principal earners. The unemployed persons in 2017–18 have more than doubled to 28.5 million from 10.8 million in 2011–12 (Fig. 8.2). Before that, the number of unemployed persons was hovering around ten million in the country in 1999–00, 2004–05 and 2009–10. The rise in unemployment in the recent years is also due to an addition of about 18 million persons in the labour force in 2017–18 over 2011–12, although only 0.5 million net jobs were added during this period. The rural and urban dynamics shows a different pic-

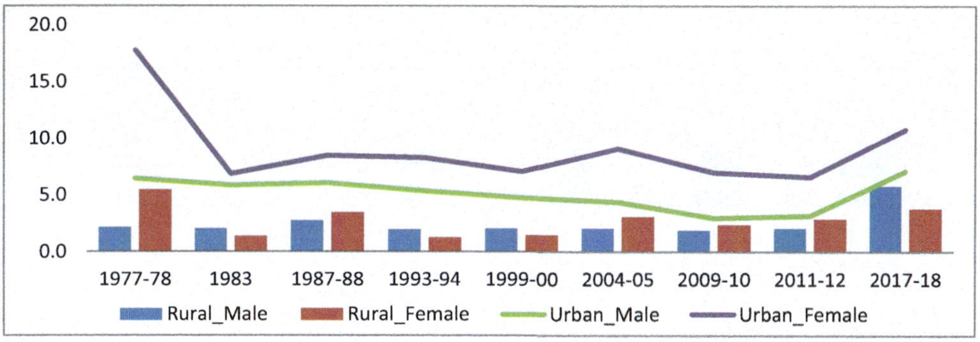

Fig. 8.1 Unemployment rate (in %): NSSO various rounds

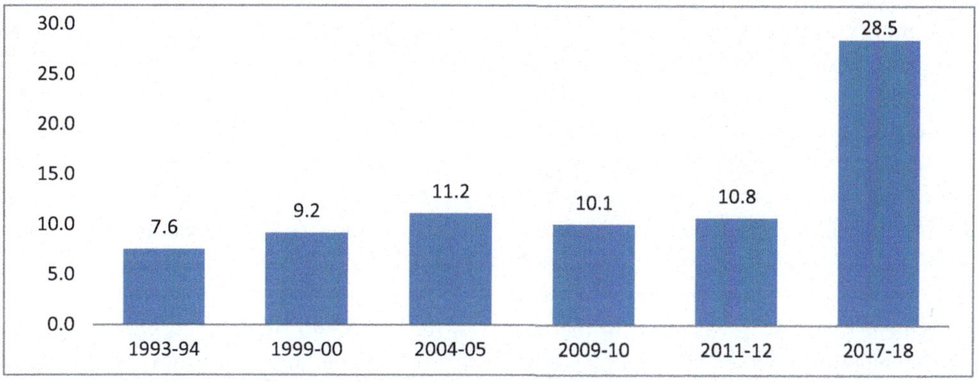

Fig. 8.2 Number of person unemployed (in million)

ture. In the rural sector, between 2011–12 and 2017–18, about 17.7 million jobs have declined while about 18.3 million have been added in the urban sector. On the labour force side, about 26.4 million persons have been added to the urban labour force whereas there is a decline of 8 million persons in the rural sector during this period. These could be the reasons for why the rural male unemployment rate which usually hovered around a low of 2% or so in the past suddenly shot up and even exceeded the female unemployment rate. Among the females, the phenomenon of 'discouraged dropouts' from the labour market is common. With the lack of rural diversification resulting in shrinking job opportunities, the rural women are not unlikely to quit the labour market. On the other hand, the PDS and a number of schemes available for the rural population might not have allowed the spillage of unemployment to cause a deterioration in the consumption poverty.

Relating to unemployment, there are two features which stand out sharply. First, the acceleration in the unemployment rate has occurred across states. There are a number of states which have crossed 5% unemployment rate. These are Goa, Manipur, Kerala, Mizoram, Nagaland, Haryana, Assam, Punjab, Jharkhand, Tamil Nadu, Uttrakhand, Bihar, Odisha, Tripura, Utter Pradesh, Arunachal Pradesh, Himachal Pradesh, J&K and Rajasthan. In fact, the unemployment rate in some of these states has reached double digits, viz., Goa, Manipur, Kerala, Mizoram and Nagaland. In comparison, in 2011–12, there were only three State/UTs, Tripura, Nagaland and Lakshadweep, which had a double-digit unemployment rate. Even an unemployment rate of 5% was not recorded in too many states or UTs (in addition to the three mentioned above four more, Kerala, Assam, Chandigarh and Andaman and Nicobar, belonged to this list in 2011–12). Further, another important feature of the regional

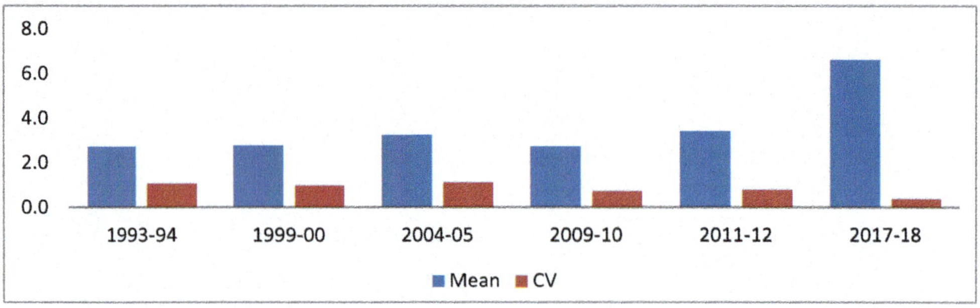

Fig. 8.3 Mean and CV of state-wise unemployment rates (%)

unemployment rate is that across the states the variation has declined since 2009–10, as reflected in the declining coefficient of variation (Fig. 8.3). This indicates that the convergence of the unemployment rate across states is in place: The deterioration on the employment front is very much evenly distributed spatially. However, there could be another side to this which need not involve a gloomy interpretation of the data. After all, in a poor region/country it is difficult to afford unemployment for long and thus the usual status of unemployment rate was always low in the past. Given the availability of the consumption support schemes particularly in the rural areas as mentioned above, the rising opportunities for the young educated and school drop-out young adults to migrate, thanks to the contractors, have raised remittance flow, which in turn may have reduced the compulsion for the non-migrants to pick up petty and low-productivity jobs. For a long time in the informal economy, which for many it is the 'employment of the last resort', the situation of excess-supplies-limited-demand was prevailing widely, leading to large-scale underemployment and a residual absorption of labour in low-productivity activities. The wages varied widely across activities, leading to a multi-modal wage distribution within this sector, though on an average low labour productivity, meagre earnings, poor hiring conditions and the lack of upward mobility add up to what is called the lack of 'decent employment'. Possibly within the informal sector, some of the jobs are fetching higher earnings and given the well-knitted support structure the desperation to strive hard is on the decline.

This note focuses more on the two questions as to why there is acceleration in the unemployment rate across states in the recent years and what explains a decline in the state-wide variation in the unemployment rate.

In spite of rising business subcontracting from the formal to the informal sector, evidence of rising real wages within the informal sector is rather scanty, which is aggravated by the existence of multi-layers of intermediaries/contractors.

The common perception about unemployment is that high and rising unemployment rates are a reflection of disequilibrium in the economy, which may further aggravate the social and economic problems. What is causing high unemployment rate is even more important to understand. As the economic literature also suggests, it may increase in response to development and structural change taking place in a country. This note is a preliminary observation focussing on the economic transitional processes which involve structural changes in economic development, labour force participation and its sex composition, educational and skill attainment and urbanisation pursuits and have the capacity to influence the unemployment rate.

8.2.1 Results and Discussion

There are some empirical studies which systematically studied the relation between unemployment rate and development. The work by Caselli (2005) plotted the unemployment rates against log per capita income, showing that contrary to common perceptions, unemployment rates are not higher in poorer countries. Feng et al. (2018) analysed household surveys or censuses of 199 country-year surveys, covering 84 countries, and spanning from 1960 to 2015 and observed that the unemployment rate is increasing in response to GDP per capita and the highly educated workers are more likely than the low-educated workers to be unemployed in poor countries.

The literature providing evidence for skill-biased technological change in the rich and poor economies also argues that the unemployment rate among the skilled workers should decline with increase in development and technological change. Banerjee et al. (2016) estimated a negative correlation between the logarithm of observed GDP per worker and the unemployment rate of skilled workers at −0.428.

Lewis (1954) and Harris and Todaro (1970), using a two-sector model, documented that labour shifts from the traditional low-productivity sector (rural) to the modern high-productivity sector (urban) in the course of development. The high unemployment rate in the urban sector prevails mainly because of migration of surplus labour from the rural areas, which accelerates with an increase in connectivity between the two sectors. This may be taken to argue that higher levels of urbanisation may correspond to higher unemployment rates. Further, the most mobile and capable labour force is likely to migrate from the rural to urban areas, which may decrease the labour force and the unemployment rate in the rural areas.

This gives us a clue to understand the new trends in unemployment in response to development indicators, such as education, skills, urbanisation and the share of traditional sector.

8.2.2 Development and Unemployment Rate

Employment is likely to increase with improvement in the level of development (indicated in terms of per capita income). The mechanism works through production processes where more and more labour is utilised to produce more output and income. However, the unemployment may increase with increase in per capita income for several reasons. First, the production process may utilise less labour with change in production technology and may not keep up the pace with increase in the labour force. Further, the income may disproportionally originate from the sector which employs a very small segment of labour on account of productivity gains. Moreover, income generation may be limited to a small section of the labour force without employing more labour.

Graph 8.1 plots the unemployment rate (axis vertical) against the log of per capita income of a state (horizontal) year-wise. It shows no concrete association between the two. The correlation coefficient is weak (except 2011–12) and statistically insignificant. The evidence can be taken to suggest that there is no concrete association between growth and unemployment rate in India. This weak association is consistent over the years. The direction of their association is also not clear. Therefore, one is not sure whether the unemployment rate is higher in poor states or whether it is lower in better developed states, vice versa.

8.2.2.1 Structural Change and Unemployment Rate

The unemployment rate is likely to increase during the structural change in the economy. There are several reasons for this. First, disguised unemployment is likely to become more visible once the activity status changes from the farm to the non-farm sector. For example, members of the households engaged in the farm sector, assisting, partly or wholly, in farm activities are more likely to admit that in the non-farm sector they do work as per principal or subsidiary status. However, once the family starts working in the non-farm sector, the labour of all other members may not be utilised as it was possible in agriculture. The family leaving the farm sector may also have their threshold wage limit and job preferences, which increase the duration of remaining unemployed with a view to finding a suitable work. Third, a better-educated person of a farming family may disassociate himself as a worker

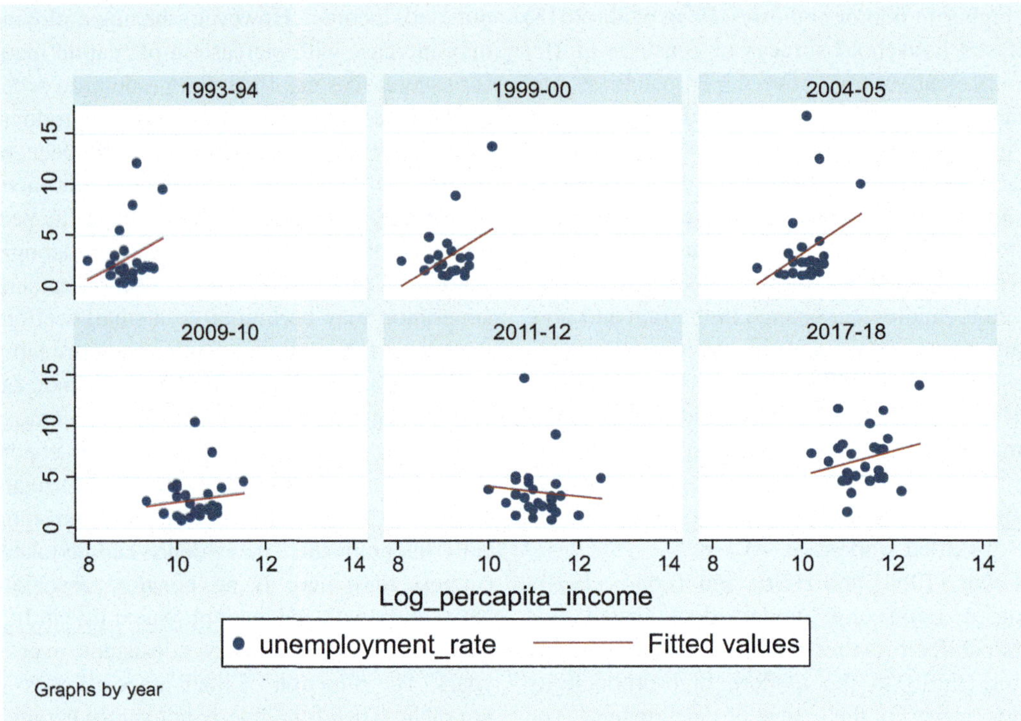

Graph 8.1 Unemployment rate % and Percapita Income of States in India

in agriculture but may still claim as part of the labour force looking for work in the non-farm sector. Fourth, the capital-intensive technology in the agriculture sector frees up labour who may not find an alternative livelihood option elsewhere.

Graph 8.2 plots the unemployment rate against the share of workforce employed in agriculture, which shows a relatively strong association between these two over the years. The statistically significant negative correlation between these two range between −0.6 and −0.8 (except for 1993–94) at different points in time from 1993–94 to 2017–18.This shows that the states where the agriculture share of the workforce is low are also the states with higher unemployment rates, and vice versa. This change in the rural sector of moving away from agriculture to non-agriculture has been strongly associated with the unemployment rates.

These trends of declining workforce in agriculture are likely to continue for several reasons. First, the operational holding size is declining over the years, making survival on land alone difficult or making the holding unviable. An NSSO survey found that about 70% of the farmer households are spending more than what they are earning. Second, the productivity in agriculture is relatively low compared to non-farm and other modern sector activities. The preference of the farmers is changing, as reflected in the NSSO surveys to suggest that about 40% of the farmers want to leave the profession given the option (Agarwal et al. 2016).[1]

8.2.2.2 Education, Skill and Unemployment Rate

Education is likely to increase unemployment for several reasons. First, education and skills raise the threshold wages of a potential worker who may prefer to remain unemployed, and keep searching for suitable jobs. Second, education

[1]Agarwal, Binaand Ankush Agrawal (2016), "To farm or not to farm? Indian farmers in transition, Global Development, Institute Working Paper Series: 001."

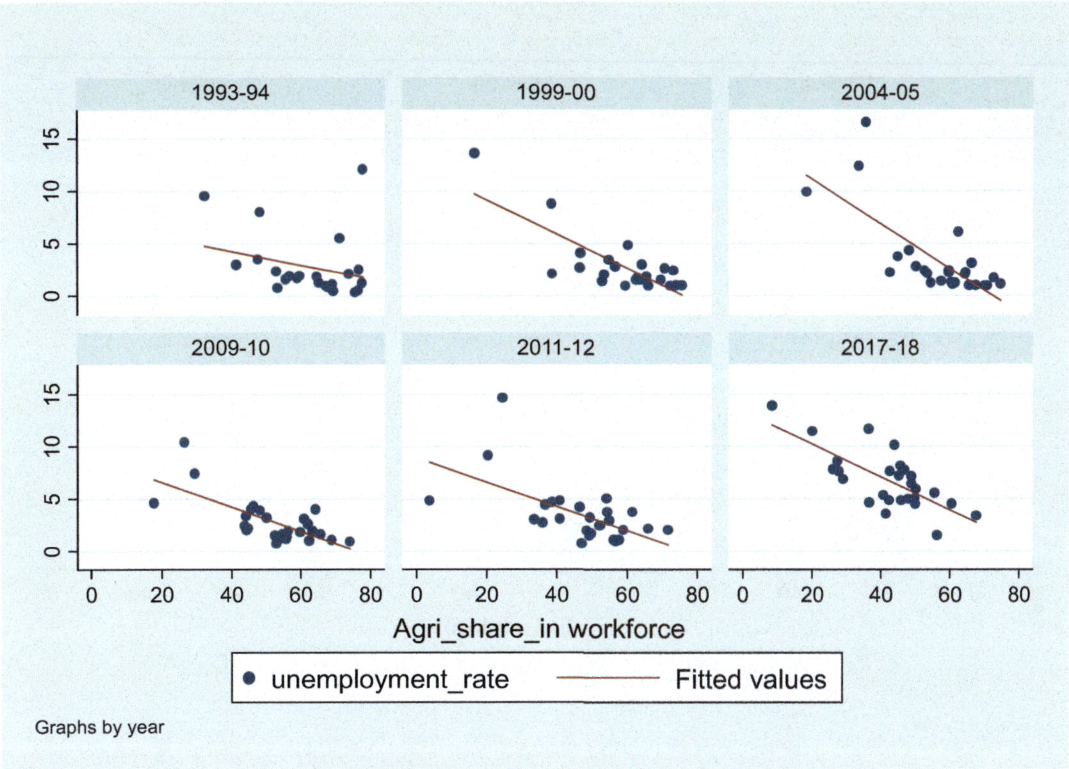

Graph 8.2 Unemployment rate (%) and Share of Agriculture in workforce (%)

and skills also have the potential to change the job preference of a potential worker, and therefore, they may prefer to remain unemployed till a suitable job is found.

On the other hand, in the short run, participation in education and skill may delay the labour force participation. The demand for new skills has risen in the recent past on account of improved premium on education and skills due to skill-based technological progress in the organised industry and services activities. The skill premium on higher education may hold millions in the educational institutional for knowledge gain/skill development but this being a temporary phenomenon, education is more likely to increase the unemployment rate in the economy.

Graph 8.3 plots the unemployment rate against the percentage of persons having 'secondary and above education'. It shows a positive association between these two. The positive correlation coefficient is found to be strong and statistically significant in 1999–00 and 2017–18 while for other years it is weak, positive though. The states with a higher percentage of population having relatively higher education and skills are more likely to observe higher rate of unemployment.

8.2.2.3 Urbanisation and Unemployment Rate

Urbanisation is an important economic transition process. States in India are witnessing variation in their level of urbanisation. Even if the official urbanisation rate is moderate, the large-scale emergence of census towns in 2011 is a strong witness to the rapid transformation that is taking place in the rural areas. The possibility of disguised unemployment in the urban labour force is low compared to the rural areas and therefore, with an increased urbanisation process the unemployment rate also becomes more visible. Besides, frictional unemployment and technological unemployment are also part of the urban labour force in the relatively skilled jobs.

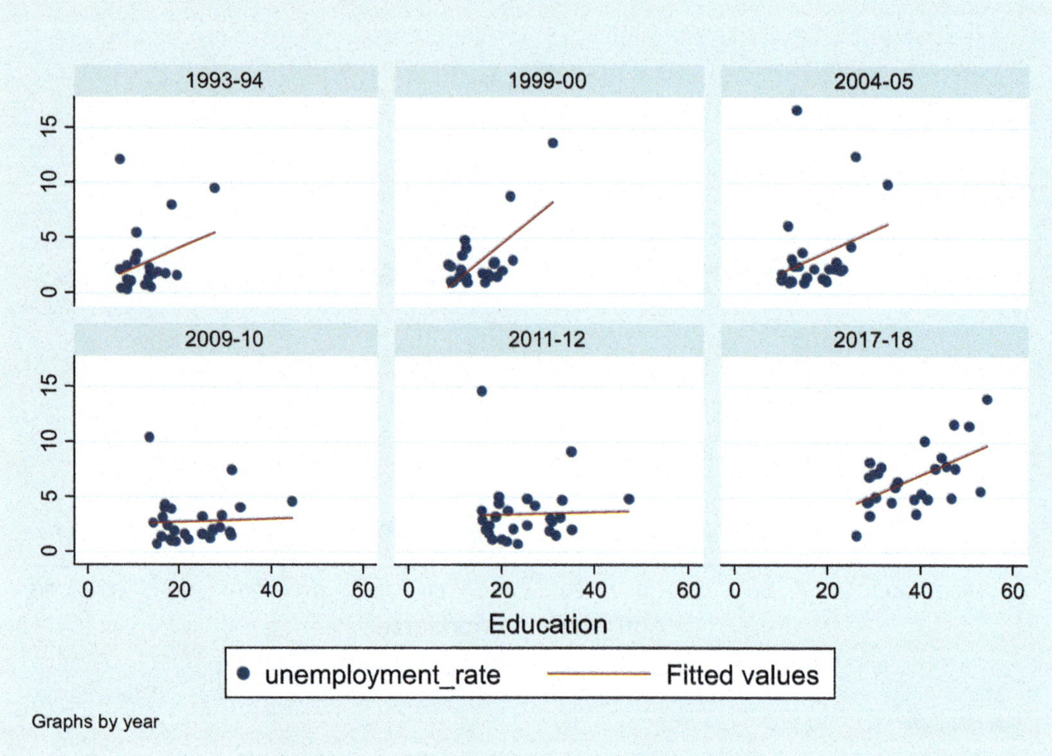

Graph 8.3 Unemployment rate (%) and Secondary and above education persons (%)

Graph 8.4 plots unemployment rates against the percentage of urban population and it important to note that no concrete association between the two is evident until the year 2017–18 for which the correlation is strong, positive and statistically significant.

8.3 Conclusions

There is acceleration in the unemployment rate in the recent years along with the convergence of unemployment rate across states. The stagnation in job creation and decline in the total employment in the manufacturing sector and the structural changes in the economy appear to be pushing the unemployment level up. Though per capita income as such is not seen to be related to unemployment rate across states and over time, the structural change and educational attainments do unravel a strong effect. The changes in the rural sector with a declining dependence on the farm sector are associated with a rising unemployment rate. Though water scarcity and low crop diversification are prevalent, new processes of contractor-led single-member migration from rural households and remittance flow, emergence of new non-farm activities and consumption support schemes are instrumental to the new transformations occurring in the rural areas and their transition to urban space. Besides, the association of the unemployment rate with educational attainments and the urbanisation–unemployment nexus observed from the recently released periodic labour force survey data are testimony to the brighter side of the development story of India.

It is also noted from the recent data that the share of the informal sector in total employment has declined compared to the earlier NSSO's enterprise survey results relating to the usual status workers. The rise in the unemployment rate as seen in the backdrop of the falling share of the informal sector employment offers space for

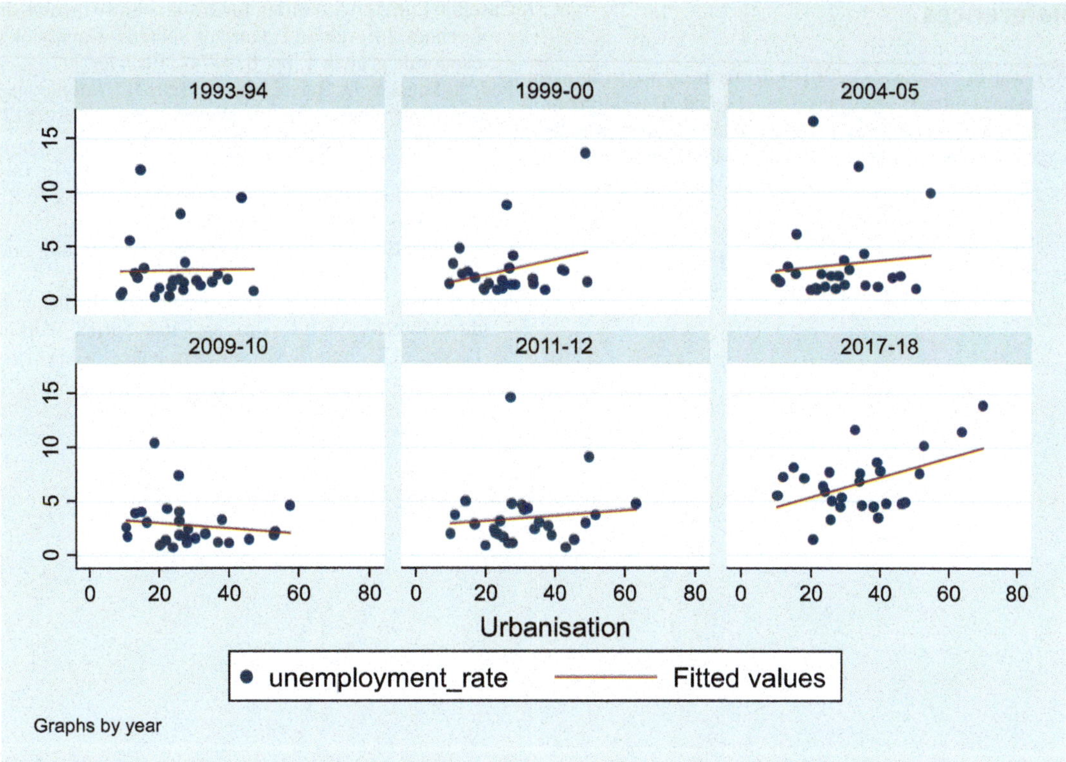

Graph 8.4 Unemployment rate (%) and Urbanisation

interpretation that negates the desperation of the workers to join the informal sector, keeping the open unemployment rate low.

While pessimism gathering around the rising unemployment rate needs to be controlled, it is equally important to pay attention to the larger issue of employment creation. Since wage employment cannot be provided on a large scale, there is need to promote entrepreneurship by creating a facilitating environment for start-ups, innovations etc. Keeping in view the skill gaps in various activities, emphasis has to be given on vocational education and training which can be integrated with the general education system. Instead of confining training to the initial entry level in jobs, up-skilling/re-skilling would be necessary to adjust to the requirement of the changing economy. It also requires availability of adequate training infrastructure and considerable improvement in the quality of these institutions. Further, interventions for the provision of on-the-job training is important as with increasing contractualisation the employers do not have any interest in providing training to the employees. The other aspect relates to the geographical location of the skill-imparting institutions. Unless the low-income households both from rural and urban areas are able to access skill, the transfer of labour from low productivity to high productivity activities would remain unrealised. However, from the demand side, it is equally important that high productivity job opportunities must grow rapidly so that labour can be motivated to acquire skills and make itself employable. Hence, the relevance of the growing business environment cannot be overlooked. The new technology with its high dependency on capital intensity is expected to pose serious challenges to employment creation both in the manufacturing and in the services sector. Without major scale effects in place, how are the dual objectives of growth with employment generation to be realised then? Often, we do not realise the importance of adequate purchasing power, the lack of which may result in economic deceleration, making growth unsustainable in the long run.

References

Agarwal B, Agrawal A (2016) To farm or not to farm? Indian farmers in transition. Global Development Institute Working Paper Series 2016-001, https://hummedia.manchester.ac.uk/institutes/gdi/publications/workingpapers/GDI/GDI_WP2016001_Agarwal_Agrawal_final.pdf

Banerjee A, Basu P, Keller E (2016) Cross-country disparities in skill premium and skill acquisition. Unpublished Working Paper, Durham University

Caselli F (2005) Accounting for cross-country income differences. In: Aghion P, Durlauf SN (eds) Handbook of economic growth 1, pp. 679–741, Elsevier

Feng Y, Lagakos D, Rauch James E (2018) Unemployment and Development. NBER Working Paper No. 25171, https://www.nber.org/papers/w25171

Harris JR, Todaro MP (1970) Migration, unemployment and development: a two-sector analysis. Am Econ Rev 60(1):126–142

Lewis WA (1954) Economic development with unlimited supplies of labor. Manch Sch 22(2):139–191

Ying F, Lagakos D, Rauch JE (2018) Unemployment and development. NBER Working Paper Series, Working Paper 25171. http://www.nber.org/papers/w25171

Avoiding Premature Deindustrialization in India: Achieving SDG9

9

Selim Raihan

9.1 Introduction

Industrialization is the process of structural transformation through which resources (especially labor) move from agriculture to industry and more specifically to manufacturing. Industrialization generates many benefits which include diversification of the economy, employment generation, technology transfer, and welfare improvement of the people. As a result of industrialization, an economy experiences an increase in the share of manufacturing in Gross Domestic Product (GDP) as the production of manufactured goods increases. Simultaneously, an economy also experiences a rise in manufacturing share in employment. In the developing countries, industrialization plays a critical role in accelerating the process of economic development. Industrialization leads to an increase in national income by ensuring the maximum use of scare resources, and raises the export of manufactured goods. Moreover, Industrialization opens up employment opportunities and thus helps in poverty reduction. It also allows the government to make long-term investments in infrastructure, skills formation, and institutional building, which contributes to the development of the economy. The manufacturing sector possesses some characteristics which can establish the necessary linkages for sustaining a virtuous circle of growth and structural transformation.

Industrialization can be described as the process by which the structure of an economy is transformed from an agricultural-base to a manufacturing-base. On the contrary, deindustrialization can be termed as a process leading to the fall in manufacturing activities, in terms of both output and employment, of an economy. In general, natural deindustrialization occurred in most of the developed economies where after achieving a certain level of per capita GDP, there was a gradual decline in the share of manufacturing value-added in GDP. This deindustrialization process coincided with the growing share of services value-added in GDP. Deindustrialization, in this regard, emerges as a natural outcome of the economic development process in those developed economies.

In contrast to the aforementioned natural process, a number of developing countries are experiencing premature deindustrialization when the shares of manufacturing in GDP and employment, already much lower than those of the early industrializers, begin to fall at a level of per capita GDP much lower than those of early industrialized economies. According to Rodrik (2016), premature industrialization in many developing

S. Raihan (✉)
Department of Economics, University of Dhaka, SANEM, Dhaka, Bangladesh

S. Hazra, A. Bhukta (eds.), *Sustainable Development Goals*, Sustainable Development Goals Series,
https://doi.org/10.1007/978-3-030-42488-6_9

countries is leading to a situation when these economies are becoming more and more service-oriented economies without going through a proper process of industrialization. Most of the Latin American countries are prime examples of such premature deindustrialization.

SDG 9 (industry, innovation, and infrastructure) should be at the heart of priorities of economic development in India as SDG9 has strong linkages with other SDGs. A recent study by UNESCAP (UNESCAP 2017) shows that the calculated infrastructural index of South Asia is much lower than the average value of the index of Asia-Pacific developing economies. South Asian countries have a much wider gap in infrastructure index when compared with the average value of the index of Asia-Pacific developed economies. All these refer to the fact that there is need for a significant volume of infrastructural investments in India from now until 2030. For the industrialization target under SDG 9, India is off the track as the country is experiencing premature deindustrialization.

Against this backdrop, this paper looks at the pattern of industrialization in India and explores the phenomenon of premature deindustrialization in India. The paper also seeks to identify the strategies for India to avoid the premature deindustrialization process.

9.2 An Overview of Industrialization of the Indian Economy

The Indian economy underwent important structural transformation over the past five decades. The share of agriculture in GDP declined drastically from as high as 56.7% in 1960 to 14.4% in 2018 (Fig. 9.1). The shares of services and industry were almost the same in 1960, of around 22%. However, by 2018, the share of the services sector increased to 54% and that of industry increased to 31.6%. It is important to mention here that the industry includes manufacturing, mining, and construction.

Despite the large decline in the share of GDP, agriculture's share in total employment declined from as high as 72% in 1960 to 47% in 2018 (Fig. 9.2). The share of industry in total employment was 11.7% in 1960, which increased gradually to 21.5% in 2018. Finally, the share of services increased from 16.4% in 1960 to 31.5% in 2018. All these indicate that agriculture still employs around half of the Indian labor force and the next major employment-generating sector is the services sector.

Since the industry sector includes manufacturing, mining, and construction, in order to explore the trend in industrialization, we analyze the trend in the manufacturing share in GDP (Fig. 9.3) and employment (Fig. 9.4). Figure 9.3 suggests that, the manufacturing share in GDP in India has shown a fluctuating trend since 1960. The share reached its peak level of 17.8% in 1979 and 1995, but in 2018, the share declined to only 15%. Since 2010, the share has a secular declining trend. Figure 9.4 suggests that the manufacturing share in employment in India demonstrated a fluctuating but a rising trend until 2002. In 1960, the share was as low as 9.6%, which by 2002 increased to 12.5%. After 2003, the share fluctuated and since 2012, the share had been showing a gradual decline. By 2018, the share stood at 11.4%.

The composition of merchandise imports in India changed over time since the beginning of the 1960s (Fig. 9.5). The share of agricultural raw materials and food declined. In 1962, the shares of agricultural raw materials and food were 9.2% and 17%, respectively, which came down to only 2% and 6%, respectively, in 2017. The share of fuel increased considerably since the early 1980s. In 2017, the share of fuel in total merchandise imports was 30%. The share of ores and metals remained around 6% throughout the period under consideration. The major component of import, with fluctuations, has been the manufactures. In 1962, the share of manufactures was 58% and by 2017, the share still remained at 54.8%.

The composition of the merchandise exports is shown in Fig. 9.6. Over the years, the shares of agricultural raw materials and food declined quite substantially. While in 1962, the shares of agricultural raw materials and food were 9.3% and 39%, respectively, the shares came down to

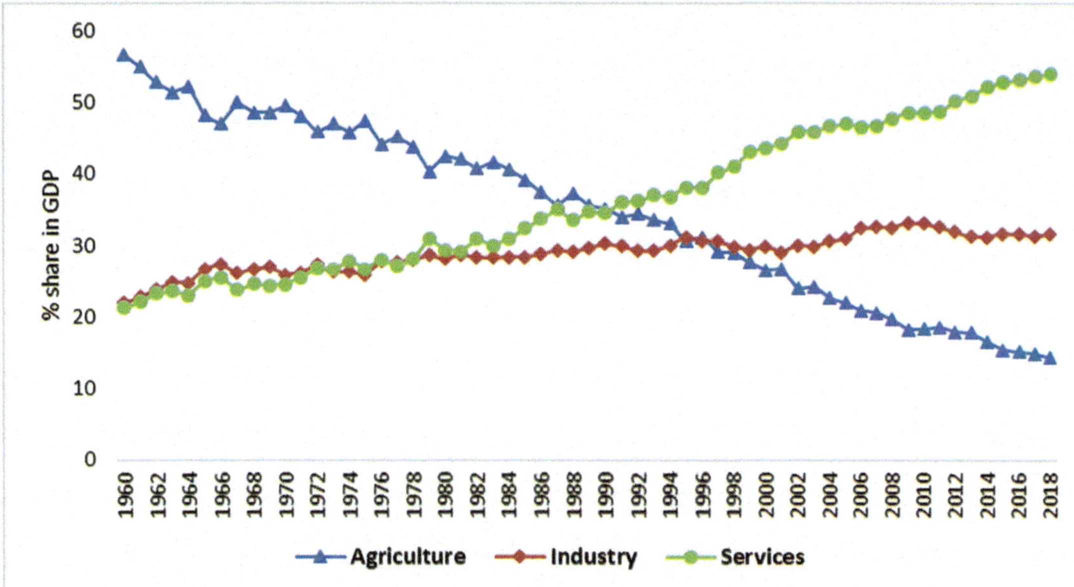

Fig. 9.1 Sectoral share of GDP (%). (Data source: World Development Indicators, World Bank)

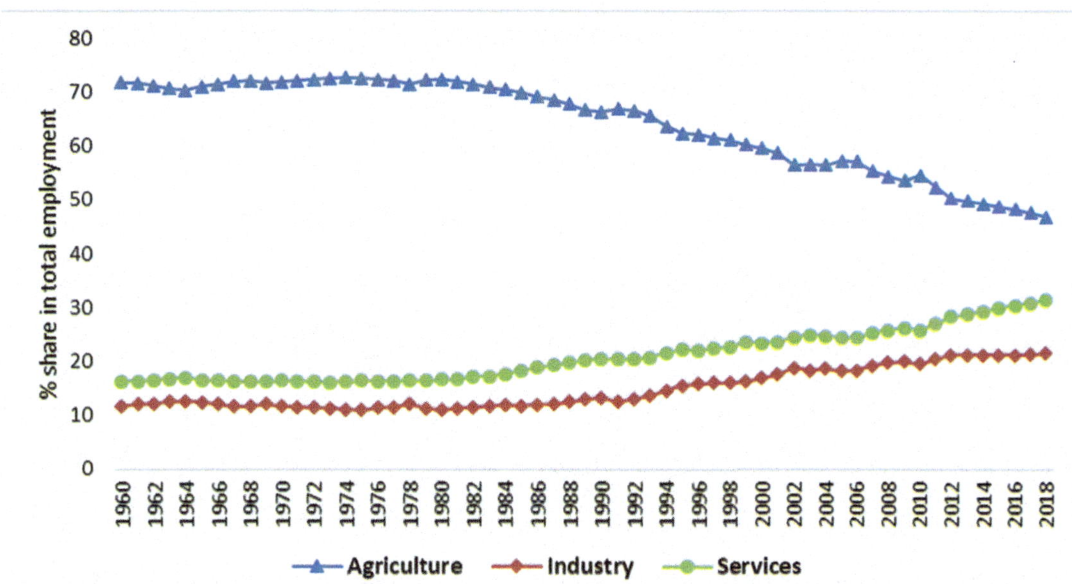

Fig. 9.2 Sectoral share of employment (%). (Data source: GGDC 10-Sector Database and World Development Indicators, World Bank)

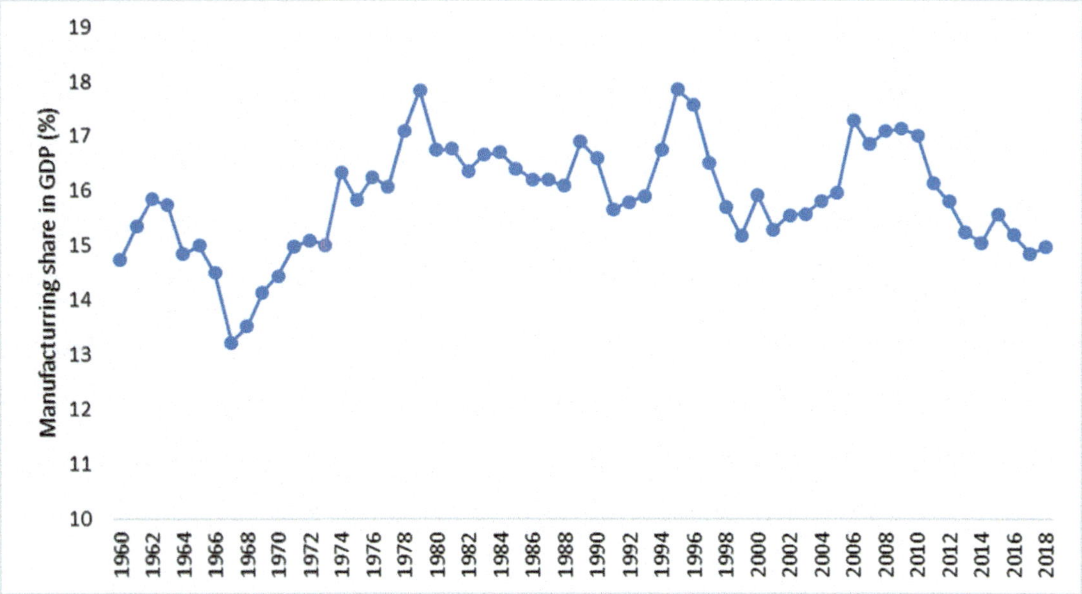

Fig. 9.3 Manufacturing share of GDP (%). (Data source: World Development Indicators, World Bank)

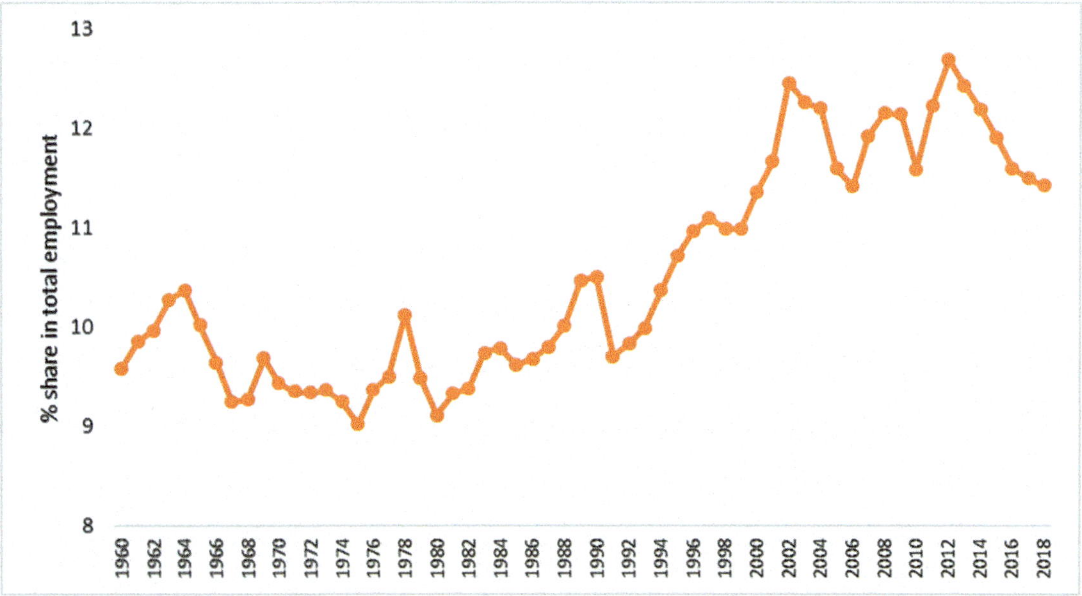

Fig. 9.4 Manufacturing share of employment (%). (Data source: GGDC 10-Sector Database and author's projection)

only 1.3% and 11.8%, respectively, in 2017. Since the early 2000s, the share of fuel increased and in 2017, the share stood at 12%. The share of ores and metals also decline from around 7% in 1962 to 3.8% in 2017. The share of manufactures increased quite substantially during the period under consideration. In 1962, the share of manufactures was 43.4%, which increased to 70.7% in 2017. Figure 9.7 suggests that, with the rise in manufactures share in merchandise exports in

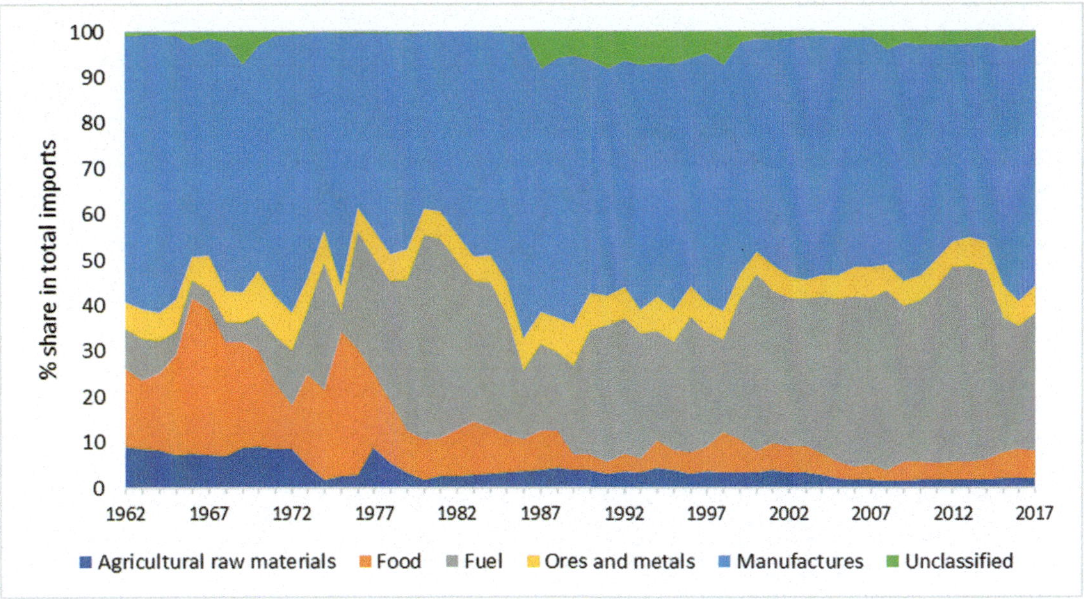

Fig. 9.5 Composition of merchandise imports (%). (Data source: GGDC 10-Sector Database)

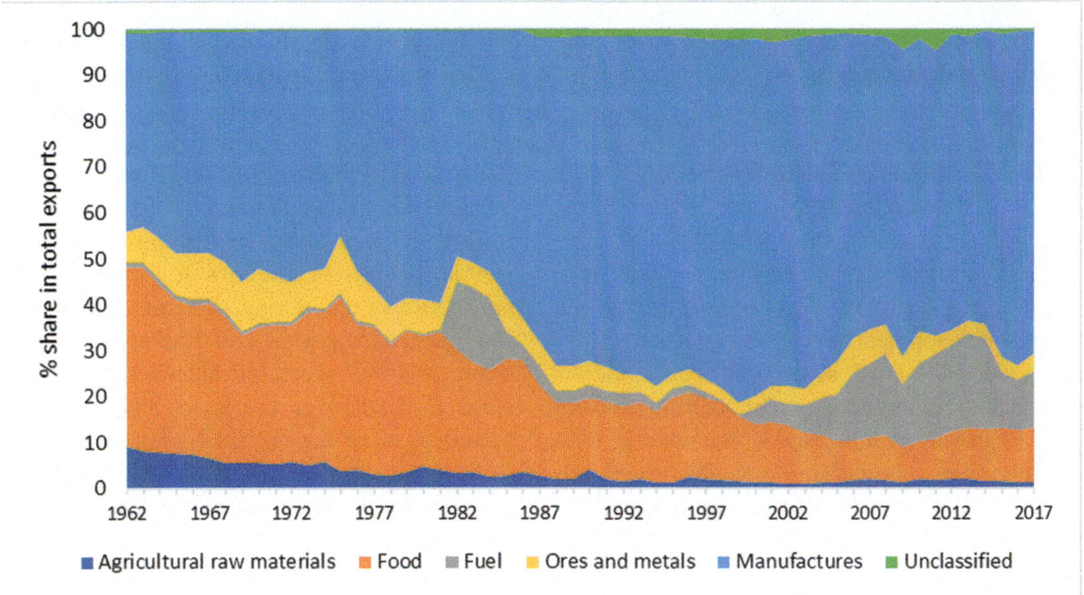

Fig. 9.6 Composition of merchandise exports (%). (Data source: GGDC 10-Sector Database)

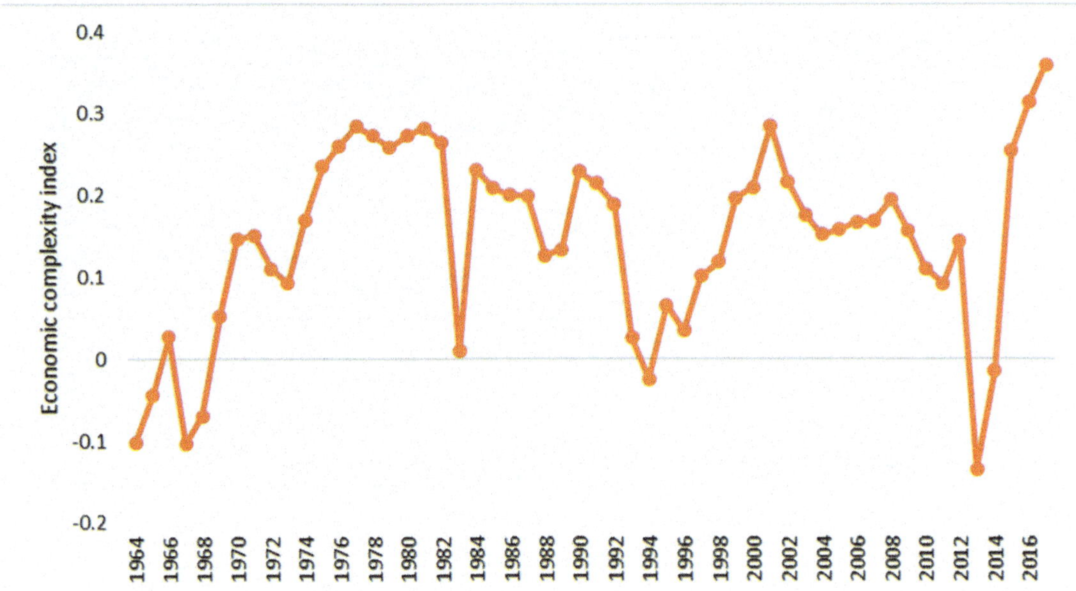

Fig. 9.7 Economic complexity index of India. (Data source: https://oec.world/en/rankings/country/eci/)

recent years, the economic complexity index also increased in India, indicating that India has been able to export more complex and higher value-added manufactures.

9.3 Is India Deindustrializing Prematurely?

There is no denying the fact that manufacturing has played a key role in the economic growth and overall development processes for many developed and advanced developing countries. As Rodrik (2016) suggests, manufacturing contributes to growth both because of the positive reallocation effect and because manufacturing tends to experience relatively stronger productivity growth over the medium to longer term, which has large positive economic and social benefits. Successful countries have changed their economic structures to benefit from manufacturing as the driver of economic growth. Therefore, premature deindustrialization as well as a low level of the manufacturing base is not good news for the South Asian countries. It thwarts the opportunities of rapid economic growth in these countries. It is very pertinent to mention here that one of the targets of the ninth Sustainable Development Goals (SDGs) is thus to "promote inclusive and sustainable industrialization and, by 2030, significantly raise industry's share of employment and gross domestic product, in line with national circumstances, and double its share in least developed countries."

Figure 9.8 presents manufacturing share in GDP in some Asian countries. It is quite interesting to observe that during 1960 and 2018, while South Korea, Malaysia, and Thailand were able to increase their shares of manufacturing value-added in GDP over the years and reached a peaked level of around 30% and then started experiencing a declining trend, for India, the share remained 15% and in recent years the share started to decline. Even Bangladesh, with a much lower share of manufacturing value-added in GDP in 1960 was able to surpass India in 2010, and by 2018, Bangladesh's share was much higher than that of India.

Is India experiencing premature deindustrialization? To answer this question, in Fig. 9.9, we present a graph using data of GDP per capita and the share of manufacturing value-added in GDP

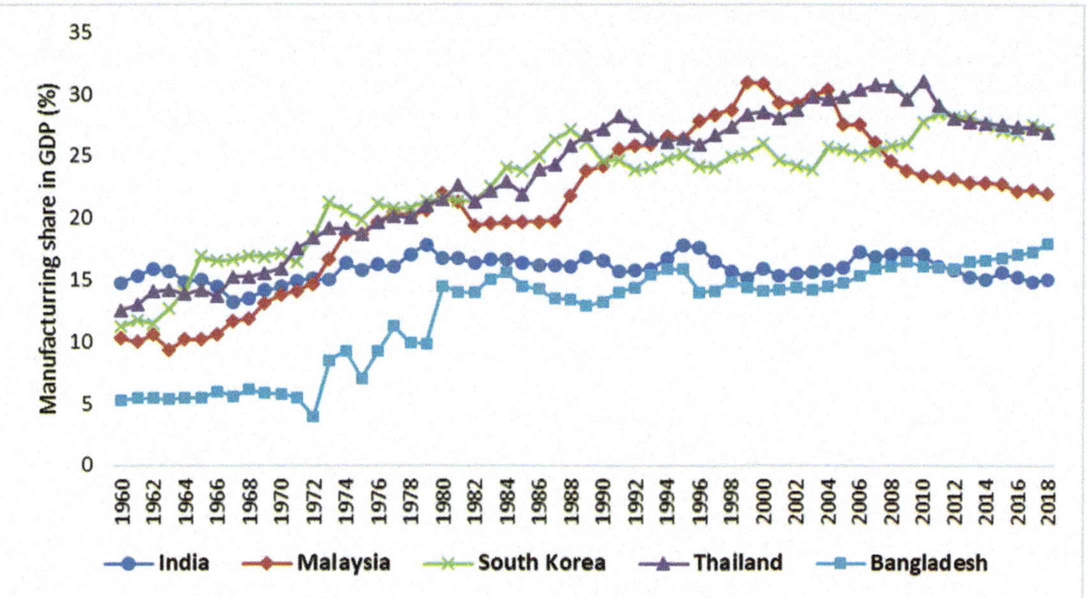

Fig. 9.8 Manufacturing share in GDP in a comparative perspective (%). (Data source: World Development Indicators, World Bank)

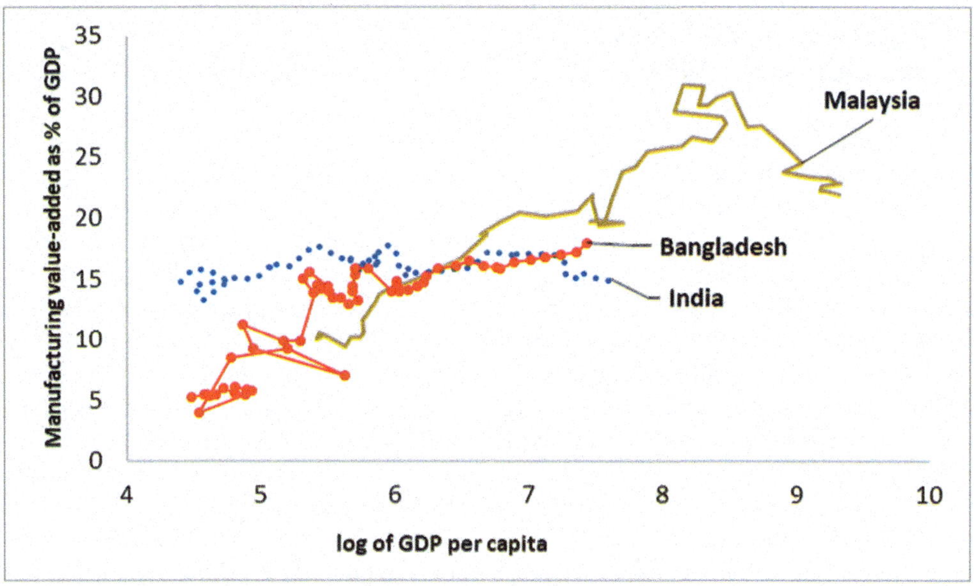

Fig. 9.9 India's declining share of manufacturing in GDP at a low level of GDP per capita. (Data source: World Development Indicators, World Bank)

for the years from 1960 to 2018 for three countries: Malaysia, India, and Bangladesh. The graph shows that Malaysia, over the years, with the rise in GDP per capita, was able to firmly increase its manufacturing value-added share in GDP from 10 to 30% (during 1960 and 2004). However, after reaching a high level of per capita GDP of around US$ 8000, the share started to decline; and this pattern follows the typical pattern of the deindustrialization process of the advanced countries. In contrast, with many fluctuations, India very slowly increased its share of manufacturing value-added in GDP from 13.7 to 18% (during 1960 and 2008). However, for India, the challenge is the declining share of manufacturing value-added in GDP since 2008 and by 2018, the share declined to 15%, which has a resemblance to premature deindustrialization. In contrast, as is shown in Fig. 9.9, Bangladesh has been experiencing a rising share of manufacturing value-added in GDP at a much faster rate than India at a similar level of per capita GDPs.

9.4 How to Avoid Premature Deindustrialization in India

9.4.1 Factors Affecting Industrialization: A Cross-Country Experience

In order to explore factors that affect the manufacturing share of GDP, a cross-country panel econometric analysis is conducted using a panel data of 107 developing countries for the years 1970–2016. The UN classification for developing countries has been used. In case of developed economies, over the years the contribution of manufacturing value-added to GDP tends to decline. Therefore, we have considered only developing countries in our analysis. All data have been obtained from World Bank's World Development indicators. The fixed effect panel regression results suggest that the size of the population, share of households with access to electricity, lower rate of applied tariff rate, domestic private sector credit as share in GDP, investment as share in GDP, labor force participation rate of younger people (age 15–24), and share of public expenditure on education in GDP have a positive and statistically significant association with higher manufacturing share in GDP.

The size of the population can be used as a proxy for the size of the internal demand. There is a significant positive relationship between manufacturing expansion and internal demand so that, other things being equal, countries with larger internal demand tend to have a higher manufacturing share. The access to electricity variable can be considered as a proxy of infrastructure, especially electricity infrastructure, of a country. Electrification is an important factor for industrialization. Furthermore, an outward-looking industrial strategy allows access to large markets and a growing demand which encourage large-scale industrialization programs. Moreover, trade liberalization helps boost the industrialization process by ensuring supply of imported inputs at free trade prices, providing access to technology and capital and by helping to establish a more competitive exchange rate. The strategies followed by economies who were successful in industrialization include the adoption of trade liberalization in conjunction with setting up of special economic zones, export processing zones, and industrial bonded zones as strategies for promoting Foreign Direct Investment (FDI) and supply export-oriented firms with duty-free, tax-free imported inputs. Trade played a crucial role in expediting structural change and industrial development in these nations. The existence of better financial institutions helped influence the industrialization process by facilitating efficient allocation of resources and ensuring larger private sector credit in proportion to GDP. The presence of an efficient banking system ensured the availability of finance to firms, especially small and medium-sized firms, and reinforced domestic entrepreneurship helped speed up the pace of industrialization. The countries with higher levels of private investments, backed by high levels of domestic savings and FDI, were also successful in the industrialization process. The increased participation of youth IN the labor market helps reap the benefits of demographic dividend.

Human capital development in the form of sufficient technically and scientifically qualified personnel can help meet the increase in demand and contribute to industrial development. The foundation of a competitive industrial sector can be developed and the appeal of investments can be raised by generating immobile national assets, through education spending in particular. Therefore, any form of industrialization demands an increase in government spending on education.

Finally, experiences from successful countries suggest that better functioning institutions, capable of guaranteeing better rule enforcement, transparency, management of corruption, and government stability could improve the business climate and stimulate the entrepreneurial spirit. On the contrary, the existence of significant governance deficiencies could render difficult the building up of a solid industrial sector and complicate the leading of a dynamic industrial policy. Therefore, reforms for improved administrative procedures and reduction of regulatory delays are critically important.

It can be summed up that in order to initiate a sustained process of strong industrialization, a boost in investments and an improvement in education are crucial; the management of trade and capital openness are also vital factors; financial sector development and the promotion of both macroeconomic stabilities in the form of lower levels of debt and high levels of political and social stability and institutional stability are essential for achieving sustained industrialization. In addition, Infrastructure development, uninterrupted access to energy and innovation can act as catalyzers in the process of nurturing industrialization.

9.4.2 Addressing Policy-Induced Challenges

There are a number of policy-induced challenges. The reform of trade and industrial policies in the 1980s and 1990s helped India achieve the current level of progress in manufacturing. However, returns from those reforms have been exhausted, and also there are now some policies in place toward the wrong directions. There is a need for strategic and dynamic industrial policies aiming at rapid expansion and diversification of manufacturing through large-scale domestic and foreign investments. Given the changes in the global and regional trade scenarios, the need for such strategic trade and industrial policies is more important now than ever.

9.4.3 Enhancing Trade Orientation

One of the worrying signs of India's pattern of trade is the falling trade orientation. From a very low base in 1960, India's trade orientation saw a gradual rise over the next four decades until 2010 (Fig. 9.10). In 1960, the import-GDP ratio was as low as 6.8%, which increased to 31.3% in 2012. However, the import-GDP ratio saw a secular decline during 2012 and 2016, and a rising trend since 2016. In 2018, the import-GDP ratio stood at 23.4%. The export GDP ratio increased from a very low share of only 4.5% in 1960 to 24.4% in 2013. The export-GDP ratio also saw a secular decline during 2013 and 2017 and a rising trend in 2018. By 2018, the export-GDP ratio stood at 19.7%. India's trade orientation needs to be increased and expansion of more export-oriented manufacturing sector needs to be encouraged.

9.4.4 Attracting Large-Scale Foreign Direct Investment (FDI)

Foreign direct investment (FDI) plays an important role in the long-run economic growth of an economy. FDI develops productive capacity through transfer in technology, enhances domestic labor skills through global managerial practices, and contributes to human capital development. FDI assists in integrating the domestic markets with the global market. Furthermore, FDI bridges the gap between domestic savings and investment and spurs economic growth, which is a powerful tool for alleviating poverty. Although developing countries understand the need for FDI to boost economic

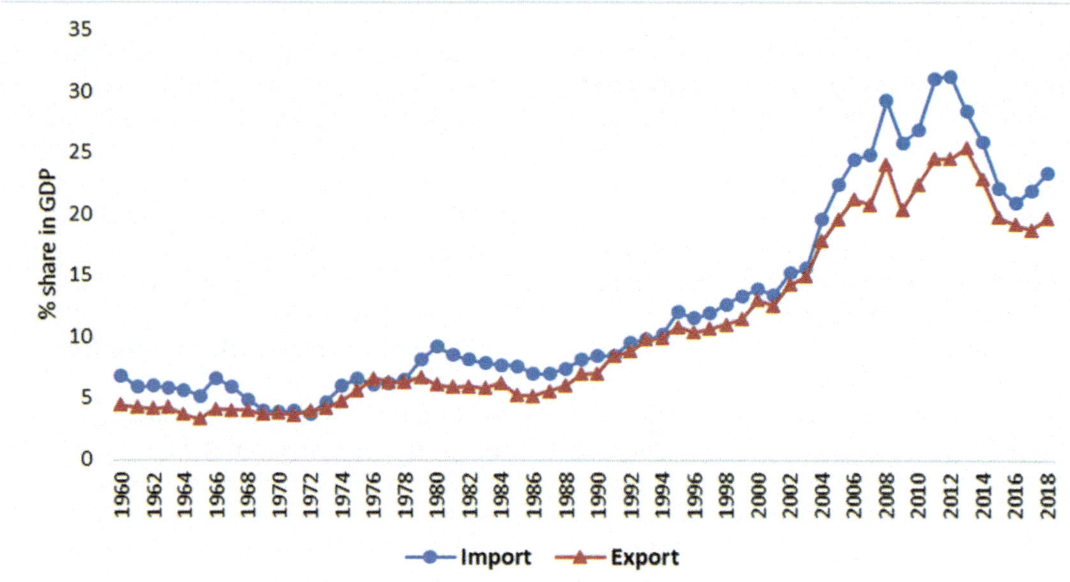

Fig. 9.10 India's falling trade orientation. (Data source: World Development Indicators, World Bank)

growth in their countries, not all countries have been successful in attracting FDI equally.

Figure 9.11 shows a comparison between India and Malaysia with respect to the share of FDI in GDP. It is clearly observed that India's FDI orientation had been much lower than Malaysia for the entire period under consideration (1975–2017), except the year 2008. India started with a very low and negative FDI-GDP ratio of 0.01% in 1975. Until 2001, India's FDI-GDP ratio remained below 1% of GDP. Between 1975 and 2001, Malaysia's average FDI-GDP ratio was 4.3% and in some years, the ratio reached 8–9%. India's FDI-GDP ratio reached its peak level of 3.6% in 2008. However, since 2009, the ratio started falling and in 2017, the ratio stood at 1.5%. Figure 9.12 suggests that in comparison to Malaysia, India is actually experiencing a declining share of FDI in GDP at a much lower level of GDP per capita.

Raihan (2017a) argued that certain factors are key to attracting FDI, and policies should be designed to take into account these factors. To attract FDI, relevant trade policy reforms leading to higher degree of openness are essential. With the increased importance of globalization, trade openness has become a key component to growth. Liberalization of trade leads to greater specialization and division of labor, leading to higher productivity and export capabilities. Furthermore, infrastructural development is needed to attract larger FDI in an economy. FDI is positively associated with the magnitude of domestic investment. Low or stagnant domestic investment may show lack of business confidence by the domestic investors, which may convey negative messages to foreign investors. Therefore, the government needs to improve the business environment, reduce the cost of doing business, and facilitate domestic investment through eliminating policy-induced and supply-side constraints.

9.4.5 Making Special Economic Zones (SEZ) Effective

SEZs in India have to deliver what they promise. Raihan (2016) argued that the standards of infrastructure and business environment within SEZs have to be up to the global marks. Delays in implementation and unsatisfactory delivery of services would make the SEZs unsuccessful. The

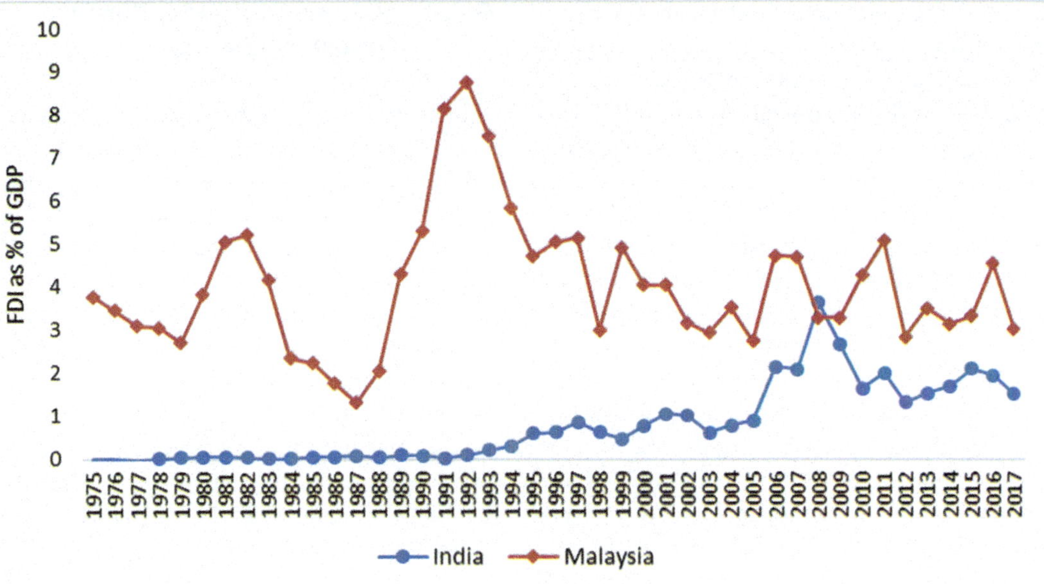

Fig. 9.11 FDI as % share of GDP in India and Malaysia. (Data source: World Development Indicators, World Bank)

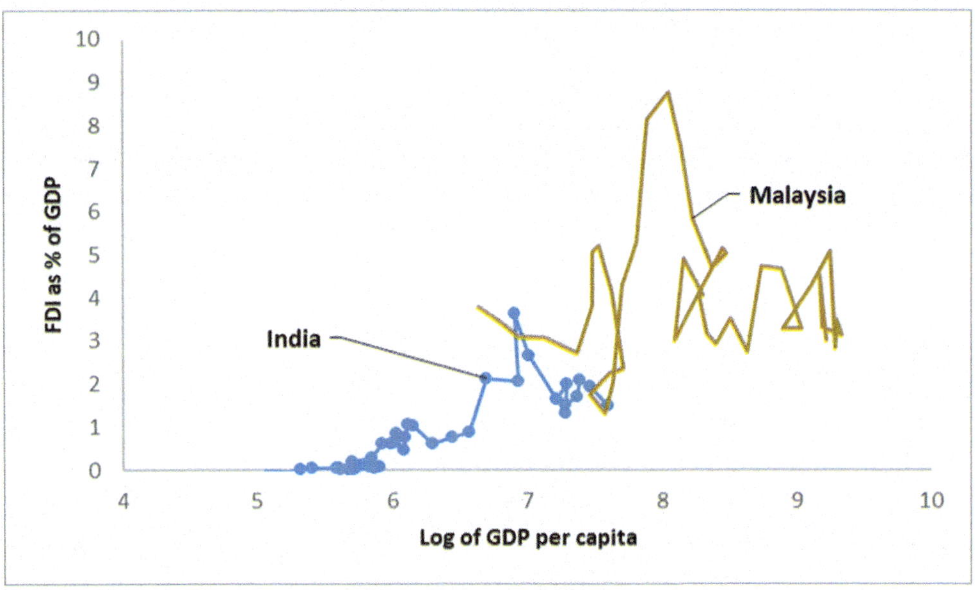

Fig. 9.12 India's declining share of FDI in GDP at a low level of GDP per capita. (Data source: World Development Indicators, World Bank)

contrasting experiences of China and India are very relevant. While China was very successful in establishing well-functioning SEZs by effectively addressing infrastructural and land issues, India, so far, in most cases failed to do the same. It is also important to understand that while SEZs are aimed at creating 'efficient' enclaves, improvements in the business environment and infrastructure of the overall economy cannot be overlooked. If there are vast differences in the quality of infrastructure and business environment between SEZs and rest of the economy, then excessive and continued external support would be needed for the survival of SEZs, which can have large financial implications. SEZs would need to be connected to 'efficient' sea and land ports. Otherwise, many of the benefits of the SEZs would be lost. Therefore, port infrastructure and its efficiency would need to be improved substantially. Furthermore, the quality of roads, connecting SEZs and ports, would need to be upgraded. In contrast to India, China's success was in establishing SEZs nearby efficient ports and developing much-improved road networks connecting SEZs with those ports. SEZs should aim for facilitating economic and export diversification, leading to progressive structural transformation of the economy. Emphasis should be on production of high value-added and diversified products. Therefore, sectors with high potentials of economic and export diversification should get the priority in the SEZs. The whole issue of the management of SEZs is very important. The gravity of institutional aspects for the good functioning of SEZs cannot be underestimated. Therefore, it has to be ensured that the institutions governing the operations of SEZs are competent enough.

9.4.6 Human Capital for Rapid Industrialization

Education is crucial for enhancing human capital in an economy, which in turn increases workers' productivity and thus contributes to economic growth. The importance of investing in human capital has been discussed in the economic literature for long. Probably, the strongest argument for investment in human capital came from the endogenous growth theory, which highlights that investment in human capital together with innovation and knowledge are important contributors to economic growth. As the global market moves toward accelerated automation, increasing the investment in human capital is now more important than ever. What often causes the difference between the ability of workers in the developed and developing countries is the poor performance of the education system in the developing countries.

Despite the fact that India made considerable progress in gross-enrolment in primary and secondary education, the country is seriously lagging behind in ensuring quality education for all. If we consider the 'average years of schooling' (Fig. 9.13) as an indication of the status of education of any country, in 2017, the 'average years of schooling' in India was only 6.4 which was

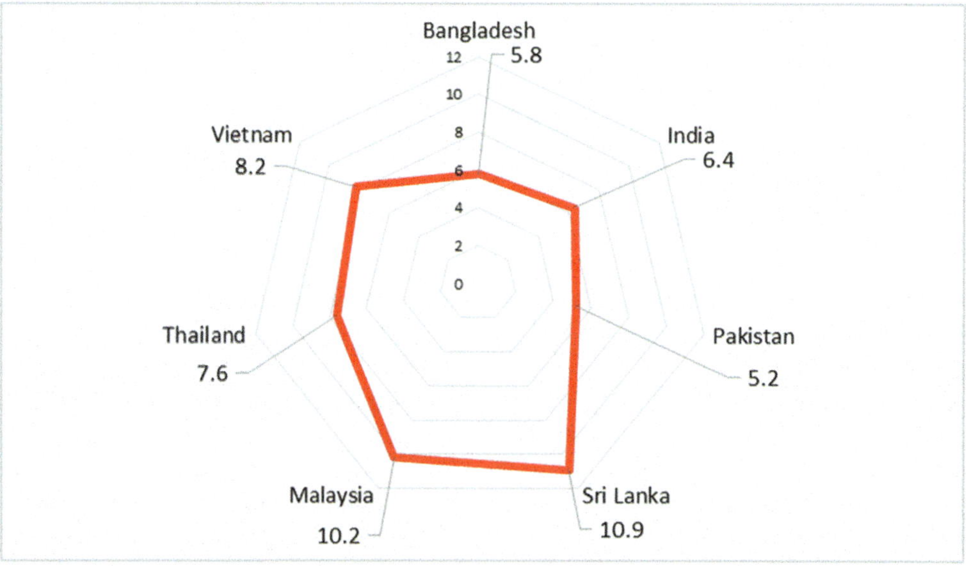

Fig. 9.13 Average years of schooling in India in a comparative perspective in 2017. (Source: UNDP)

higher than Pakistan (5.2) and Bangladesh (5.8). However, India was far behind Sri Lanka (10.9) and some of the leading Southeast Asian countries like Malaysia (10.2), Thailand (7.6), and Vietnam (8.2).

Therefore, as India is lagging behind in educational infrastructure and outcome, the government must consider using the public expenditure on education as a critical tool to achieve the targets. India should re-evaluate its prioritization of public spending, and reorient such spending more toward social sectors like education and health. It should also be kept in mind that the increase in the ratio of public expenditure on education to GDP should coincide with the improvement in the quality of institutional arrangements in the education systems.

9.5 Conclusion

How to substantially increase the manufacturing value-added share in GDP and thus promote manufacturing-led economic growth in India? As Raihan (2017b) argued, India has to adopt the right kind of policies and programs which can trigger much faster rate of growth of the manufacturing sector compared to those of agricultural and services sectors. The experiences of the successful countries show that human capital has made a major difference. In India, compared to the East and Southeast Asian countries, both the quantity and quality of human capital are at much lower levels. Therefore, policies and programs should be targeted at the rapid enhancement of human capital in India. There is also a need for pro-active trade and industrial policies in terms of providing effective incentives to domestic investors, setting up special economic zones and attracting foreign direct investment (FDI) for diversified manufacturing industries. Such policies should also be aimed at integration of the domestic manufacturing industries with the global value chains.

References

Raihan (2016) How to make Special Economic Zones successful. The Daily Star. https://www.thedailystar.net/op-ed/economics/how-make-special-economic-zones-successful-1325266

Raihan (2017a) Ways to attract investment through SEZs. The Daily Star. https://www.thedailystar.net/drivers-economy/ways-attract-investment-through-sezs-1364833

Raihan (2017b) Are South Asian countries 'deindustrialising' prematurely? The Daily Star. https://www.thedailystar.net/op-ed/are-south-asian-countries-deindustrialising-prematurely-1400050

Rodrik D (2016) Premature deindustrialization. J Econ Growth 21:1–33. https://doi.org/10.1007/s10887-015-9122-3

UNESCAP (2017) Asia-Pacific countries with special needs development report 2017. Bangkok

10 How Is India Flaring in Achieving SDG 10 on Reduced Inequality?

Zubayer Hossen and Bazlul Haque Khondker

10.1 Introduction

Economic inequality refers to the unequal distribution of income and opportunity between different groups in society. Inequality can take many forms. Many leading scholars have argued that inequality is multifaceted in nature, as inequality is associated with many high-priority indicators (Scheidel 2017; Piketty 2014; Milanović 2016). Inequality is present in income structure and also can be observed in health status, educational attainment, employment, access to food and water, access to social security, opportunity, choice and many more. Interestingly, aforementioned aspects of inequality are intertwined such as improvement in water and sanitation can mitigate inequality in health status or income inequality can be marginalised by obtaining parity in educational activities.

For the past few decades, inequality has been on the rise across the globe and thus, it has been a concern for almost all countries around the world. Though some countries have reduced the numbers of people living in extreme poverty, economic gaps have continued to grow as the richest amass unprecedented levels of wealth. Since 1980, the share of national income going to the top 1% has increased rapidly in North America (defined here as the United States and Canada), China, India and Russia and more moderately in Europe. On the contrary, the Middle East, sub-Saharan Africa, and Brazil, have had relatively stable, but extremely high levels of inequality. Though rapid economic growth in Asia, particularly in China and India, has lifted many people out of extreme poverty, the richest 1% of the people have been mainly benefitted by the economic gains. Their share of global income is still much higher than their 16% share in 1980, although it has declined at more than 20% since the 2008 financial crisis (World Inequality Report 2018).

There are many research works showing that inequality affects the economic growth. That is why, the relationship between aggregate output and the distribution of income is an important topic in macroeconomics (Galor 2011). In recent years, the impact of income inequality on economic growth has received significant attention in policy circles. The different discourses show that the changes in income inequality have an effect on GDP per capita. However, the degree of effect may differ between rich and poor countries.

Z. Hossen (✉)
South Asian Network on Economic Modeling (SANEM), Dhaka, Bangladesh

B. H. Khondker
South Asian Network on Economic Modeling (SANEM), Dhaka, Bangladesh

Department of Economics, University of Dhaka, Dhaka, Bangladesh

Policy Research Institute (PRI), Dhaka, Bangladesh

S. Hazra, A. Bhukta (eds.), *Sustainable Development Goals*, Sustainable Development Goals Series,
https://doi.org/10.1007/978-3-030-42488-6_10

There is a major concern regarding poverty and inequality among the policy-makers in India, though the economy has been going through steady growth in the recent years. Ali and Son (2007) mentioned that indeed sustained growth is a necessary condition for reducing poverty, but it is not the sufficient condition because of the presence of inequality. Chandrasekhar and Ghosh (2015) revealed that inequality in India has not shown any nosedive (considering income and consumption) since her independence, even though growth has been commendable. On the other hand, according to the statistics of RIS (2016), although there has been an upward trend in overall inequality, state-level disaggregation shows some states with negative or stagnated inequality during 2000–2010. Such heterogeneity calls for rigorous research to understand the phenomenon more clearly, which can lead to construct better policy framework.

Sustainable Development Goal (SDG) on reduced inequality (SDG 10) recognises inequality within and among nations as a major concern despite progress in and efforts at narrowing disparities of opportunity, income and power. Therefore, SDG 10 sets the targets to reduce the inequality and determines the indicators to measure the gap for each target. Keeping the SDG 10 in the background, Government of India has taken initiatives to reduce inequality. Different programmes have been undertaken as per the comprehensive strategy of social inclusion, women empowerment and access to better social security. In some areas, the progress is visible. Gini Coefficient of income has fallen from 36.8 to 33.6 during the years 2010–2015. The advancement is also noticed at the state level. Some states are already on track to achieve the SDG 10 by 2030. Among 29 states, 3 states are categorised as achievers; 20 states are identified as front runners and 2 states are defined as aspirants (NITI Aayog 2018). However, though there are successes, India will need to make a long jump to achieve the targets by 2030.

Against this backdrop, this chapter attempts to explore the targets and indicators under SDG 10, identify the challenges in implementing the goal, assess the current progress achieved by India at the national level and state level, conduct a cross-country analysis to understand where India actually stands, discuss the lessons that India can take from other better performing countries, and suggest way forward for an inclusive society.

10.2 Synopsis of SDG 10

SDG 10 echoes the necessity of establishing a society where each person will enjoy the benefits of the economic progress equally. Thus, it accentuates on progressively reducing not only income inequalities but also inequalities of outcome. This is possible only by ensuring access to equal opportunities and promoting social, economic and political inclusion of all, irrespective of age, sex, disability, race, ethnicity, religion or other status relevant within a society. Since the goal also stresses on reducing inequality among countries, it asks for enhancing representation and voice for developing countries in decision-making in international institutions. With this aspiration, the following targets and indicators were developed by the Inter-Agency and Expert Group on SDG Indicators (IAEG-SDGs) and agreed to as a practical starting point at the 47th session of the UN Statistical Commission held in March 2016.

Though SDG 10 correctly identifies increasing inequality across borders as a threat to inclusive development, there are two potential challenges in implementing the goal: (1) equivocal conceptualisation and (2) absence of required data. The targets under SDG 10 cover very broad areas and thus, they do not specifically mention about how to reduce inequality within country or across countries. Besides, like the other SDGs, the unavailability of data does not always allow to assess the current situation and estimate the needs in order to achieve the goal.

Inequality takes on many forms in a large and diverse country like India. There are inequalities in income and consumption; structural inequalities which take the form of inequalities based on gender, religion, caste and social groups as well as regional inequalities, all of which manifest in inequalities of opportunities and access. India has

Box 10.1 SDG 10 in the Global Context[a]

Targets	Indicators
10.1 By 2030, progressively achieve and sustain income growth of the bottom 40% of the population at a rate higher than the national average	10.1.1 Growth rates of household expenditure or income per capita among the bottom 40% of the population and the total population
10.2 By 2030, empower and promote the social, economic and political inclusion of all, irrespective of age, sex, disability, race, ethnicity, origin, religion or economic or other status	10.2.1 Proportion of people living below 50% of median income, by age, sex and persons with disabilities
10.3 Ensure equal opportunity and reduce inequalities of outcome, including by eliminating discriminatory laws, policies and practices and promoting appropriate legislation, policies and action in this regard	10.3.1 Proportion of the population reporting having personally felt discriminated against or harassed within the previous 12 months on the basis of a ground of discrimination prohibited under international human rights law
10.4 Adopt policies, especially fiscal, wage and social protection policies, and progressively achieve greater equality	10.4.1 Labour share of GDP, comprising wages and social protection transfers
10.5 Improve the regulation and monitoring of global financial markets and institutions and strengthen the implementation of such regulations	10.5.1 Financial Soundness Indicators
10.6 Ensure enhanced representation and voice for developing countries in decision-making in global international economic and financial institutions in order to deliver more effective, credible, accountable and legitimate institutions	10.6.1 Proportion of members and voting rights of developing countries in international organizations
10.7 Facilitate orderly, safe, regular and responsible migration and mobility of people, including through the implementation of planned and well-managed migration policies	10.7.1 Recruitment cost borne by employee as a proportion of yearly income earned in country of destination 10.7.2 Number of countries that have implemented well-managed migration policies
10.A Implement the principle of special and differential treatment for developing countries, in particular least developed countries, in accordance with World Trade Organization agreements	10.A.1 Proportion of tariff lines applied to imports from least developed countries and developing countries with zero-tariff
10.B Encourage official development assistance and financial flows, including foreign direct investment, to States where the need is greatest, in particular least developed countries, African countries, small island developing States and landlocked developing countries, in accordance with their national plans and programmes	10.B.1 Total resource flows for development, by recipient and donor countries and type of flow (e.g. official development assistance, foreign direct investment and other flows)
10.C By 2030, reduce to less than 3% the transaction costs of migrant remittances and eliminate remittance corridors with costs higher than 5%	10.C.1 Remittance costs as a proportion of the amount remitted

Source: SDG Knowledge Platform, United Nations
[a]https://sustainabledevelopment.un.org/sdg10

a number of legislations and programmes for empowerment and socio-economic development of women and different social groups. There are several programmes in place which aim to provide equal opportunity in education to girls and children from vulnerable sections of the society. The Government of India's emphasis on the JAM trinity (Jan Dhan–Aadhaar–Mobile) presents a broad strategy of inclusion, financial security and social empowerment. India has several national level schemes like Pradhan Mantri Jan Dhan Yojana, Deen Dayal Upadhyay Grameen Kaushal Yojana and employment schemes like Prime Minister Employment Generation Programme, which are aimed at reducing social, economic and political gaps and progressively achieving greater equality in the country.[1]

[1]NITI Aayog (2018). *SDG India Index: Baseline Report 2018*. New Delhi, India: NITI Aayog.

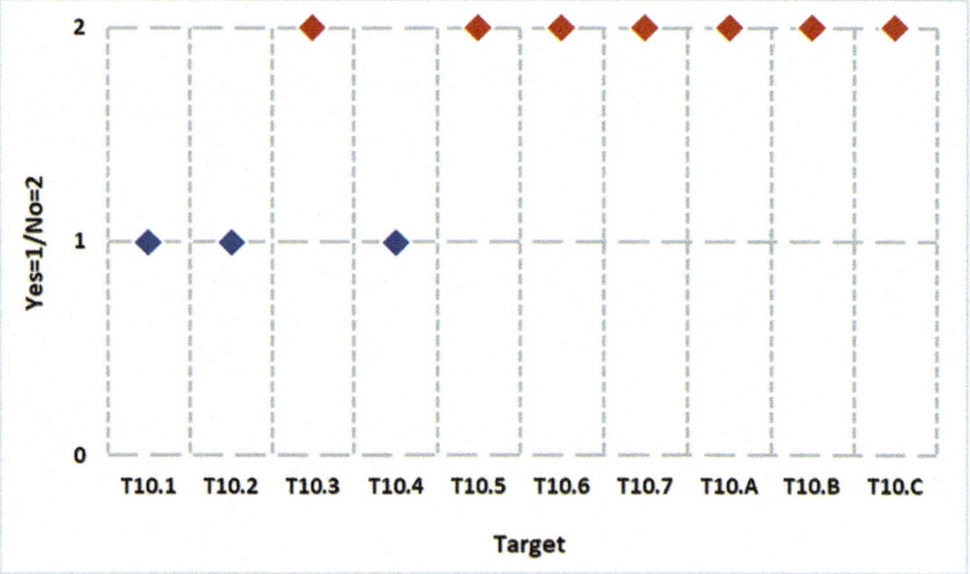

Fig. 10.1 Targets accepted by India. (Data Source: NITI Aayog (2018))

Box 10.2 Indicators Adopted by India

SDG global target	Indicator selected for SDG India index	National target value for 2030
10.1 By 2030, progressively achieve and sustain income growth of the bottom 40% of the population at a rate higher than the national average	1. Palma ratio of household expenditure in urban India[a]	1
	2. Palma ratio of household expenditure in rural India[2]	1
10.2 By 2030, empower and promote the social, economic and political inclusion of all, irrespective of age, sex, disability, race, ethnicity, origin, religion or economic or other status	3. Ratio of transgender labour force participation rate to male labour force participation rate	1
10.4 Adopt policies, especially fiscal, wage and social protection policies, and progressively achieve greater equality	4. Percentage of scheduled caste sub plan fund utilisation	100
	5. Percentage of tribal sub plan fund utilised	100

Data Source: NITI Aayog (2018)

[a]Measured as the ratio of the monthly consumption expenditure of the top 10% households to the monthly consumption expenditure of the bottom 40% households

To measure India's performance towards reduced inequality, five national-level indicators have been identified, which capture three out of the ten SDG targets for 2030 outlined under SDG 10. These indicators have been selected based on availability of data at the national level and to ensure comparability across States and Union Territories (UTs) (Fig. 10.1).

10.3 India's Headway Analysis

The international community, especially the vulnerable nations—the least developed countries, the landlocked developing countries and the small island developing states—has made significant steps towards lifting people out of poverty. However, disparities among people are still prominent within these countries. Inequality within countries has risen, while income inequality between countries may have been reduced. Inequality impedes the progress through depriving people of opportunity and subjecting many to conditions of extreme poverty. Rising inequalities adversely impact human development. According to the inequality-adjusted human development index (HDI), sub-Saharan Africa loses 33% of its HDI to inequality and South Asia by 25%. That is why, special effort is required to make the economic growth inclusive and reach out to everyone with the benefits of growth.

To achieve SDG 10, higher income growth of the bottom 40% of the population, universality in principle policies, attention to the needs of disadvantaged and marginalised populations, inclusion of all in social as well as political spheres, etc. will be fundamental. To achieve greater equality and promote social, economic and political inclusion of all by 2030, the Government of India has emphasised on the three-pronged Jan Dhan–Aadhaar–Mobile programmes aiming at a comprehensive strategy of inclusion, financial empowerment and social security. As a consequence of cumulative effort, at the national level, progress is observed for the selected indicators under SDG 10. The data indicate that India has shown better performance in all the indicators except indicator 1—Palma ratio of household expenditure in urban India. The ratio is 1.41 for India, where the target is 1.0 (Table 10.1).

In case of performance analysis at the regional level, the progress rate varies across states and Union Territories (UTs). The states and UTs have been given an SDG index score based on their performance in the indicators. The SDG index score for SDG 10 ranges between 38 and 100 for states and between 52 and 100 for UTs. Based on the score, the States and UTs have been divided into achiever (100), front runner (65–99), performer (50–64), and aspirant (0–49). Among the states, Meghalaya, Mizoram and Telangana are the achievers, while Dadra and Nagar Haveli, Daman and DIU and Lakshadweep are the achievers among the UTs. Totally 29 states and UTs are front runners and two states are aspirants (NITI Aayog 2018). The majority of the less progressed states and UTs have an issue in common. They have performed poorly in indicator 1 and as a result, have received a lower score (Fig. 10.2).

10.4 Where Does India Stand? A Cross-Country Analysis

The twenty-first century has witnessed economic development. Also, it has countersigned the concentration of wealth in a few hands as its necessary condition. In the year 2000, top 1% Indians enjoyed 74,935 USD per capita of wealth where the remaining 99% of individuals had only 1300 USD per capita of wealth. Over time, this gap between the richest and the poorest has proliferated. In 2000, the top 1% of population enjoyed 58 times the wealth of the rest of the population and this gap has widened to 95 times in 2014. Nevertheless, India is not the only country that is facing the challenge of increasing income inequality. During the last four decades, with the increasing economic development and growth, majority of the developing nations have experienced rise in

Table 10.1 Performance of India on indicators for SDG 10

No	Indicators	Raw Data		Index Score	
		India	Target	India	Target
1	Palma ratio of household expenditure in urban India	1.41	1	50	100
2	Palma ratio of household expenditure in rural India	0.92	1	100	100
3	Ratio of transgender labour force participation rate to male labour force participation rate	0.64	1	61	100
4	Percentage of scheduled caste sub plan fund utilisation	77.67	100	68	100
5	Percentage of tribal sub plan fund utilised	82.98	100	76	100

Data Source: Household Consumer Expenditure Report 2011, Census 2011, Labour Bureau Report, and States Reports to NITI Aayog

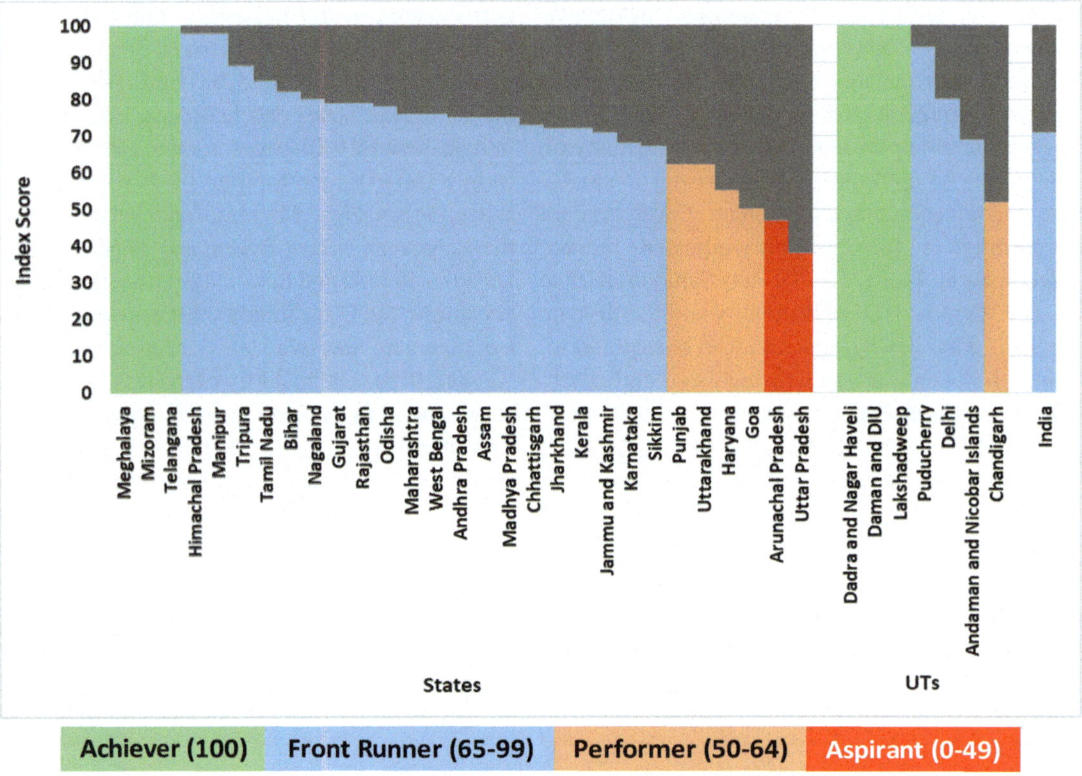

Fig. 10.2 Index score of States and UTs on SDG 10. (Source: NITI Aayog (2018))

income inequality. For instance, in most OECD countries, income inequality has widened during the past two or three decades. From the mid-1980s to the late 2000s, according to Gini coefficient, income inequality rose by 10%, while the ratio of top 10% to bottom 10% has reached its highest level in 30 years. The rise in income inequality has been far from uniform between countries. A decline has even been observed in some countries. The OECD area experienced a sort of 'inequality convergence' from the mid-1990s until the late 2000s since inequality increased in countries such as Sweden, Denmark and Finland and fell in countries such as Turkey, Mexico and Chile. However, on the aspect of income inequality, the Palma ratio[2] for the latest available year endorses the existence of income inequality across different countries. While countries like Iceland, Norway, Sweden, Denmark, and Belgium have a ratio less than or equal to 1, most of the countries have a ratio greater than 1. There are countries such as South Africa, Namibia, Botswana and Zambia with very high inequality. Besides, the list includes several developed or high-income countries with the presence of income inequality. India has a Palma ratio of 1.5, which confirms the concentration of wealth in the hands of the richest people (Fig. 10.3).

In India, the analysis of the performance of development expenditure illustrates that though the poor should have been the mostly benefitted, the development spending has benefited the rich more effectively and thereby raised the inequality within country, between states and within states. At the national level, inequality is broadly found to have risen in India during the period 2000–2015. In 2000, the share of top 10% in national income was 39.9% and it went up to 56.1%. In

[2]The Palma ratio is a measure of inequality. It is the ratio of the richest 10% of the population's share of gross national income (GNI) divided by the poorest 40% share.

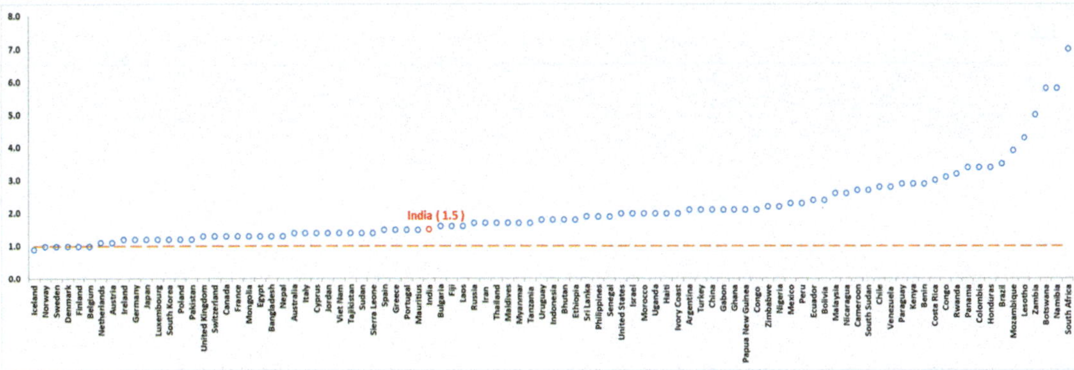

Fig. 10.3 Palma ratio for the latest available year. (Data Source: Human Development Indices and Indicators, 2018 Statistical Update, UNDP)

the same period, the national income distribution data show that bottom 50% of people had an average share of 17.3%. The share for bottom 50% was 20.6% in 2000 and it declined to 14.7% in 2015. Therefore, the share of bottom 50% decreased by 5.9 percentage point in this period. Moreover, the share of middle 40% also declined by 10.3 percentage point (Fig. 10.4).

The national income share of top 1% population has been stable with a slight fluctuation in recent years, however. For India, though the share of top 1% has not increased much, the share is still very high. During the period 2010–2015, the share was estimated to be, on an average, 21.3% for India, where the world average was 20.7%. For the same period, the average share of the richest 1% was lower for both Europe and Asia (excluding Middle East countries). They were 10.1% and 16.5%, respectively. However, it is evident from the analysis that like many other developing countries, the nature of economic growth for India has been inequality-enhancing. The available statistics affirms that there is inequality between income groups and it is in upsurge. The existing interventions have remained unsuccessful at reducing the inequality at an expected level. While poverty has fallen, most of those who have escaped poverty continue to face a high risk of falling back into it. Moreover, those who remain poor are increasingly chronically poor, and may be particularly difficult to reach through the introduction or expansion of safety nets. As a consequence, the income gap between lower and higher income groups is rising (Fig. 10.5).

Despite having short-term and long-term poverty reduction strategies, India has been unsuccessful in reducing inequality at the expected level. One of the potential explanations of this can be the Kuznets inverted-U hypothesis of income inequality. According to Kuznets (1955), as an economy develops, market forces first increase and then decrease economic inequality. Hence, as per capita national income of a country increases, in the initial stages of growth, inequality in income distribution rises and after reaching the highest degree in the intermediate level, the income inequality falls. Since India is in its development stage, inequality is showing an increasing trend. In India, the Kuznets' ratio and national per capita income data show that the inequality rate has still been increasing. Since the rate is yet to reach the maximum point, it will continue to escalate in near future. While India has a rising Kuznets' ratio, it has been stable at 1.46 for Sri Lanka for last two decades. Of course, compared to Sri Lanka, India, as a bigger economy and is more diversified. India has a large number of population and there are so many religions, castes and languages. Therefore, India faces more challenges in reducing inequality. However, the Kuznets' ratio has, on an average, also slumped for high-income countries and lower middle-income countries (Fig. 10.6).

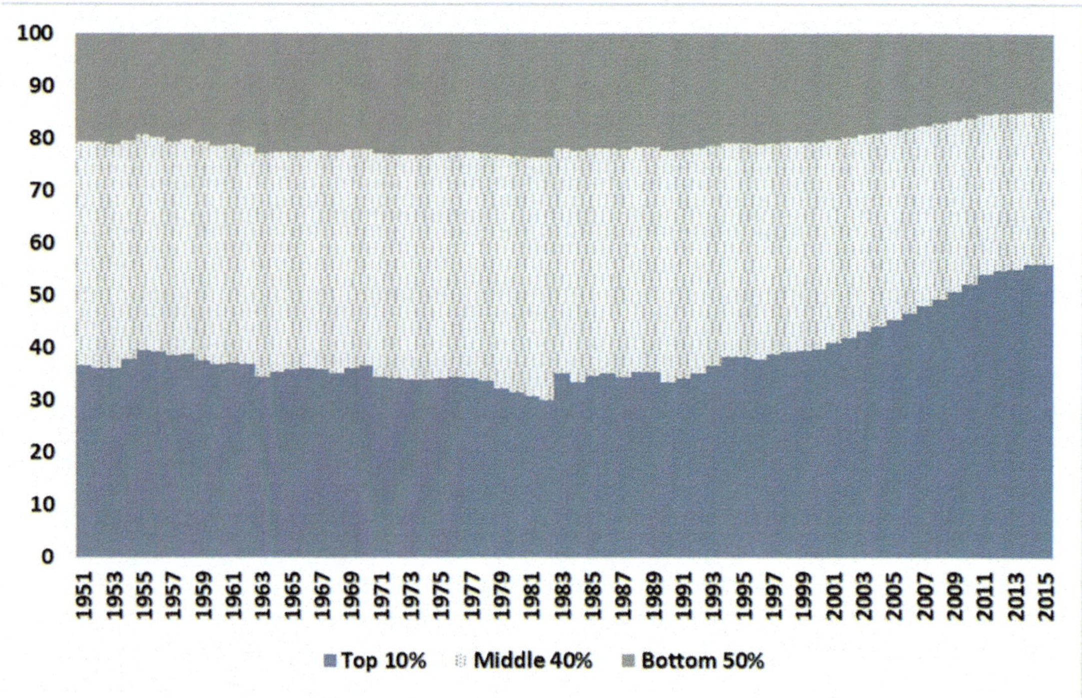

Fig. 10.4 National income distribution for India (%). (Data Source: World Inequality Database)

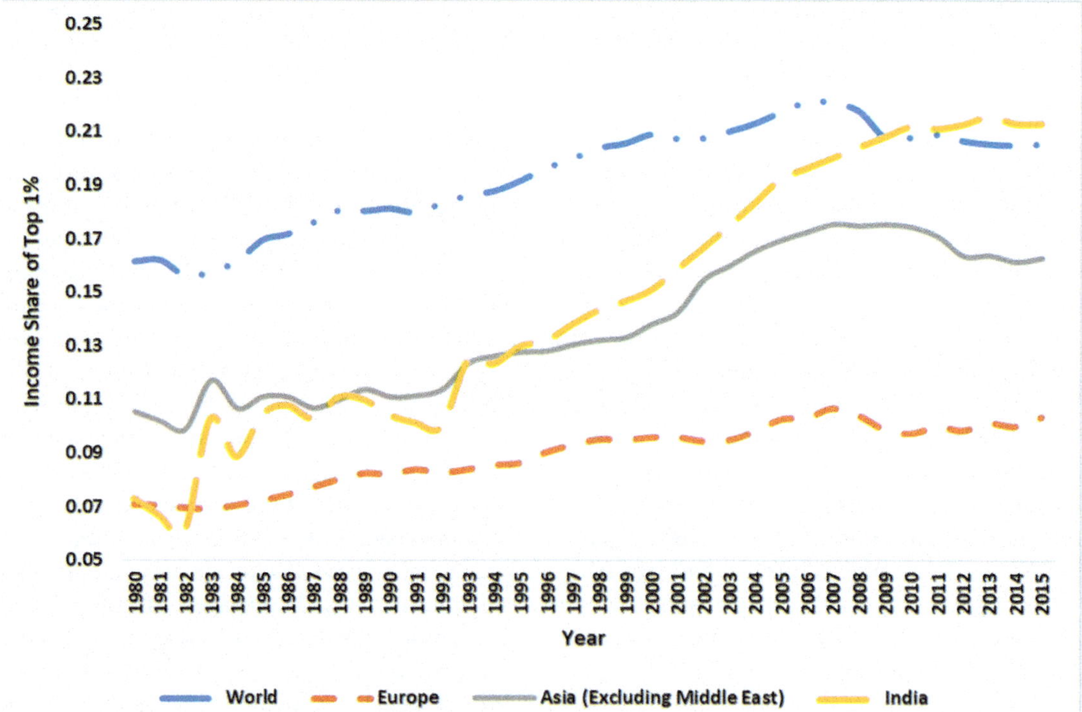

Fig. 10.5 National income share of top 1%. (Data Source: World Inequality Database)

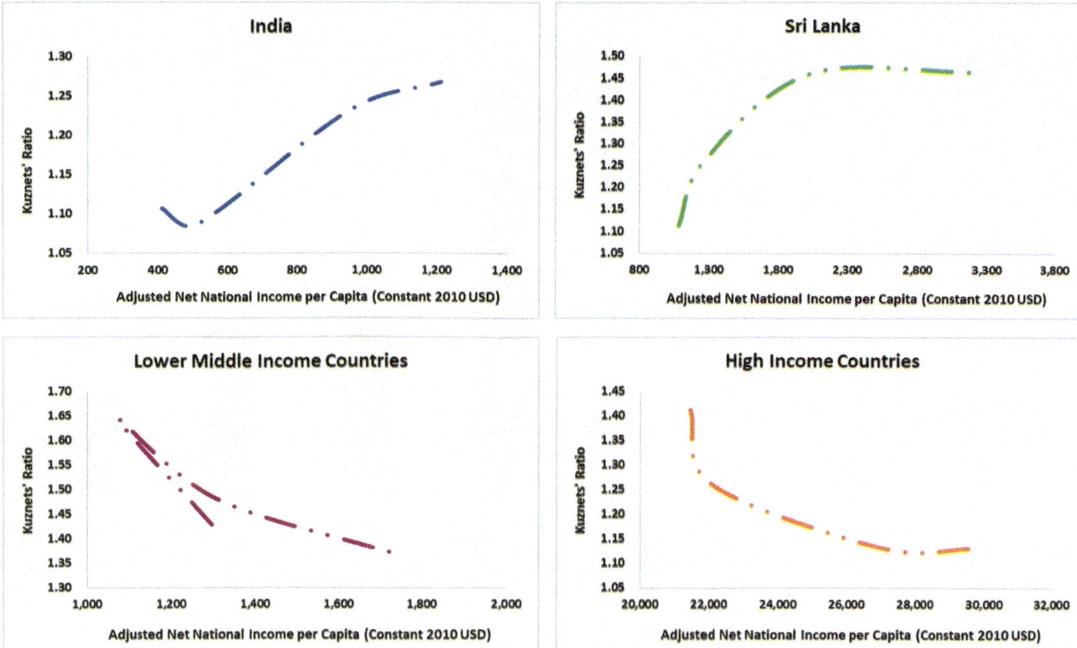

Fig. 10.6 Kuznets' ratio and recent trend of income inequality (The Kuznets' ratio is the income share of the top 20% of the population to the bottom 60% of the population.) Note: The data period for this figure is 1981–2017. Since the income distribution data are not available for each year, the authors have divided the period into four different periods: 1981–1990, 1991–2000, 2001–2010 and 2011–2017. For each period, the average has been calculated and used to construct the figure. (Data Source: World Development Indicators and Poverty and Equity, the World Bank)

10.5 Lessons from Other Countries

The current growth model pursued in India (perhaps in other emerging and developing economies as well) will tend to increase inequality and thereby dent the growth impacts on poverty. The main resource of a poor person is labour; ensuring a decent and productive employment is thus a key prerequisite for inclusive economic growth for reducing poverty and improving inequality. According to Beyer (2018)[3] economic growth in South Asia (including India) has failed to generate enough employment. They termed this outcome as 'Jobless' growth. However, an attempt to change this growth process may not be easy with the advent of automation and artificial intelligence. A better approach may be to use fiscal instruments namely tax financed social protection to smooth consumption of poor, help develop their skills, and address their life cycle risks with the aim to reduce poverty and inequality.

Social protection instrument emerged long ago during 1880s and 1990s. Some of the European countries developed these instruments to tackle rising poverty against the backdrop of industrialisation and rural to urban migration. The programmes were known as 'poor relief' programme and countries spent more than 1% of their GDP on these programmes. Gradually the 'poor relief' programmes have been replaced with well-designed inclusive social protection system based on the risks associated with life cycle of a citizen. An inclusive life cycle social protection system generally covers risks of various stages of life, such as pregnancy and early childhood; school children; youth; working age and the elderly. The European countries now allocate around 25% of their GDP on inclusive life cycle-based social protection system (Fig. 10.7).

[3]Beyer, Robert Carl Michael (2018), "*Jobless growth? (English)*. South Asia economic focus." Washington, D.C.: World Bank Group. http://documents.worldbank.org/curated/en/825921524822907777/Jobless-growth.

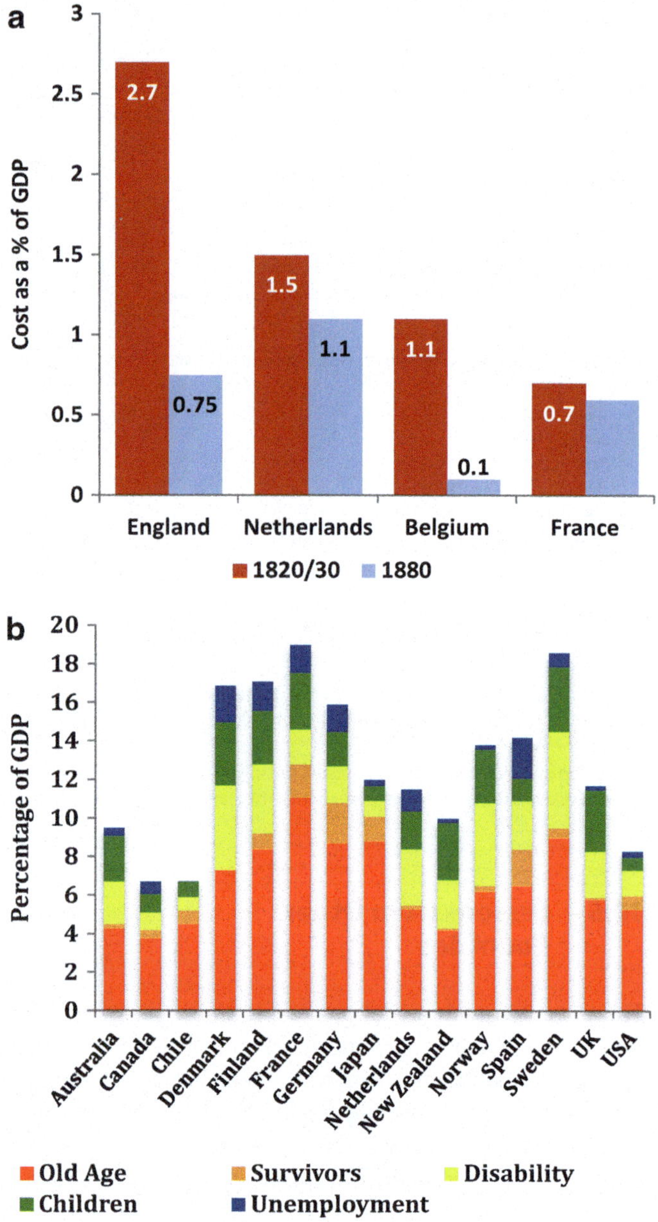

Fig. 10.7 (**a**) Poor relief budgets in the nineteenth century: 18320/0 and 1880 (% of GDP). Source: Lindert (2004, Growing Public: Social Spending and Economic Growth since the Eighteenth Century; Volume 1: The Story. Cambridge University Press: New York). (**b**) Spending on social protection schemes in developed countries (% of GDP). (Data Source: Social Expenditure Database, OECD)

Some of the emerging (Brazil, South Africa, Argentina, and Mexico etc.) and developing economies (Bangladesh) are also following the experiences of the European countries (particularly the experiences of the OECD countries) by embracing an inclusive life cycle-based social protection system. The figure below clearly shows that emerging economies are covering a larger segment of their citizen with social protection allocation of reasonable amount of resources. In comparison, India has embraced a 'poor relief' social protection system spending around 2.5% of their GDP covering only 19% of their population. It is now time for India to discard the 'poor relief' approach and embrace the inclusive life cycle-based social protection system (Table 10.2).

Table 10.2 Beneficiary coverage, spending on social protection and Gini index for selected countries

Country/region	Beneficiary coverage (% of population)	Spending (% of GDP)	Gini
India	**19.0**	**2.7**	**33.9**
Argentina	67.0	11.2	44.5
Brazil	59.8	18.3	56.7
South Africa	48.0	10.1	63.1
China	63.0	6.3	42.1
Russia	90.4	15.5	40.1
Bangladesh	34.3	2.4	32.1
Sri Lanka	30.4	6.5	36.4
Eastern Europe	85.0	17.6	30.0
Northern, Southern and Western Europe	95.0	25	28.3

Data Source: ILO and UNDP

10.6 Vision for India's Future: Way Forward for an Inclusive Society

1. As mentioned above, India should adopt an inclusive life cycle social protection system in place of their 'poor relief' programme over the medium term. At the same time, beneficiary coverage must be increased to at least cover the vulnerable population (as opposed to only poor or extreme poor)—perhaps reasonable approximation is 40–50% of the population with flexibility of variations by regions. The coverage expansion may be justified by two reasons: (1) low-level beneficiary coverage is associated with large underestimation of poor (i.e. known as exclusion errors); and (2) use of low national poverty line excludes large number of genuine poor and vulnerable persons from poverty count. India must also increase spending on inclusive life cycle social protection system to 3% of GDP by 2025 and 6% by 2030 to cover more beneficiaries as well as to increase the size of transfer payments. By 2030, India may aim to adopt a universal child grant and universal social pension for age 65+ population with a provision for self-exclusion.
2. Every year, India has been creating 5–5.5 million jobs. But with an average labour force participation rate of 50–55%, India may need to generate 6–6.5 million jobs per year over the medium to longer run (i.e. 2030). This envisages that, every year, one million new entrants to the labour market cannot find productive work. Furthermore, according to an estimate in Envisioning India 2030,[4] another one million employment per year is needed to absorb some of the workforce who could not find suitable work over the last decade or so. At the same time, India must also try to raise female labour force participation to at least 50% from the current level of 27% for their full use. Adding it all up implies that India must increase employment by 8.5–9 million people per year over the medium to longer run (i.e. 2030).
3. Job market is changing fast in India (like other countries). In line with this changing job dynamic, India must create mechanism to train the workforce on 'future skills'. New courses and training programmes must need to be developed to impart youth training to meet the future demand. Furthermore, 'new-age learning labs' in educational institutions for hands-on training on advanced technologies should also be set up.
4. Social sector (i.e. education and health) spending in India is low at around 4% of GDP—with around 1% on health and 3% on Education, while East Asian economies spend far more to build the human capital founda-

[4]FICCI (2018), "Envisioning India 2030", Federation of Indian Chambers of Commerce and Industry, Tansen Marg, New Delhi – 110001.

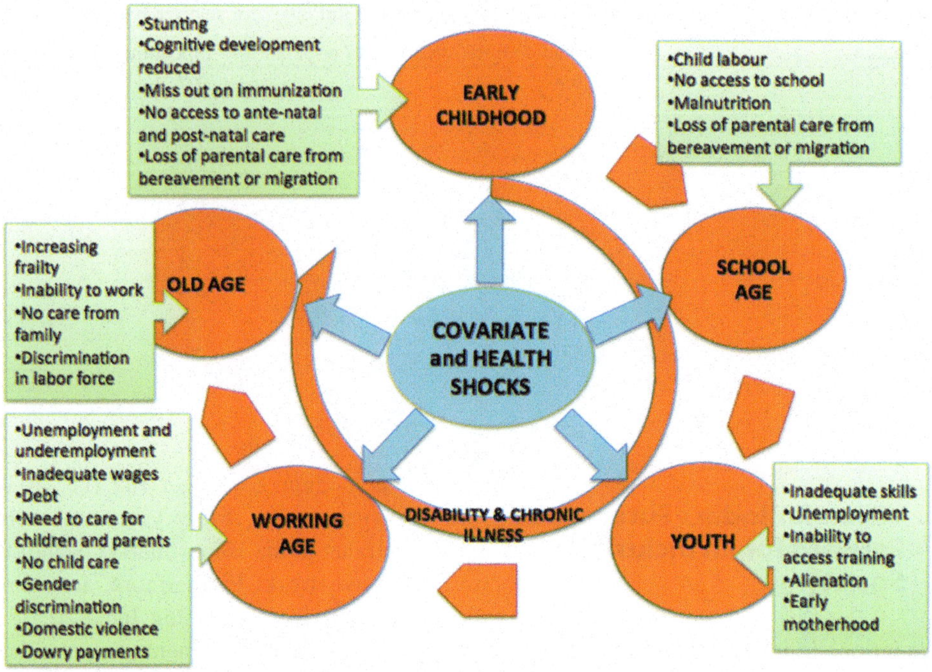

Fig. 10.8 Summary of the main life cycle risks and challenges. (Source: Kidd et al. (2017))

tion for growth. Thus, India must increase social sector spending, focusing on improving the pupil–teacher ratio by hiring more teachers. India has two million unfilled jobs in the public sector. This huge vacancy may be used to hire more teachers and health workers.

5. India is a large populous country with large variations in economic and social development, resulting in divergence between poverty rates and inequality. Special attention must thus be attached to the relatively less well-off (or lagging) regions. India may create a lagging region funds to propel economic and social development of these lagging regions (Fig. 10.8).

10.7 Conclusion

Growing inequality across nations and within nations is a major concern depressing economic growth and growth impacts on poverty reduction, and causing social unrest and conflict. Given the importance to abate the rising inequality trend, SDG 10 attaches special emphasis on ways to reduce the inequality. Like many countries, inequality is increasing in India. Data show that in India, there are both income inequalities and inequalities of access to opportunities by gender, religion, caste and region. Although in recent years income inequality has fallen, India has a long way to go to achieve SDG 10 target for 2030. India must embark on polices and strategies to reduce inequality to attain SDG 10 by 2030. They include diverse sets of policies and strategies encompassing inclusive social protection system, expanding social sector spending and innovative employment-generating schemes.

References

Ali I, Son H (2007) Defining and measuring inclusive growth: application to the Philippines. ERD Working Paper Series No. 98. ADB, Manila

Beyer RCM (2018) Jobless growth? (English). South Asia economic focus. World Bank Group, Washington, DC

Chandrasekhar CP, Ghosh J (2015) Growth, industrialisation and inequality in India. J Asia Pac Econ 20(1):42–56

Galor O (2011) Inequality, Human Capital Formation, and the Process of Development. Handbook of the Economics of Education, North Holland

FICCI (2018) Envisioning India 2030. Federation of Indian Chambers of Commerce and Industry, Tansen Marg, New Delhi-110001

Kidd S, Freeland N, Khondker B (2017) International best practice in social protection: implications for Bangladesh. In: Alam S (ed) Background research papers for preparing the national social security strategy of Bangladesh. General Economic Division (GED), Planning Commission, Government of the People's Republic of Bangladesh, Sher-e-Bangla Nagar, Dhaka

Kuznets S (1955) Economic growth and income inequality. Am Econ Rev 45(1):1–28

Lindert PH (2004) Growing public: social spending and economic growth since the eighteenth century. Cambridge University Press, New York

Milanović B (2016) Global inequality: a new approach for the age of globalization. Harvard University Press, Cambridge

NITI Aayog (2018) SDG India index: baseline report 2018. NITI Aayog, New Delhi

Piketty T (2014) Capital in the twenty-first century. Belknap Press, Cambridge

RIS (2016) India and sustainable development goals: the way forward. New Delhi

Scheidel W (2017) The great leveler: violence and the history of inequality from the stone age to the twenty-first century. Princeton University Press, Princeton

UNDP (2018) Human development indices and indicators: 2018 statistical update. New York

World Inequality Report (2018) World Inequality Lab, Paris School of Economics

Sustainability in Indian Cities: A Developing Country Perspective

11

Hippu Salk Kristle Nathan

11.1 Why Cities?

Large-scale urbanization is a phenomenon of the twentieth century. As per United Nations (2018a), in 1950, only three-fourths of a billion people in the world lived in urban areas, which constituted 29.6% of the inhabitants. The share of the urban population increased overtime and in 2008, it touched and went beyond 50%, i.e. for the first time in human history, majority of the population started living in urban areas (PRB 2008) (Fig. 11.1). At present, 4.2 billion people live in urban areas, which is 55% of the total population and by 2030, this will grow to approximately 5.2 billion i.e. 60% of the world's total population, and by 2050, to 6.7 billion, i.e. more than two-thirds of the total (United Nations 2018a).

The United Nations (2018a) estimate shows during 1950–2020, the population growth in urban area is at a compound annual growth rate (CAGR) of 2.6%, whereas the CAGR for the entire population is at 1.6%. In the next three decades, 2020–2050, the CAGR for urban area will be at 1.4%, whereas the total population is expected to grow at a rate of 0.76% (United Nations 2018a). The growth rate of population worldwide in rural areas started falling since 1970s and currently it is becoming negative (Fig. 11.2). However, the urban population will be on increasing growth rate and will taper around 2040.

The nature, impact and scale of the changes resulting from this urbanization process are more important than the statistics themselves. Firstly, urbanization is 'inevitable'. External factors and internal policies may briefly increase or decrease the pace of urbanization, but in the long run, there is hardly any change in the course of urbanization (Tannerfeldt and Ljung 2006). Secondly, the process of urbanization is neither balanced nor geographically uniform, but tends to cause 'polarization'. Industrialization, migration and economic globalization have driven the phenomenon of progressive concentration of population in urban agglomerations. This global phenomenon will ultimately lead to the single complex, diverse reality, i.e. world with densely populated 'crystallized cities' and each city crystal having not so densely populated clouds of sub-urban or rural areas linked to it physically, socially, and economically.

Though rapid urbanization is arguably the most complex and important socio-economic phenomenon of the last century, it is acknowledged that the impact of urbanization will continue to bring about major global and local changes well into the current century (Allen 2009). The impact of urbanization goes beyond urban boundary.

H. S. K. Nathan (✉)
Institute of Rural Management, Anand, Gujarat, India
e-mail: hippu@irma.ac.in

S. Hazra, A. Bhukta (eds.), *Sustainable Development Goals*, Sustainable Development Goals Series,
https://doi.org/10.1007/978-3-030-42488-6_11

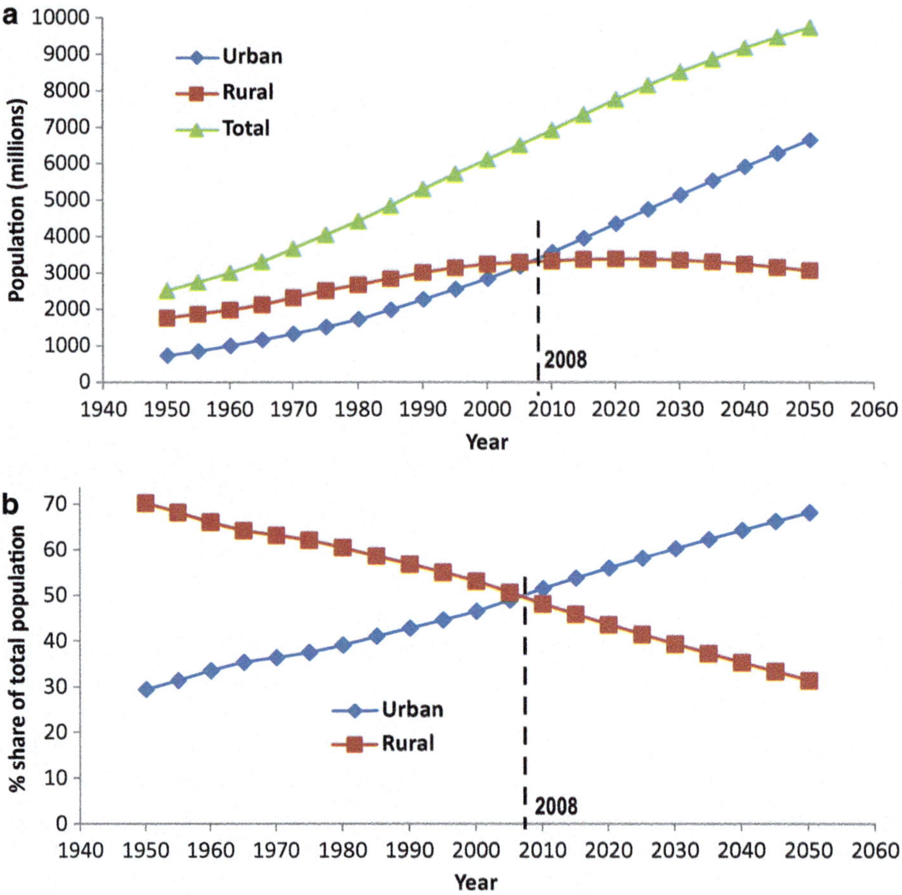

Fig. 11.1 (**a**) Populations in the world: urban, rural, total: 1950–2050. (**b**) Share of urban and rural populations of the world: 1950–2050. (*Data source*: United Nations (2018a))

Fig. 11.2 Average annual additions to urban and rural population: 1950–2050

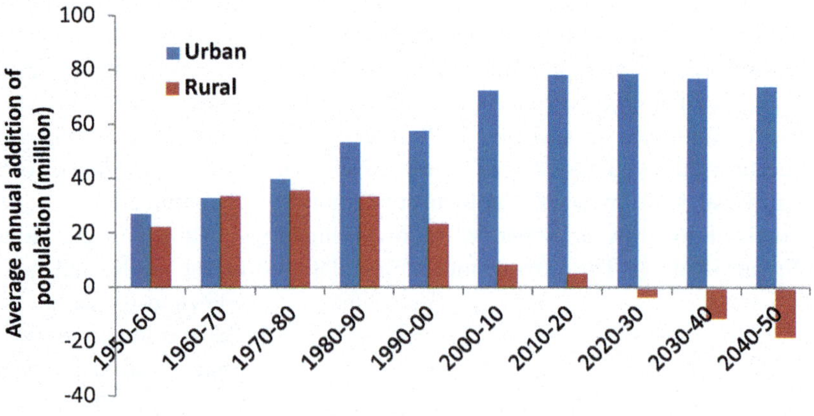

Because of the information flow, mass media, and communication and transportation the rural–urban divide has blurred and there are fewer and fewer places on the planet that are unaffected by the dynamics of cities (UNFPA 1999; Kenneth 2005). The ecological and sociological footprints of cities have spread over wider areas and cities will continue to have a significant impact on the global carrying capacity of the earth (UNFPA 1999; Montgomery et al. 2004; Allen 2009).

Urbanization and socio-economic development are closely linked. Countries with high GNP per capita are in most cases more urbanized (World Bank 2000). Urban economies in developing countries as well as in industrialized countries are related to economies of scale and agglomeration. They are characterized by proximity of labour, capital, technology—which are specialized and diversified; and by markets—which are large and networked (SIDA 1995). National economies are largely influenced by cities; two-thirds of the GDP of many countries is already generated in major urban centres, and 80% of future anticipated economic growth will be in cities. As per World Bank (2000), there is a weak but clear relationship between the degree of urbanization and the level of quality of life which citizens enjoy. Appropriate urban development does not only benefit cities but enhances rural development too; they are not alternative, but mutually reinforcing. Urban areas play an important role as providers of markets, services, and employment for rural surroundings (BMBF 2004; SIDA 2006).

In the paper entitled 'Why research on mega-urban regions matters!', Heinrichs (2005) has reasoned the socio-cultural significance of urbanization. The author argues that cities are the locus where social conflicts are played out and creative change takes place. They are laboratories of change that absorb local, national and global changes and impacts, which continuously combine and condense into new 'opportunities' and 'risks'. Cities influence global sustainable development because of their speed, scale and scope of global connectedness. Urban areas are engines of economic, scientific, and cultural development and centres of knowledge, information, and other social resources that foster democratic movements (SIDA 2006). They provide the future with more choices (SIDA 2008). So, cities have more positives than negatives; and the problems surrounding cities are because of the failure of effective governance within cities rather than inherent characteristics of cities themselves (Williams et al. 1999; Jenks 2000).

11.2 Sustainability of Cities

Urbanization is understood as a shift from a predominantly rural to a predominantly urban society with major and irreversible changes in production and consumption and the way people interact with nature; and it is time to examine cities through 'sustainability' lens (Allen 2009). The ecological footprint of cities is far bigger than their actual territory.[1] Cities occupy just 3% of land area, but consume 70% of energy and are responsible for three-fourths of the carbon emissions (United Nations 2019). Such concentration of resources and its disparity across urban populace result in sustainability risks in all dimensions—economic, social, and ecological.

Sustainability of resources in urban area has multiple dividends—both economic and non-economic. Urban areas are fulcrums of economic activities and urbanization is considered both as a consequence of and as a requirement for economic development. Increasingly, urban growth is influencing and is being influenced by continued global economic integration and the struggle for countries—and cities—to be competitive in the global marketplace (Cohen 2004). Cities are also centres of modern living, where the participation of female labor force is greater and where indicators of general health and well-being, literacy, women's status, and social mobility are typically higher (Cohen 2006). However, urbanization must be judged not only from a perspective of prosperity, but also from the point of view of maintaining a natural balance. Urbanization is not necessarily bad, but the rapid change, as

[1] 'Ecological footprint' is an accounting tool that enables us to estimate the resource consumption and waste assimilation requirements of defined human population or economy in terms of productive land area (Wackernagel and Rees 1996).

experienced by developing countries, brings a lot of socio-economic and environmental problems (SIDA 1995), which puts a question mark on their sustainability.

The Goal 11— 'Make cities and human settlements inclusive, safe, resilient and sustainable'—of Sustainable Development Goals addresses the above-mentioned concern. This goal is spelt out in terms of seven targets concerning, (1) safe and universal housing with basic services, (2) sustainable disabled-friendly transportation, (3) participatory urban planning and management, (4) safeguarding natural heritage, (5) protection and resilience against disaster, (6) waste management, and (7) access to green and public spaces (UNSC 2016). The SDG 11 has also additional targets relating to positive linkages between urban, peri-urban, and rural areas, climate adaption and mobilization of local resources (UNSC 2016). All these targets are further expressed under 15 indicators as given in Table 11.1 (UNSC 2016).

11.3 Cause of Continuous Cities

The world has urbanized even faster than originally predicted by 1972 Malthusian report, *Limits of Growth* (Davis 2004). Urbanization has two primary components, viz. the natural increase and net migration. The age structure of the urban population explains the boom in natural increase. Urbanites, being predominantly young, are in reproductive age with a wish to have children, which contribute to higher natural increases (Devas and Rakodi 1993; SIDA 1995; Kamete

Table 11.1 Indicators of goal 11

Sl. no.	Indicators
1	Proportion of urban population living in slums, informal settlements or inadequate housing
2	Proportion of population that has convenient access to public transport, by sex, age and persons with disabilities
3	Ratio of land consumption rate to population growth rate
4	Proportion of cities with a direct participation structure of civil society in urban planning and management that operate regularly and democratically
5	Total expenditure (public and private) per capita spent on the preservation, protection and conservation of all cultural and natural heritage, by type of heritage, level of government (national, regional and local/municipal), type of expenditure, and type of private funding
6	Number of deaths, missing persons and persons affected by disaster per 100,000 people
7	Direct disaster economic loss in relation to global GDP, including disaster damage to critical infrastructure and disruption of basic services
8	Proportion of urban solid waste regularly collected and with adequate final discharge out of total urban solid waste generated, by cities
9	Annual mean levels of fine particulate matter (e.g. PM2.5 and PM10) in cities (population weighted)
10	Average share of the built-up area of cities that is open space for public use for all, by sex, age and persons with disabilities
11	Proportion of persons victim of physical or sexual harassment, by sex, age, disability status and place of occurrence, in the previous 12 months
12	Proportion of population living in cities that implement urban and regional development plans integrating population projections and resource needs, by size of city
13	Proportion of local governments that adopt and implement local disaster risk reduction strategies in line with the Sendai Framework for Disaster Risk Reduction 2015–2030
14	Number of countries with national and local disaster risk reduction strategies
15	Proportion of financial support to the least developed countries that is allocated to the construction and retrofitting of sustainable, resilient and resource-efficient buildings utilizing local materials

et al. 2001).[2] Migration is identified to be an important cause of urban population growth, particularly in the early stages of urbanization. Much of urban migration is driven by rural populations' desire to avail the advantages that urban areas offer on the one hand and to escape from the disadvantages of rural society on the other hand. The World Bank (1995) maintains that people in rural areas, being drained by poverty and natural disasters, make a rational choice to move to urban areas, as urban areas provide much better livelihood opportunity, fewer social restrictions and an expectation of a better life that has largely been met. In urbanization literature, these factors are classified as standard 'push-pull' factors (see Table 11.2).

The 'push' factors are poverty-stricken rural economy (which primarily depends on low-yielding subsistence agriculture), lack of livelihood options (minimal or no presence of industry and service sectors), natural disasters like floods and draughts and social restrictions (to the disadvantaged caste, creed or gender). The 'pull' factors are greater access to livelihood opportunities (portfolio diversification), education, health care, energy options, and services such as entertainment. The pull factors also include perceived higher wage, greater infrastructure, and general attraction towards faster and economically more vibrant life and lifestyle. In general, the urban poor have fewer opportunities than the urban non-poor, but still they are better off than rural populations.

Table 11.2 Push and pull factors in rural–urban migration

Push factors	Pull factors
Poverty	Prosperity and wealth
De-pesantization (agrarian crisis)	Livelihood portfolio diversification
Natural disaster, resource scarcity	Infrastructure, market
Social restrictions (caste, gender)	Education, health care and other services

11.4 Developed–Developing Divide

Both in terms of rate and extent of urbanization and cause and implication of it, there is a marked difference between developed and developing economies. However, the phenomenon of large-scale urbanization and growth of mega and million-plus cities is global in nature, and more so in the developing world. In the last 60 years, there has been an accelerated movement of people towards cities in developing countries and the trend will continue in future decades. The increasing number of megacities in developing countries is a testimony to this.[3] In 1950, there was only one megacity, i.e. New York. In 1975, there were 4, and in 2000, they were 18. In 2018, there are 33 such cities of which 27 are located in developing countries (United Nations 2018b). It is important to note that a substantial portion of future population growth will not be in huge agglomerations since they have saturated and will be in smaller cities and towns, where the challenge to accommodate so many people is not small though (World Bank 2000; UN-Habitat 2003). Jenks (2000) reports most aggressive growth to be occurring in the million plus cities, i.e. cities with population 1–10 million. There were 86 million-plus cities in 1950 that grew nearly to 400 by the turn of the century (United Nations 2002), and by 2018, there are 548 of them, and by 2030, there will be more than 700 such cities (United Nations 2018b).

Global Urban Observatory (2003) has noted that 95% of the build-up of humanity during 2000–2030 will occur in the urban areas of developing countries. The rapidness of urbanization in developing region is evident from the fact that in 2000, for every one person living in cities in developed countries, there are two in the cities of

[2] Natural growth accounts for nearly 60% of urban growth globally (Devas and Rakodi 1993).

[3] Cities with more than ten million inhabitants are called megacities (United Nations 2018b).

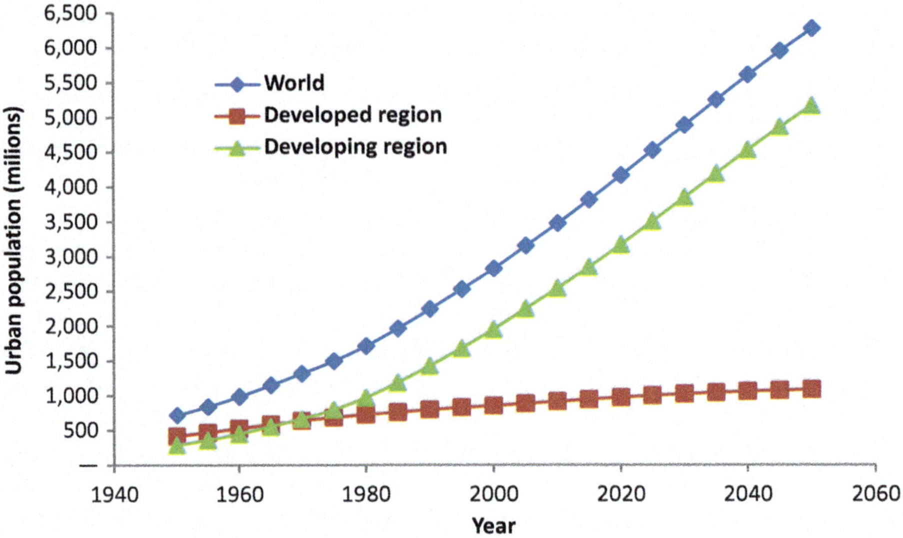

Fig. 11.3 Urban population in developed and developing regions: 1950–2050

the developing world. Within 30 years, this proportion is predicted to rise to 1:4 (Jenks 2000). As 120,000 people in developing countries move to cities every day (Cities Alliance 2008), urban areas have become centres of high resource consumption. Urban populations consume much more food, energy, and durable goods than their rural counterparts (Parikh and Gokran 1991). Cities have been seen as the cause of environmental degradation and resource depletion, casting an ecological footprint across the globe, far beyond their immediate regions (Girardet 1996; Wackernagel et al. 1997). The greater demand for the resources in urban areas is largely driven by high income and consumption of the rich class. So, at a country level, there has been a polarization of resources between urban and rural areas.

Region-wise,[4] developed countries were highly urbanized by the 1950s and the pace of urban growth slowed down in these countries. The developing countries, on the other hand, were just beginning an accelerating urbanization process during the same time. From Fig. 11.3, it is apparent that all most all the future growth in world urban population will be due to population growth in the urban areas of developing countries, averaging 2.57% per year during 2010–2050. In contrast, the urban population of the more developed regions is expected to increase very slowly, from 0.93 billion in 2010 to 1.10 billion in 2050 (United Nations 2009).

Studying the urban share of population for developed and developing regions in 1950, 2000 and 2030 (projected) one can see that the share of urban–rural population in developed regions in 1950 almost resembles that of less developed region for 2030 (Fig. 11.4).[5] Less developed regions face a time lag of approximately 80 years with respect to the developed world in the urbanization process.

The push and pull factors, described in the previous section, acted in different intensities in developed and developing countries. Urbanization in developed countries followed European model that is characterized by law of advancing human-

[4]The regional division follows United Nations (2009) classification. The 'Developed region' comprises Europe, Northern America, Australia/New Zealand and Japan. 'Developing region' is the rest of the world.

[5]This categorization of developed and less developed does not mean that cities in one region are all alike. There are similarities, as well as significant differences, in cities across this divide (Jenks 2000). However, this oversimplification is useful in macro-sense.

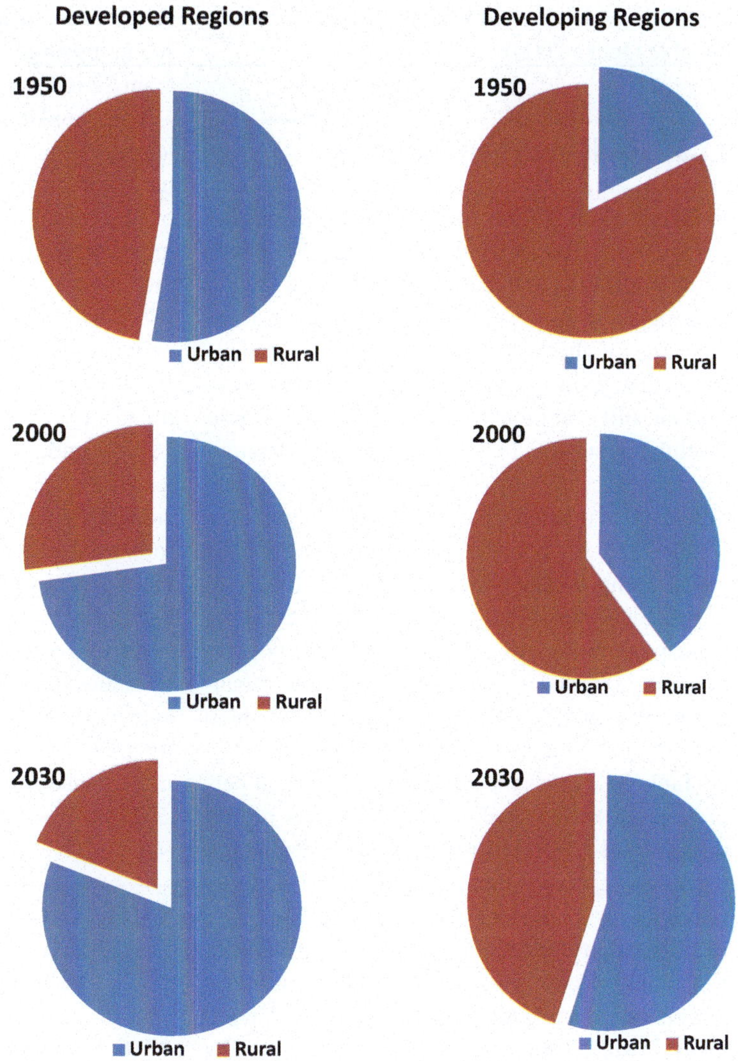

Fig. 11.4 Snapshots of urban–rural population shares of developed and developing regions: 1950, 2000 and 2030

ity through advancing technology wherein 'pull' factors of urban areas were more significant. This urbanization is explained through people migrating to urban areas to get absorbed into a growing urban economy. But, the second-generation urbanization, which followed in developing region, has occurred with very limited, stagnant or negative urban economic growth. The urbanization in sub-Saharan Africa, Latin America, the Middle East, and parts of Asia are examples of such kind. This urbanization process, which is at times referred to as 'over-urbanization' (Gugler 1982; Simon 1995), is largely explained by 'push' factors like de-peasantization and lack of livelihood options and services in rural areas.[6] Hence, there is an accelerating movement of surplus rural labour to urban slums even as cities ceased to be job machines (Davis 2004). This is interpreted as transfer of poverty from rural to urban areas, termed as 'urbanization of poverty' in literature. In this context, urban areas acted as a refuge from an impoverished and restricted

[6]Relationship between urbanization and industrialization is weak for developing countries (Burgess 2000).

rural life, which has turned economically unsustainable (Jenks 2000).

11.5 Indian Urban Surge

India is the second most populous country in the world after China and with more than 17% of the world population and 2.3% land area, it is second to Bangladesh in population density among large countries of the world (UNSD 2010).[7,8] India's population growth has been accompanied by a gradual shift over time from a largely rural agrarian dwelling population to one that lives in urban non-agricultural centres (MGI 2010). The growth rate difference in urban population and total population justifies the same (Table 11.3, Fig. 11.5). The annual compound growth rates (ACGR) of the total population during 1951–2011 is 2.0%, whereas the corresponding figures for urban and rural population are 3.0% and 1.7%, respectively. The urbanization rate in 1951 is slightly above 17%, which has increased to more than 30% in 2011.

Table 11.4 gives the number of towns and percentage of urban population by size class of city during 1901–2001.The million-plus cities within the Class I cities, though only 35 in number accounts for 38.6% of the population. The share of population in different city sizes is plotted in Fig. 11.6, which shows continuous concentration of population and activities in large cities (Kundu 1983; Datta 2006). MGI (2010) estimates that by 2030, India's urban population will rise to 590 million, which is twice the population size of the United States today. As per the same estimates, India will have 68 million plus cities by 2030, whereas entire Europe has 35 such cities today. Seventy per cent of the new employment would be generated in cities and the urban economy will see an upsurge in urban middle-class population from 22 to 91 million.

The rapid urbanization comes with tough challenges to conquer. In the past, India has witnessed how rapid urbanization has led to massive growth of slum followed by misery, poverty, unemployment, exploitation, inequalities, and overall degradation in the quality of urban life (Datta 2006).[9] Also, like other developing countries, India is a party to urbanization primarily instrumented by 'rural push' rather than 'urban pull' and transfer of poverty from rural to urban areas (Mukherji 1993; Kundu et al. 2001; Datta 2006). India's cities are critical for inclusive growth; hence, provision of better quality life to city dwellers is an important challenge sustainable urban development policy of the country will face. City's sustainability must consider both global and local perspectives. Sustainable urban forms will only be achievable if they are underpinned by a policy background which commits to global sustainability goals, but leaves room for local formation and implementation of solutions (Williams et al. 2000).

11.6 India's City Sustainability Challenges: *A Developing Country Syndrome*

11.6.1 Incomplete Realization of Merits of Urbanization

Urbanization is a compaction or densification process, which will result in reductions in travel distances and thus vehicle emissions, and that the high densities can create greater viability for service provision, public transport, waste disposal, health care and education, among other things

[7]Large countries mean countries having area of 100,000 km² or more.

[8]It is interesting to note that India's population by 2009 itself was more than one sixth of the world and is more than 1.5 times of that of entire Europe (World Bank 2011). India is the most populous country in the world after China, but before 2030, it will cross China and get the distinction, which it will almost certainly never lose (Haub and Sharma 2006).

[9]India's urbanization has been criticized in literature for concentration of population in a few large cities without a corresponding increase in their economic base, and has been termed as 'pseudo urbanization' (Breese 1969) or 'dysfunctional urbanization' (Raza and Kundu 1978) or 'involuted urbanization' (Mukherji 1993).

Table 11.3 Population and urbanization in India: 1951–2011

Year	Total population (crore)	Population growth rate[a] (%)	Urban population (million)	Urbanization rate[b] (%)	Urban population growth rate[a] (%)
1951	36.1	NA	6.2	17.29	NA
1961	43.9	1.98	7.9	17.97	2.37
1971	54.8	2.24	10.9	19.91	3.29
1981	68.3	2.23	15.9	23.34	3.87
1991	84.6	2.16	21.8	25.71	3.16
2001	102.9	1.97	28.6	27.82	2.77
2011[c]	120.4	1.58	36.3	30.13	2.40

Source: Census of India (2001)

[a]Growth rate is calculated as Annual Compound Growth Rate (ACGR). ACGR for 1961–1971 is given against 1971 year

[b]Urbanization rate is defined as the ratio of urban population to total population

[c]Population data for 2011 is from PRB (2007). ACGR for urban population during 2001–2011is 2.4% (HPEC 2011)

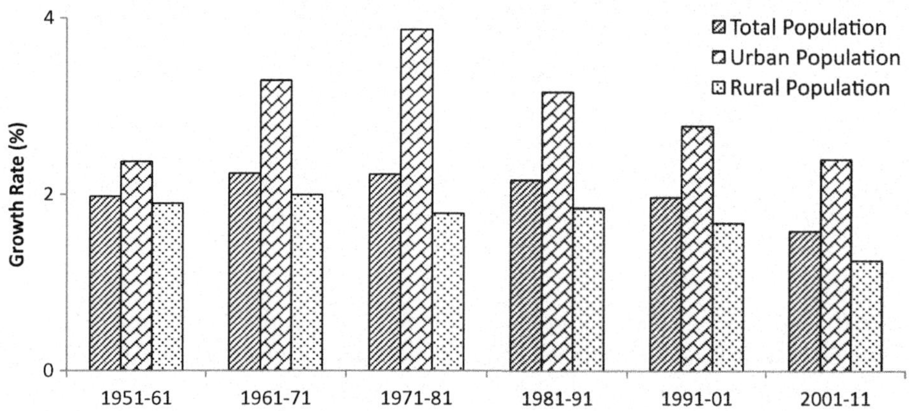

Fig. 11.5 Growth rate in population (total, urban, and rural) in India: 1951–2011

Table 11.4 Number and population proportion in cities by size class: 1901–2011

	No. of towns/cities by size class[a]						Percentage of urban population by size class					
Year	I	II	III	IV	V	VI	I	II	III	IV	V	VI
1901	24	43	130	391	744	479	26.0	11.3	15.6	20.8	20.1	6.1
1911	23	40	135	364	707	485	27.5	10.5	16.4	19.7	19.3	6.6
1921	29	45	145	370	734	571	29.7	10.4	15.9	18.3	18.7	7.0
1931	35	56	183	434	800	509	31.2	11.7	16.8	18.0	17.1	5.2
1941	49	74	242	498	920	407	38.2	11.4	16.4	15.8	15.1	3.1
1951	76	91	327	608	1124	569	44.6	10.0	15.7	13.6	13.0	3.1
1961	102	129	437	719	711	172	51.4	11.2	16.9	12.8	6.9	0.8
1971	148	173	558	827	623	147	57.2	10.9	16.0	10.9	4.5	0.4
1981	218	270	743	1059	758	253	60.4	11.6	14.3	9.5	3.6	0.5
1991	300	345	947	1167	740	197	65.2	11.0	13.2	7.8	2.6	0.3
2001	393	401	1151	1344	888	191	68.7	9.7	12.2	6.8	2.4	0.2
2011	468	474	1373	1686	1784	424	70.2	8.5	11.1	6.4	3.4	0.5

Source: Nallathiga et al. (2018)

[a]Population range for the size class of the cities is as follows: Class I: >1,00,000, Class II: 50,000–1,00,000, Class III: 20,000–50,000, Class IV: 10,000–20,000, Class V: 5000–10,000, Class VI: <5000

Fig. 11.6 Growth of cities by size class: 1901–2011. (*Source*: Nallathiga et al. (2018))

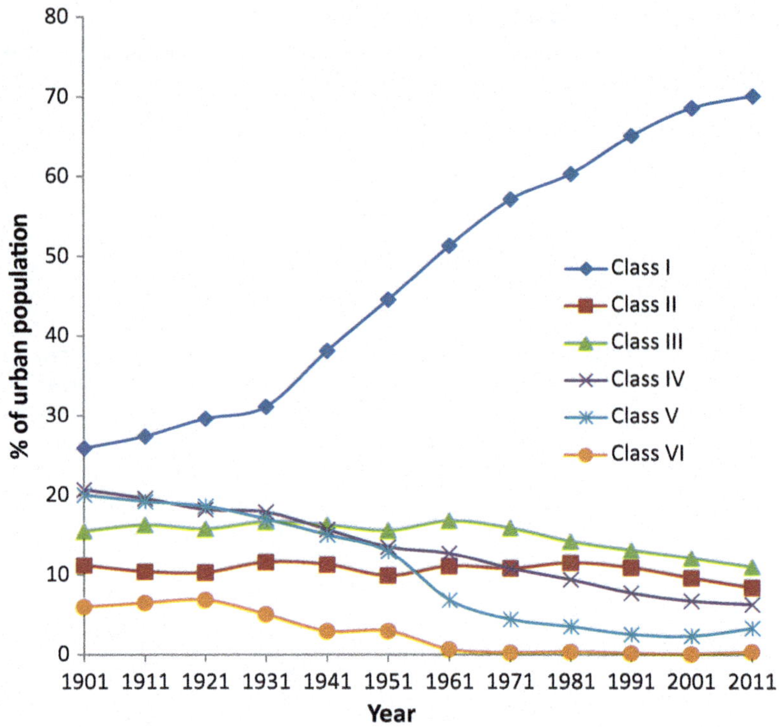

(Burgess 2000). However, the merits of urbanization are more realized in case of developed economies. Less developed economies like India experience high demographic growth, low-levels of economic development, high income inequalities, small urban budgets, and shortages of environmental infrastructure, shelter, and basic services. For these, the merits of urbanization are not fully realized and often they result in counterproductive unsustainable risks in terms of housing and sanitation crisis (in informal settlements or slums), infrastructure overload, overcrowding, congestion, air pollution, environmental degradation, health hazards, and lack of public and green space.

11.6.2 Genesis of Slums

The genesis of slums is an outcome where the push factor dominates in the process of urbanization. In many developing economies like India, because of the unsustainable rural economy, surplus labor is pushed out to look for avenues in cities. These people settle themselves in pavements or informal dwellings (slums) and earn and live at a subsistence level. An estimated 70% of the urban population in the least developed countries lives in slums without proper shelter, water, sanitation, electricity, transport and other infrastructure and services (SIDA 2006). In India, the latest census shows that there is 22.4% of the urban population living in slums and in absolute terms, there is a 25% rise in the slum population between 2001 and 2011 (GoI 2019).

11.6.3 Demand Outstripping the Supply

Densification is preferred in the developed world because higher densities in cities will lead to cheaper infrastructure costs and absorption of spare capacity. However, the same is not true for cities in developing countries like India, as there is no spare capacity and the existing capacity is being overused. Demand overtaking supply in cities can be exemplified taking Mumbai's case. Mumbai's suburban trains are world's most overcrowded (Railway Gazette 2010), and run at least three times more than the rated capacity during peak hours (CEE–WR 2011). Also, high scarcity coupled with high demand for land in Mumbai makes its rent for commercial office the highest in the world (Burgess 2000).[10]

11.6.4 Polarization Within

In India, like many developing economies, there exists a kind of polarization in resource use other than the rural–urban divide. This polarization is within the urban areas. This is attributed to over-urbanization or urbanization of poverty as explained in Sect. 11.4. Urban areas of developing world are characterized by extreme differences in income and living conditions across population groups and often have high rates of poverty and low rates of service delivery for the poorest. At the other end, affluent lifestyles and profligate use of land by the rich class result in a disproportionate use of resources and urban forms that are often unsustainable (Jenks et al. 1996). The urban rich of developing countries consume as much as, if not more than, those in developed countries as they ape their lifestyles (Richardson et al. 2000). Thus, there are resource polarizations at two levels: one, between urban and rural areas and the other, between the rich and the poor within urban areas.

11.6.5 Informality of Economy

The structural characteristics of urban economy for India like many developing countries are different from those of developed ones. It is characterized by high share of employment and output from informal sector, small-scale workshops dispersed throughout the low-income settlements and street businesses (Burgess 2000). These characteristics have helped to address the high unemployment issue, but these have also created major social and environmental externalities including congestion, waste disposal, fire and health risks, and other unsustainable trends.

11.7 A Strategy for a Sustainable City Solution

Most of the India's urban problems are pegged to the fact that roughly one in four persons is residing in informal settlements. These settlements are characterized by unsafe housing and inadequate infrastructure in terms of water, sanitation, electricity, and internal roads. So, these are the areas which are subjected to poor transportation facility. Also, the slum dwellers are the ones who became victims of weather extremities because of lack of capacity—financial, technological, and resource-wise. Being unplanned, the waste management in slums remains a huge challenge. Also, slum areas being one of the highly dense areas in terms of population,[11] are devoid of open and green spaces. So, slum population concerns five of the seven targets (excepting the one on participatory urban planning and management and other on safeguarding natural heritage) of SDG 11.

A strategy of sustainable city solution in India can be to find a solution for the housing issues of slum dwellers. Once this issue is addressed, issues of internal roads, disaster preparedness, waste management, and public and green spaces can be addressed. Let us understand the issue of housing (or lack of it) in the first place. Let us discus this issue in the context of any Class I city of India, *say* Mumbai.

Let us assume that a private builder has constructed a housing society having 100 odd flats for higher-middle-class families in one decent residential localities of Mumbai. The families inhabiting this building do not stay in isolation with city. Rather, they get integrated to the city life through different agents. These agents are

[10]The increase in rent may have causalities other than supply demand mismatch. However, reasons for real estate price rise are out of the scope of this study.

[11]For instance, the density of a central neighbourhood of Dharavi slum, i.e. Chamda Bazar is 336,643 people/km^2 and this compares with 29,500/km^2 for Mumbai as a whole, which is the most densely populated city in the world, and 55,077/km^2 for Kwun Tong, the most densely populated area of Hong Kong (Vasudevan 2009).

none other than the watchmen, the gardeners, the housemaids, the cooks, the drivers, the newspaper delivery men, the milk men, the garbage collectors, etc. Without them, a life in a housing society can never be imagined. They provide the requisite support services that smooth the otherwise difficult city life. And the list of these people go beyond the immediate visible. In includes the person who made a duplicate key when one family accidentally forgot the key in the room and locked the door from outside. In also includes the small boy who sets the choke right when a highly qualified engineer of the housing society could not make out the reason for his two wheelers failing to start and hence dragged the vehicle to the other side of the road. No doubt, the housing society gets valuable services from these people. So, the next question comes—where all these many people live?

The answer to this question is that they stay on footpaths, under the bridge, at railway platforms, or in slums. Though their services are very valuable, they are underpaid. Because of availability of excess labor, that too unorganized, and as these service providers are replaceable, they cannot demand high price. In short, the families in the housing society do not pay enough to these service providers for them to have a standard of life anywhere close to themselves. So, possibility of one of these service providers as one of the neighbours in the same housing society does not arise. Their income is at a sustenance level where they cannot afford even a house at a fraction of the cost of the housing societies for which they work. So, there is no surprise they end up in slums.

So, now the more important question is: who takes responsibility of providing affordable housing to these service providers—the government, the builder, or the families who got their valuable services and underpaid them? The answer is all—each stakeholder has its distinct role to provide affordable housing to the under privileged. The role of the builders is primary in this regard. It is well known; builders provide houses for middle-class, higher-middle-class, rich, and ultra-rich people and make considerable profits. Hence, it must be made mandatory for every builder that, for certain share of commercial houses, they are to construct social housing and sell them at an affordable price. This is in similar lines to the concept of Corporate Social Responsibility (CSR).[12] Like corporates, the real estate giants, in the present situation of high disparity in housing condition in society, need to contribute towards building a future, which is more just and equitable. The role of government is very critical. First and foremost, it has to make the required law for the real estate sector for construction of low-cost housing and enforce the same on all builders and facilitate space for the same. The role of the larger society is to debate and discuss the issue to create an ambience, which will persuade government to enact laws in favour the urban poor. The proposed model for city housing is given in Fig. 11.7.

The communication infrastructure is the very next thing which has a direct relation to housing. People, who cannot afford houses in the heart of city, move to the periphery where the housing is affordable. But these houses are preferred only when the communication network is strong between the centre and periphery. So, the low-cost housing can be built at the periphery, with a public transport system that is very fast and efficient.

Working people hostels in big cities added to the above framework can make us get rid of slums. It is usually one person of the family who migrate to city first looking for a livelihood option. These hostels can be either government or private run, which will have dormitories which only act as an ad hoc arrangement made for migrants till they settle with city life and bring their family to stay in one of the low-cost housings described above. In this new housing model, slum is replaced by 'hostels and low-cost buildings'. This will solve all infrastructure and sanitation related issues.

However, most of the efforts to improve the situation might get offset with continuous popu-

[12]The World Business Council for Sustainable Development (1999) defines CSR as 'the continuing commitment by business to behave ethically and contribute to economic development while improving the quality of life of the work force and their families as well as of the local community and society at large'.

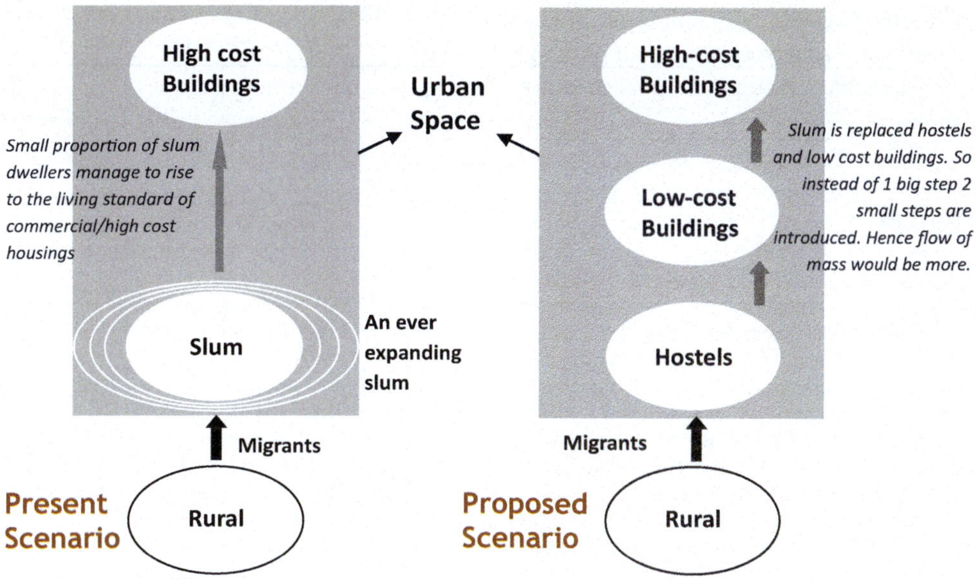

Fig. 11.7 A model to build 'city without slums'

lation rise. India's population has quadrupled since independence and the ever-increasing population remains the underlining cause behind issues behind congestion, strains in housing, infrastructure, and unsustainable risks for any resource and service. Therefore, effective family planning programme will go a long way in ensuring sustainability of both urban and rural society.

11.8 Concluding Remarks

India's urbanization is at crossroads. Like other developing countries, rapid growth of cities is seriously outstripping their capacity to provide adequate services to the citizens. In order to improve sustainability of urban areas, it is worthwhile to understand the cause and pattern of urbanization and their implications on quality of life. This study shows how the level, rate, and consequences of urbanization are different for developed and developing region of the world. Being 'rural push' driven, urbanization in developing region throws multitude of challenges in terms of growing informal sector and settlements, infrastructural and environmental overloads, and high incidence of poverty and inequality. India's urbanization, which shows progressive concentration of population towards large cities, faces similar challenges. Urbanization is inevitable and is a positive sign for growth, but in order to tap its merit, cities need to be well managed. As Cohen (2006) ascertained, if well managed, cities can offer important opportunities for economic and social development.

This study has proposed a housing model to achieve most of the targets of SDG 11, which advocates for inclusive, safe, resilient and sustainable cities. The study proposes a step-wise housing scenario for the city where working people hostel is the first shelter for a rural migrant followed by low-cost housing, which are located at the periphery of the city, but connected with the core with efficient, affordable, rapid, and environment-friendly transport system. A corporate social responsibility model can be invoked by the government for the builders to mandate low-cost housing, which can get cross-subsidized by the residents of high-cost housing. The civil society can facilitate an ecosystem for the implementation of such a system. 'A city without slum' is realizable with different stakeholders—government, builders, non-slum dwellers, and civil society. Chalking out their roles and movement towards slum-free society coupled with effective family planning programme is instrumental in achieving almost all of the targets of sustainable development goal meant for urban areas.

References

Allen A (2009) "Sustainable cities or sustainable urbanization", palette—the grand challenge of sustainable cities. UCL's J Sustain Cities (Summer Issue):3.18–3.20

BMBF (Federal Ministry of Education & Research) (2004) The urban transition: research for the sustainable development of the megacities of tomorrow. Division 622, Global Change. BMBF, Germany

Breese G (1969) Urbanisation in newly developing countries. Prentice Hall, New Delhi

Burgess R (2000) The compact city debate—a global perspective. In: Jenks M, Burgess R (eds) Compact cities—sustainable urban forms for developing countries. Spon Press, Taylor & Francis Group, London and New York

CEE–WR (Chief Electric Engineer—Western Railway) (2011) Information obtained from personal interaction with Sr. Section Engineer, EMU services, Electrical Dept., CEE Office, Churchgate, Western Railway, Mumbai

Census of India (2001) Census data online—2001. New Delhi. http://censusindia.gov.in/2011-common/censusdataonline.html

Cities Alliance (2008) Cities alliance in action. Annual Report

Cohen B (2004) Urban growth in developing countries: a review of current trends and a caution regarding existing forecasts. World Dev 32(1):23–51

Cohen B (2006) Urbanization in developing countries: current trends, future projections, and key challenges for sustainability. Technol Soc 28(1–2):63–80

Datta P (2006) Urbanization in India. Paper prepared for European Population Conference, 21–24 June. www.infostat.sk/vdc/epc2006/papers/epc2006s60134.pdf. Accessed 15 Jan 2011

Davis M (2004) Planet of slums. New Left Rev (26):5–34

Devas N, Rakodi C (eds) (1993) Managing fast growing cities: new approaches to urban planning and management in the developing world. Longman, Harlow

Girardet H (1996) *The Gaia Atlas of Cities: New Directions for Sustainable Urban Living*, Gaia Books Ltd., London

Global Urban Observatory (2003) Slums of the World: the face of urban poverty in the new millennium? Working Paper, United Nations Human Settlements Programme, UN-Habitat, Nairobi, Kenya

GoI (Govt. of India) (2019) Handbook of urban statistics 2019. Ministry of Housing and Urban Affairs, GoI

Gugler J (1982) Overurbanization Reconsidered *Economic Development and Cultural Change*, Vol. 31, No. 1, pp. 173–189

Haub C, Sharma OP (2006) India's population reality: reconciling change and tradition. Popul Bull 61(3):1–20. Population Reference Bureau (PRB)

Heinrichs D (2005) "Why research on mega urban regions matters!", Department Urban and Environmental Sociology. UFZ Centre for Environmental Research, Leipzig–Halle

HPEC (The High Powered Expert Committee) (2011) Report on Indian urban infrastructure and services. Ministry of Urban Development, Govt. of India, March

Jenks M (2000) Sustainable urban form in developing countries? Introduction chapter. In: Jenks M, Burgess R (eds) Compact cities—sustainable urban forms for developing countries. Spon Press, Taylor & Francis Group, London and New York

Jenks M, Burton E, Williams K (eds) (1996) The compact city: a sustainable urban form? E & FN Spon, London

Kamete AY, Tostensen A, Tvedten I (2001) From global village to urban globe, urbanisation and poverty in Africa: implications for Norwegian aid policy. Report R 2001: 2, Development Studies and Human Rights, Chr. Michelsen Institute

Kenneth L (2005) *Rural-Urban Interaction in the Developing World*, Routledge, London

Kundu A (1983) Theories of city size distribution and Indian urban structure—a reappraisal. Econ Polit Wkly 18(3):1361–1368

Kundu A, Lalitha N, Arora SL (2001) Growth dynamics of informal manufacturing sector in urban India: an analysis of interdependencies. In: Kundu A, Sharma AN (eds) Informal sector in India perspectives and policies. Institute for Human Development and Institute of Applied Manpower Research, New Delhi, pp 93–123

MGI (McKinsey Global Institute) (2010) India's urban awakening: building inclusive cities, sustaining economic growth, MGI, April

Montgomery M, Stren R, Cohen B, Reed HE (eds) (2004) Cities transformed. Demographic change and its implications in the developing world. Earthscan, London

Mukherji S (1993) Poverty induced migration and urban involution in India: cause and consequences. International Institute for Population Sciences, Mumbai, pp 1–91

Nallathiga R, Taneja S, Gupta A, Gangal B (2018) Sustainability of urban fringe development and management in NCT-Delhi: a case study. In: Mukherjee J (ed) Sustainable urbanization in India: challenges and opportunities. Springer, Singapore

Parikh J, Gokran S (1991) Consumption patterns: the driving force of environmental stress. Working Paper, Indira Gandhi Institute of Development Research (IGIDR), Mumbai

PRB (Population Reference Bureau) (2007) The future population of India: a long range demographic view. Population Foundation of India, August

PRB (2008) World population highlights, key findings from PRB's 2008 world population data sheet. Popul Bull 63(3), p 9

Railway Gazette (2010) Loan to relieve world's most overcrowded trains. Railway Gazette News, 06 Jul. http://www.railwaygazette.com/news/single-view/view/loan-to-relieve-worlds-most-overcrowded-trains.html. Accessed 31 Dec 2010

Raza M, Kundu A (1978) Some aspects of disfunctional characteristics of urbanisation-socio-economic development problems in South and South East Asia. Popular Prakashan, Bombay

Richardson HW, Bae C-HC, Baxamusa MH (2000) Compact cities in developing countries—assessment and implication. In: Jenks M, Burgess R (eds) Compact cities—sustainable urban forms for developing countries. Spon Press, Taylor & Francis Group, London and New York

SIDA (Swedish International Development Cooperation Agency) (1995) Towards an urban world: urbanization and development assistance, Stockholm

SIDA (2006) Fighting poverty in an urban world: support to urban development. SIDA Policy Info, Department for Infrastructure and Economic Cooperation, Division for Urban Development, SIDA

SIDA (2008) The world goes to town—Sweden in the Urban World. Department for Environment, Climate Change and Sustainable Services, SIDA, October

Simon D (1995) Urbanization, globalization and economic crisis in Africa. In: Rakodi C (ed) The urban challenge in Africa: growth and management of its large cities. United Nations University Press, Tokyo, New York, pp 74–110

Tannerfeldt G, Ljung P (2006) More Urban Less Poor—an introduction to urban development and management. Swedish International Development Cooperation Agency (SIDA), London, SIDA & Earthscan

UNFPA (United Nations Fund for Population Division) (1999) State of World Population: six billion, a time for choice. United Nations Population Fund, New York

UN-Habitat (2003) The challenge of the slums: global report on human settlements 2003. London

United Nations (2002) World urbanization prospects, the 2001 revision. Population Division, Department of Economic and Social Affairs, United Nations

United Nations (2009) Data online, world urbanization prospects: the 2009 revision population database. Population Division, Department of Economic and Social Affairs, United Nations

United Nations (2018a) Download center. World urbanization prospects: the 2018 revision. Population Division, Department of Economic and Social Affairs, United Nations. https://population.un.org/wup/Download/. Accessed 13 Aug 2019

United Nations (2018b) The world's cities in 2018—data booklet. Population Division, Department of Economic and Social Affairs, United Nations. https://www.un.org/en/events/citiesday/assets/pdf/the_worlds_cities_in_2018_data_booklet.pdf. Accessed 15 Aug 2019

United Nations (2019) Goal 11-sustainable cities and communities, facts and figures at https://www.un.org/sustainabledevelopment/cities/. United Nations. https://developmentperspectives.ie/SDGChallenge/InformationPacks/SDG%2011%20Sustainable%20Cities%20and%20Communities%20.pdf. Accessed 15 Aug 2019

UNSC (United Nations Statistical Commission) (2016) Final list of proposed sustainable development goal indicators, Annex IV. Report of the Inter-Agency and Expert Group on Sustainable Development Goal Indicators. https://sustainabledevelopment.un.org/content/documents/11803Official-List-of-Proposed-SDG-Indicators.pdf. Accessed 15 Aug 2015

UNSD (United Nations Statistics Division) (2010) United Nations demographic yearbook 2008. UNSD, Department of Economic and Social Affairs, United Nations. http://unstats.un.org/unsd/demographic/products/dyb/dyb2008.htm

Vasudevan S (2009) Sustainability of business correspondent model for financial inclusion in slums in Mumbai. http://www.iibf.org.in/documents/reseach-report/Report-20.pdf. Accessed 20 Aug 2019

Wackernagel M, Rees WE (1996) Our ecological footprint: reducing human impact on earth. New Society Publishers, Gabriola Island

Wackernagel M, Onisto L, Linares AC, Falfan ISL, Garcia JM, Guenero AIS, Guenero GS (1997) Ecological footprint of nations: how much nature do they use?—how much nature do they have? Universidad Anahuac de Xalpa, Mexico

Williams K, Jenks M, Burton E (1999) How much is too much? Urban intensification, social capacity and sustainable development. Open House Int 24(1):17–25

Williams K, Burton E, Jenks M (eds) (2000) Achieving sustainable urban form. E & FN Spon, London

World Bank (1995) Better urban services: finding the right incentives. Development in Practice Series, Washington, DC

World Bank (2000) Entering the 21st century—world development report 1999/2000. Washington, DC

World Bank (2011) Online data, millennium development goals. World Bank. http://data.worldbank.org/about/millennium-development-goals. Accessed 31 Mar 2011

World Business Council for Sustainable Development (1999) Corporate social responsibility: meeting changing expectations. In: Crane A, Matten D, Spense IJ (eds) (2008) Readings and cases in global context. Routledge-Cavendish, Oxford; retrieved from Okoye A (2017) Legal approaches and corporate social responsibility. Routledge, London

Sustainable Consumption Pattern in India

12

Sanchita Daripa and Soumyananda Dinda

12.1 Introduction

The economic development agenda in the twenty-first century begins with the Millennium Development Goals (MDGs) for the global economy. The MDGs is the most significant global development programme with targeted goals, which has shifted the focus on sustainable development from unsustainable individualistic development approach. Truly, the MDG is a new global development doctrine, which initially focuses on poverty and human development, highlighting certain defined and targeted goals for achieving development. After five decades of economic development practices, it is realized that traditional development doctrines have failed to ensure development over time with limited resources on the Earth, and alternatively sustainable development approach emerges for sustaining human civilization (Dinda 2017). The UN has set up universal goals in the framework of Sustainable Development Goals (SDGs) initiated through the MDGs in the early twenty-first century. The UN goals are relevant for both developing and developed nations. The MDGs mainly focus on development of human capital, while SDGs sets a wide coverage on human capital formation and its protection for sustaining humankind. Now, the global human society feels an urgent need for interconnections among economic, social and environmental dimensions of development. Recently, the MDGs are set to be achieved integrating *Sustainable Consumption and Production (SCP)* patterns in the framework of new development goals. The core of sustainable development goals is to make sustainable consumption and production (Akenji and Bengtsson 2014). As consumption and production are interdependent, question arises on the operationalization of the SCP within the market system (Xu et al. 2018). Here is a need to develop the mechanics for promotion of sustainable consumption. Sustainable consumption should focus on non-polluting consumption, and consumption process should not generate pollution (Banerjee et al. 2016; Dinda 2014, 2016, Goswami et al. 2017; Nayak et al. 2015; Raman et al. 2014) and avoid resource waste. The above said are the initial indicators to assess the sustainable consumption. Any development suggests concomitant changes in environmental impacts, which can be attributed to the consumption households (Scherer et al. 2018). The indicators of sustainable consumption are directly or indirectly connected with energy and material consumption. Is there any evidence of sustainable consumption in India? This chapter

S. Daripa · S. Dinda (✉)
Department of Economics, University of Burdwan, Burdwan, West Bengal, India
e-mail: sdinda@eco.buruniv.ac.in

S. Hazra, A. Bhukta (eds.), *Sustainable Development Goals*, Sustainable Development Goals Series,
https://doi.org/10.1007/978-3-030-42488-6_12

attempts to provide the fuel consumption pattern and its distribution across states, and highlights some indicators like material consumption footprint per capita and its intensity in India.

For the said study purpose, this chapter is organized as follows: Sect. 12.2 describes and highlights sustainable consumption and production pattern for achieving sustainable development goals. Section 12.3 provides the energy consumption pattern in India. Section 12.4 reports with analytical focus on material consumption footprint in India. Final Sect. 12.5 concludes with remarks.

12.2 Concept of SCP and Its Relevance

The concept of sustainable consumption and production has emerged and is gaining importance in development policy meetings. Recently, it has been realized that sustainable consumption and production pattern is one of the most important global discourses, which is included in the MDGs to achieve the global development agenda in post-2015. Truly, consumption and production process are the pivotal or focal point of achieving sustainable development goals (SDGs). Truly, SCP is embedded in the SDGs. Now, the UN merges the MDGs with environmental agenda for achieving sustainable development. It is urgently necessary to integrate the production system and consumption patterns with sustainable development for its establishment with predominant goals. Currently, sustainable consumption and production patterns have gained importance in policy dialogues at the international arena (Gasper et al. 2019; Ma et al. 2019; Xu et al. 2018; Scherer et al. 2018). The Rio + 20 (the UN Conference on Sustainable Development) pointed out the mistake of cornering SCP in the MDGs and suggested a proposal for 10-Year Framework of Programme on SCP patterns. Research on SCP should address the integration of economic growth, environmental protection and social inclusiveness from both consumption and production sides.

12.2.1 SDG 12 Ensures SCP

Sustainable Development Goal 12 (SDG 12), ensuring sustainable consumption and production pattern, promotes increased human well-being while decoupling economic growth from resource use and environmental degradation. Goal 12 sets the targeted goal on sustainable consumption and production, emphasizing on '*doing more with less*' and ensures that the needs of the present generation are fulfilled without compromising that of the future generation. Goal 12 ensures sustainable consumption and production process, which is interconnected with other goals. Entire socio-economic development evolves over production and consumption pattern, which should ensure sustainability of this consumption and production process over time unless and until improving resource efficiency and/or reducing its degradation.

SCP encourages socio-economic development within possible market ecosystem and its carrying capacity in the global limits. Sustainable consumption reveals the consumer's choices of goods and services focusing on basic needs like food, shelter, clothing, leisure, etc (Heltberg 2005). This is the demand-side economy for quality consumption to sustain in the long run. Sustainable consumption suggests to fulfil the basic needs and improve quality of life without harming the environment, economy and society overall. So, the present choice of consumption should not be traded-off with damaging possible production (Ouedraogo 2006; Masera et al. 2000).

Sustainable production focuses on resource efficiency, highlighting minimizing risk on environment and human society. It is the supply-side economy assessing the impacts of production processes on environment, economy, and society. Sustainable production certainly refers to the optimum use of resources at all stages of production cycle, which aims to reduce its ecological footprint without any burden shifting between different stages of product life cycle. The global need is to adopt policy to reduce the footprint of each and every country. In this context, how do we gauge such complex issues? What are the targets and corresponding measureable indicators? The next sub-section discusses the main targets and potential indicators.

12.2.2 The Targets and Potential Indicators

This section describes possible ideological background of Goal 12 associated with targets and technical architecture for monitoring them and suggests some potential indicators. The goal on Sustainable Consumption and Production has 11 major targets and its progress is monitored by defined indicators. Now, we discuss on the guideline of implementation programme framework as mentioned as Implement the 10-Year Framework of Programme on sustainable consumption and production (10 YFP). All countries will take action; the developed countries will take the lead and help the developing countries with improving their capabilities and accountability. 10 YFP is more of an implementation framework and covers all other targets. Major targets related to achieve SCP by 2030 are (see SDG 12 in the UN website: http//in.one.un.org/sdg-12) as follows:

1. Achieve the sustainable management and efficient use of natural resources.
2. Reduce food losses along production and supply chains, including post-harvest losses. By 2030, per capita global food waste should be reduced to half, both at the retail and consumer levels.
3. Achieve the environmentally sound management of chemicals and all wastes throughout their life cycle, in accordance with agreed international frameworks, and significantly reduce their release into air, water and soil in order to minimize their adverse effects on human health and natural environment.
4. Reduce waste generation through reduction, recycling, reuse and prevention.
5. Encourage companies, especially large and transnational companies, to adopt sustainable practices and to integrate sustainability information into their reporting cycle.
6. Promote public procurement practices that are sustainable, in accordance with national policies and priorities.
7. Ensure that people everywhere have the relevant information and awareness for sustainable development and lifestyles in harmony with nature.
8. Support developing countries to strengthen their scientific and technological capacity to move towards more sustainable patterns of consumption and production.
9. Develop and implement tools to monitor sustainable development impacts for sustainable tourism that creates jobs and promotes local culture and products.
10. Rationalize inefficient fossil-fuel subsidies that encourage wasteful consumption by removing market distortions, in accordance with national circumstances, including by restructuring taxation and phasing out those harmful subsidies, where they exist, to reflect their environmental impacts, taking fully into account the specific needs and conditions of developing countries and minimizing the possible adverse impacts on their development in a manner that protects the poor and the affected communities.

Relevant goals and their corresponding indicators are presented in Table 12.1, which briefly summarizes SDG Goal 12.

Table 12.1 describes briefly the relevant targets of Goal 12 and their corresponding possible indicators. SDG 12 ensures sustainable consumption and production patterns through eight specific targets, 12.1–12.8 and three target-related means of implementation (12.a–12.c). The sustainability is viewed in the lens of production efficiency in relation to use of natural resources (12.2), food production and supply-related losses (12.3), management of chemical and wastes (12.4), reduce waste generation (12.5), reporting of sustainable corporate practice (12.6) and public procurement (12.7) and ensure universal access to information for sustaining lifestyles (12.8). The target for rationalizing fossil fuel subsidies (12.c) and the least waste generation (12.5)

Table 12.1 Goal 12 ensures sustainable consumption and production patterns: targets and indicators

Targets	Indicators
12.1 Implement the 10-Year Framework of Programmes on Sustainable Consumption and Production Patterns, all countries taking action, with developed countries taking the lead, taking into account the development and capabilities of developing countries	12.1.1 Number of countries with sustainable consumption and production (SCP) national action plans or SCP mainstreamed as a priority or a target into national policies
12.2 By 2030, achieve the sustainable management and efficient use of natural resources	12.2.1 Material footprint, material footprint per capita, and material footprint per GDP
	12.2.2 Domestic material consumption, domestic material consumption per capita, and domestic material consumption per GDP
12.3 By 2030, halve per capita global food waste at the retail and consumer levels and reduce food losses along production and supply chains, including post-harvest losses	12.3.1 Global food loss index
12.4 By 2020, achieve the environmentally sound management of chemicals and all wastes throughout their life cycle, in accordance with agreed international frameworks, and significantly reduce their release to air, water and soil in order to minimize their adverse impacts on human health and the environment	12.4.1 Number of parties to international multilateral environmental agreements on hazardous waste, and other chemicals that meet their commitments and obligations in transmitting information as required by each relevant agreement
	12.4.2 Hazardous waste generated per capita and proportion of hazardous waste treated, by type of treatment
12.5 By 2030, substantially reduce waste generation through prevention, reduction, recycling and reuse	12.5.1 National recycling rate, tons of material recycled
12.6 Encourage companies, especially large and transnational companies, to adopt sustainable practices and to integrate sustainability information into their reporting cycle	12.6.1 Number of companies publishing sustainability reports
12.7 Promote public procurement practices that are sustainable, in accordance with national policies and priorities	12.7.1 Number of countries implementing sustainable public procurement policies and action plans
12.8 By 2030, ensure that people everywhere have the relevant information and awareness for sustainable development and lifestyles in harmony with nature	12.8.1 Extent to which (1) global citizenship education and (2) education for sustainable development (including climate change education) are mainstreamed in (a) national education policies; (b) curricula; (c) teacher education; and (d) student assessment
12.a Support developing countries to strengthen their scientific and technological capacity to move towards more sustainable patterns of consumption and production	12.a.1 Amount of support to developing countries on research and development for sustainable consumption and production and environmentally sound technologies
12.b Develop and implement tools to monitor sustainable development impacts for sustainable tourism that creates jobs and promotes local culture and products	12.b.1 Number of sustainable tourism strategies or policies and implemented action plans with agreed monitoring and evaluation tools
12.c Rationalize inefficient fossil-fuel subsidies that encourage wasteful consumption by removing market distortions, in accordance with national circumstances, including by restructuring taxation and phasing out those harmful subsidies, where they exist, to reflect their environmental impacts, taking fully into account the specific needs and conditions of developing countries and minimizing the possible adverse impacts on their development in a manner that protects the poor and the affected communities	12.c.1 Amount of fossil-fuel subsidies per unit of GDP (production and consumption) and as a proportion of total national expenditure on fossil fuels

Source: The UNEP

might capture production and consumption, which could lead towards sustainability.

10 year framework of programmes on sustainable consumption and production should be implemented; all countries should take action, with developed countries taking the lead and taking into account the development and capabilities of developing countries. Is there any mismatch between targets and identified indicators? How do we implement Goal 12 in our daily life? Is there any evidence? In the next section, we attempt to answer these questions focusing on consumption pattern in India with empirical evidence.

12.3 State-Wise Fuel Consumption Pattern in India

Initially, this section describes the fuel consumption pattern at the household level in India, and also attempts how far it is consistent with SCP. The last part provides forecasting on material footprint in India by 2030.

12.3.1 Energy Consumption and Development

Energy consumption is essential for economic activities, and energy requirements are increasing at a high rate along with economic development. In a developing economy, domestic energy consumption forms a large part of total energy consumption. Household energy demand comes basically from the fuels they consume. Fuel forms an indispensable part of daily consumption basket of households. Household consumes fuel to serve three basic domestic purposes—cooking, lighting and space heating (Van Der Kroon et al. 2014). These purposes are served by a variety of sources of fuels namely coal, petroleum, LPG, electricity, gas, fuel oil, charcoal, dung cakes, firewood, etc. Some of these fuels are traditional fuels like coal, charcoal, dung cakes, crop residues, etc. The efficiency of these fuels is lower than that of other fuels like LPG, electricity, etc. Apart from the efficiency standards, combustion of traditional fuels creates pollution, and they are also termed as *dirty fuels*. Their combustion also leads to several health hazards. Modern fuels like petroleum, LPG, biogas, electricity, etc. are termed as *clean fuels*, which are more efficient than traditional fuels (Adrianjen 2013; Takama et al. 2012).

Forest is often considered as a renewable source of energy (Couture et al. 2012), while firewood is considered as a dirty fuel since it adds to CO_2 emissions upon combustion. In developing countries, a large part of supply of firewood comes from deforestation, which also acts as a threat to environment (Barnes et al. 2010; Davis 1998; Gebreegziabher et al. 2012; Hosier and Dowd 1987; Couture et al. 2012; Arnold et al. 2006; Brouwer and Falcao 2004; Campbell et al. 2003; Kebede et al. 2010; Leach 1992). Developing economies are ushered into a world of uncertainty with unsustainable development. How do we ensure sustainable development? Obviously, fuel choice of households and their actual consumption decision are important research issues for achieving sustainable development goals (SDG) within 2030 (Jingchao and Kotani 2012; Hanna and Oliva 2015; Jumbe and Angelsen 2011). Household's fuel choice and amount of fuel consumption decision may certainly help to control the risk of uncertainty associated with climate change, which could be a potential threat to sustainable development in emerging economy like India (Gupta and Kohlin 2006; Gundimeda and Kohlin 2008; Pachauri and Jiang 2008). In this context, we have to study the recent fuel consumption pattern in India at the household level that might help to reformulate and redesign fuel consumption and subsidy policy focusing on targeted SDGs. To make effective energy policy design for achieving SDGs, we have to study the fuel consumption pattern and its distribution. Household fuel choice and consumption decision in any economy is an important aspect of study, which has recently adopted a new approach for sustainable consumption in the sustainable development goals (SDGs).

12.3.2 Data

The data used for this study are secondary data, which are collected and published by National Sample Survey Office (NSSO). The NSSO conducts nationwide sample surveys relating to various socio-economic topics to collect data for planning and formulation of policies (Gundimeda and Kohlin 2008). The NSSO 68th round Socio-Economic Survey data, which were collected during the period July 2011–June 2012, are used in the current study. These data were published in September 2015. From the data, it is observed that in India, individuals are faced with ten major alternatives for cooking purposes, namely coke or coal, firewood, LPG, gobar gas, dung cakes, charcoal, kerosene, electricity, others and no cooking arrangements.

12.3.3 Preliminary Results

Using Stata software, we have performed a tabulation, which reflects that out of these ten types of cooking fuel, rural households tend to prefer firewood as the most preferred fuel and urban households are inclined more towards LPG (i.e. LPG is the most preferred fuel urban India). Table 12.2 shows that firewood (62.68%) is the most preferred fuel of rural India, while LPG (68.9%) is the most preferred fuel in urban India. Since firewood is a dirty fuel, its consumption should not be encouraged for cooking purposes.

Firewood consumed in rural India comes from various sources. Table 12.3 describes the distribution of sources of firewood in rural India. From Table 12.3, it is clear that the main source of firewood in rural India is free collection (34.32%) followed by purchase (30.05%) and home-grown (27.54%).

This study is mainly focused on fuel selection and their consumption in India. For in-depth analysis, we also examine it in rural and urban India. We consider a brief overview about the nature of the data. From the data, it is observed that in India, individuals are faced with ten major alternative sources of fuels for their energy consumption purposes, namely coke or coal, firewood, LPG, Gobar gas, dung cakes, charcoal, kerosene, electricity, others and no cooking arrangements.

12.3.4 Analysis

Our preliminary analysis begins with state-wise choice of major fuels in India and attempts to identify fuel choice preferences of each state. We have performed a state-wise tabulation of the major fuels consumed in India (Table 12.4). Table 12.4 describes the state-wise distributional pattern of household fuel consumption in India.

Table 12.2 Distribution of households as per sources of cooking fuels in India

	Rural			Urban		
Cooking code	Freq.	Percentage	Cum.	Freq.	Percentage	Cum.
Coke, coal	657	1.1	1.1	914	2.18	2.18
Firewood and chips	37,410	**62.68**	63.78	7483	17.83	20.01
LPG	14,562	24.4	88.18	28,912	**68.9**	88.91
Gobar gas	141	0.24	88.42	6	0.01	88.93
Dung cake	4203	7.04	95.46	520	1.24	90.17
Charcoal	25	0.04	95.5	115	0.27	90.44
Kerosene	619	1.04	96.54	1907	4.54	94.99
Electricity	77	0.13	96.67	245	0.58	95.57
Others	1669	2.8	99.46	313	0.75	96.32
No cooking arrangement	320	0.54	100	1546	3.68	100
Total	59,683	100		41,961	100	

Source: *NSSO 68th Round*

Table 12.3 Distribution of sources of firewood in rural India

Sources of firewood	Frequency	Percentage
Only purchase	14,562	30.05
Only home-grown stock	13,347	27.54
Both purchase and home-grown stock	1580	3.26
Only free collection	16,631	34.32
Only exchange of goods and services	93	0.19
Only gifts/charities	89	0.18
Others	2160	4.46
Total	48,462	100

Source: *NSSO 68th Round*

From Table 12.4, we observe the consumption of each fuel in each of the 35 states of India. The row totals reflect the total fuel consumption of each state while the column totals reflects the total consumption of each fuel in India. Along the rows, we observe the share of each fuel in the total fuel consumption of the particular state, while along the columns, we observe the share of each state in the total consumption of a particular fuel in India.

It is observed from Table 12.4 that 49.05% of the total fuel consumption of Jammu & Kashmir comes from LPG, which is followed by firewood, comprising of 42.76% of total fuel consumption of the state. Other important fuels are kerosene consumed by 3.63% of people and electricity by 2.66% of people. It is quite clear from Table 12.4 that LPG and firewood are the two main fuels consumed in Jammu & Kashmir. However, LPG consumption of this state consists of only 3.82% of total LPG consumption of India and firewood consumption is 3.22% of total consumption of India. ***But the share ofelectricity consumptionof this state in total electricity consumption of India is highesti.e.27.95%.***

The results show that 59.61% of people of Himachal Pradesh consume firewood and 36.27% consume LPG. Among the other fuels, 1.86% of people consume kerosene, 1.42% of people have no cooking arrangements and 0.44% of people of the state consume electricity. It is also observed that though firewood and LPG are the prime sources of fuel for this state, they contribute a share of only 2.71% and 1.7% of total India's consumption. The most important fuel is LPG consumed by 57.83% of the people of Punjab and followed by firewood, which is consumed by 18.58% of the people and dung cake consumed by 15.92% of the people. However, the share of these fuels in total India's consumption is 1.29% for firewood, 4.15% for LPG and 10.5% for dung cake. In Chandigarh, the most important fuel is LPG consumed by 71.15% of people followed by ***kerosene consumed by 16.03%*** of people and firewood consumed by 5.77% of people of the state. Though LPG is consumed by most people of the state, the total consumption of LPG comprises 0.51% of total LPG consumption of India. In Haryana, the most important fuel is LPG consumed by 51.87% of people of the state, followed by firewood consumed by 27.81% of the people and ***dung cake*** consumed by 16.03% of people. But the share of Haryana in total LPG consumption is 3.09%, share in firewood consumption is 1.6% and dung cake consumption is 8.79% of total India's consumption. In Delhi, 81.38% people consume LPG, 2.65% people consume firewood, 1.69% people consume kerosene and ***9.21% have no cooking arrangements***. Also, the LPG consumption of people of Delhi comprises only 1.77% of total LPG consumption of India. 62.02% people of Rajasthan consume firewood and 34.71% of people consume LPG. Firewood consumption of the people of Rajasthan comprises 5.71% of total firewood consumption of India. 50.67% of Uttaranchal people consume LPG followed by 45.57% people consuming firewood, which are the two most important fuels consumed in the state followed by kerosene and dung cake. However, the share of LPG is 2.08% and firewood is 1.81% in total LPG and firewood consumption of India. In Uttar Pradesh, 43.8% of people consume firewood, 27.48% of people consume LPG and ***24.75% of people consume dung cakes***, which are the three major sources of fuel for the state. But the consumption of this fuel comprises a share of 8.79% for firewood, 5.7% for LPG and 47.24% for ***dung cakes*** out of the total consumption of these fuels in India. Of these, the share of firewood and dung cakes in total India's consumption is highest. 44.09% people of Bihar consume firewood, 25.33% of peo-

Table 12.4 State-wise consumption of different fuels in India

State	Coke, coal	Firewood	LPG	Gobar gas	Dung cake	Charcoal	Kerosene	Electricity	Others	No cooking arrangements	Total
Jammu & Kashmir	1	1447	1660	0	34	7	123	90	2	20	3384
	0.03	42.76	49.05	0	1	0.21	3.63	2.66	0.06	0.59	100
	0.06	3.22	3.82	0	0.72	5	4.87	27.95	0.1	1.07	3.33
Himachal Pradesh	1	1216	740	0	5	2	38	9	0	29	2040
	0.05	59.61	36.27	0	0.25	0.1	1.86	0.44	0	1.42	100
	0.06	2.71	1.7	0	0.11	1.43	1.5	2.8	0	1.55	2.01
Punjab	1	579	1802	22	496	0	130	10	48	28	3116
	0.03	18.58	57.83	0.71	15.9	0	4.17	0.32	1.54	0.9	100
	0.06	1.29	4.15	14.97	10.5	0	5.15	3.11	2.42	1.5	3.07
Chandigarh	0	18	222	0	4	0	50	0	0	18	312
	0	5.77	71.15	0	1.28	0	16.03	0	0	5.77	100
	0	0.04	0.51	0	0.08	0	1.98	0	0	0.96	0.31
Uttaranchal	1	812	903	2	5	0	37	0	0	22	1782
	0.06	45.57	50.67	0.11	0.28	0	2.08	0	0	1.23	100
	0.06	1.81	2.08	1.36	0.11	0	1.46	0	0	1.18	1.75
Haryana	0	720	1343	0	415	2	23	4	65	17	2589
	0	27.81	51.87	0	16	0.08	0.89	0.15	2.51	0.66	100
	0	1.6	3.09	0	8.79	1.43	0.91	1.24	3.28	0.91	2.55
Delhi	1	25	769	0	0	1	16	3	43	87	945
	0.11	2.65	81.38	0	0	0.11	1.69	0.32	4.55	9.21	100
	0.06	0.06	1.77	0	0	0.71	0.63	0.93	2.17	4.66	0.93
Rajasthan	5	2562	1434	0	24	1	48	3	4	50	4131
	0.12	62.02	34.71	0	0.58	0.02	1.16	0.07	0.1	1.21	100
	0.32	5.71	3.3	0	0.51	0.71	1.9	0.93	0.2	2.68	4.06
Uttar Pradesh	36	3948	2477	0	2231	3	56	14	196	53	9014
	0.4	43.8	27.48	0	24.8	0.03	0.62	0.16	2.17	0.59	100
	2.29	8.79	5.7	0	47.2	2.14	2.22	4.35	9.89	2.84	8.87
Bihar	75	2019	1160	0	836	2	22	3	445	17	4579
	1.64	44.09	25.33	0	18.3	0.04	0.48	0.07	9.72	0.37	100
	4.77	4.5	2.67	0	17.7	1.43	0.87	0.93	22.45	0.91	4.5
Sikkim	1	167	554	1	0	0	11	0	0	34	768
	0.13	21.74	72.14	0.13	0	0	1.43	0	0	4.43	100
	0.06	0.37	1.27	0.68	0	0	0.44	0	0	1.82	0.76

Arunachal Pradesh	3	800	826	0	1	0	21	17	1	5	1674
	0.18	47.79	49.34	0	0.06	0	1.25	1.02	0.06	0.3	100
	0.19	1.78	1.9	0	0.02	0	0.83	5.28	0.05	0.27	1.65
Nagaland	0	300	723	0	0	0	1	0	0	0	1024
	0	29.3	70.61	0	0	0	0.1	0	0	0	100
	0	0.67	1.66	0	0	0	0.04	0	0	0	1.01
Manipur	3	1115	1353	0	2	71	7	3	0	6	2560
	0.12	43.55	52.85	0	0.08	2.77	0.27	0.12	0	0.23	100
	0.19	2.48	3.11	0	0.04	50.71	0.28	0.93	0	0.32	2.52
Mizoram	1	324	1197	0	0	1	12	1	0	0	1536
	0.07	21.09	77.93	0	0	0.07	0.78	0.07	0	0	100
	0.06	0.72	2.75	0	0	0.71	0.48	0.31	0	0	1.51
Tripura	0	1360	459	1	0	0	26	0	0	10	1856
	0	73.28	24.73	0.05	0	0	1.4	0	0	0.54	100
	0	3.03	1.06	0.68	0	0	1.03	0	0	0.54	1.83
Meghalaya	0	850	309	0	2	16	30	39	0	14	1260
	0	67.46	24.52	0	0.16	1.27	2.38	3.1	0	1.11	100
	0	1.89	0.71	0	0.04	11.43	1.19	12.11	0	0.75	1.24
Assam	3	1994	1345	7	3	1	38	0	10	38	3439
	0.09	57.98	39.11	0.2	0.09	0.03	1.1	0	0.29	1.1	100
	0.19	4.44	3.09	4.76	0.06	0.71	1.5	0	0.5	2.04	3.38
West Bengal	656	2494	2055	3	201	0	217	4	534	148	6312
	10.39	39.51	32.56	0.05	3.18	0	3.44	0.06	8.46	2.34	100
	41.76	5.56	4.73	2.04	4.26	0	8.59	1.24	26.94	7.93	6.21
Jharkhand	596	1404	591	0	57	1	17	6	1	64	2737
	21.78	51.3	21.59	0	2.08	0.04	0.62	0.22	0.04	2.34	100
	37.94	3.13	1.36	0	1.21	0.71	0.67	1.86	0.05	3.43	2.69
Orissa	59	2914	672	6	51	17	48	25	154	80	4026
	1.47	72.38	16.69	0.15	1.27	0.42	1.19	0.62	3.83	1.99	100
	3.76	6.49	1.55	4.08	1.08	12.14	1.9	7.76	7.77	4.29	3.96

(continued)

Table 12.4 (continued)

State	Coke, coal	Firewood	LPG	Gobar gas	Dung cake	Charcoal	Kerosene	Electricity	Others	No cooking arrangements	Total
Chhattisgarh	67	1553	412	5	47	2	34	12	5	37	2174
	3.08	71.44	18.95	0.23	2.16	0.09	1.56	0.55	0.23	1.7	100
	4.26	3.46	0.95	3.4	1	1.43	1.35	3.73	0.25	1.98	2.14
Madhya Pradesh	28	2726	1537	21	281	0	76	5	1	40	4715
	0.59	57.82	32.6	0.45	5.96	0	1.61	0.11	0.02	0.85	100
	1.78	6.07	3.54	14.29	5.95	0	3.01	1.55	0.05	2.14	4.64
Gujarat	7	1493	1529	10	17	0	218	2	94	59	3429
	0.2	43.54	44.59	0.29	0.5	0	6.36	0.06	2.74	1.72	100
	0.45	3.33	3.52	6.8	0.36	0	8.63	0.62	4.74	3.16	3.37
Daman & Diu	0	26	79	0	0	0	18	0	0	5	128
	0	20.31	61.72	0	0	0	14.06	0	0	3.91	100
	0	0.06	0.18	0	0	0	0.71	0	0	0.27	0.13
D&N Haveli	0	79	88	0	0	0	17	0	0	6	190
	0	41.58	46.32	0	0	0	8.95	0	0	3.16	100
	0	0.18	0.2	0	0	0	0.67	0	0	0.32	0.19
Maharashtra	11	2535	4467	26	7	4	384	7	374	229	8044
	0.14	31.51	55.53	0.32	0.09	0.05	4.77	0.09	4.65	2.85	100
	0.7	5.65	10.28	17.69	0.15	2.86	15.2	2.17	18.87	12.27	7.91
Andhra Pradesh	12	2735	3833	6	3	9	108	20	2	168	6896
	0.17	39.66	55.58	0.09	0.04	0.13	1.57	0.29	0.03	2.44	100
	0.76	6.09	8.82	4.08	0.06	6.43	4.28	6.21	0.1	9	6.78
Karnataka	0	1983	1763	21	0	0	169	15	0	145	4096
	0	48.41	43.04	0.51	0	0	4.13	0.37	0	3.54	100
	0	4.42	4.06	14.29	0	0	6.69	4.66	0	7.77	4.03
Goa	0	36	382	2	0	0	22	0	0	5	447
	0	8.05	85.46	0.45	0	0	4.92	0	0	1.12	100
	0	0.08	0.88	1.36	0	0	0.87	0	0	0.27	0.44
Lakshadweep	0	89	56	0	0	0	22	14	0	10	191
	0	46.6	29.32	0	0	0	11.52	7.33	0	5.24	100
	0	0.2	0.13	0	0	0	0.87	4.35	0	0.54	0.19

Kerala	3	2274	2025	14	0	0	16	9	2	119	4462
	0.07	50.96	45.38	0.31	0	0	0.36	0.2	0.04	2.67	100
	0.19	5.07	4.66	9.52	0	0	0.63	2.8	0.1	6.38	4.39
Tamil Nadu	0	2098	3962	0	0	0	343	6	0	237	6646
	0	31.57	59.61	0	0	0	5.16	0.09	0	3.57	100
	0	4.67	9.11	0	0	0	13.58	1.86	0	12.7	6.54
Pondicherry	0	83	426	0	1	0	18	0	0	48	576
	0	14.41	73.96	0	0.17	0	3.13	0	0	8.33	100
	0	0.18	0.98	0	0.02	0	0.71	0	0	2.57	0.57
A&N Islands	0	115	321	0	0	0	110	1	1	18	566
	0	20.32	56.71	0	0	0	19.43	0.18	0.18	3.18	100
	0	0.26	0.74	0	0	0	4.35	0.31	0.05	0.96	0.56
Total	1571	44,893	43,474	147	4723	140	2526	322	1982	1866	1,01,644
	1.55	44.17	42.77	0.14	4.65	0.14	2.49	0.32	1.95	1.84	100
	100	100	100	100	100	100	100	100	100	100	100

Source: Author's calculation

ple consume LPG and ***18.26% of people consume dung cakes***. But the share of firewood is 4.5%, LPG is 2.67% and that of dung cake is 17.7% out of total India's consumption of these fuels.

In Sikkim, 72.14% of people consume LPG and 21.74% of people consume firewood. The share of these fuels in total consumption of the fuels in India is given by 1.27% and 0.37% respectively. 49.34% of people of Arunachal Pradesh consume LPG and 47.79% of people consume firewood. The share of LPG and firewood in the total consumption of LPG and firewood in India is 1.9% and 1.78%, respectively. In Nagaland, 70.61% of people consume LPG and 29.3% of people consume firewood and their respective shares in India's total consumption of LPG and firewood are 1.66% and 0.67%, respectively. 52.85% of Manipuri people consume LPG and 43.55% of people consume firewood and their respective shares in India's total consumption of LPG and firewood are 3.11% and 2.48%. In Mizoram, 77.93% of people consume LPG and 21.09% of people consume firewood, while their respective shares in India's total consumption of LPG and firewood are 2.75% and 0.72%. 73.28% of people of Tripura consume firewood and 24.73% of people consume LPG and their respective share in India's total consumption of firewood and LPG is 3.03% and 1.06%. In Meghalaya, 67.46% of people consume firewood and 24.52% of people consume LPG and their respective share in India's total consumption of firewood and LPG is 1.89% and 0.71%. In Assam, 57.98% of people consume firewood and 39.11% of people consume LPG and their respective share in India's total consumption of firewood and LPG is 4.44% and 3.09%, respectively.

In West Bengal, 39.51% of people consume firewood, 32.56% of people consume LPG and ***10.39% of people consume coke or coal*** as their major source of fuel. The respective shares in India's total consumption of these fuels are 5.56%, 4.73% and 41.76%. Among these, the share of coke or coal consumption of West Bengal has the highest share in India's total coke or coal consumption. 51.3% of people of Jharkhand consume firewood, 21.59% of people consume LPG and ***21.78% of people consume coke or coal*** as their major source of fuel. The respective shares in India's total consumption of these fuels are 3.13%, 1.36% and 37.94%.

72.38% of people of Orissa consume firewood, 16.69% people consume LPG and their respective shares in total India's consumption of firewood and LPG are 6.49% and 1.55%. In Chhattisgarh, 71.44% of people consume firewood, 18.95% of people consume LPG and their respective shares in total India's consumption of firewood and LPG are 3.46% and 0.95%. In Madhya Pradesh, 57.82% of people consume firewood, 32.6% of people consume LPG and their respective shares in total India's consumption of firewood and LPG are 6.07% and 3.54%, whereas in Gujarat, 44.59% of people consume LPG, 43.54% people consume firewood and their respective shares in total India's consumption of LPG and firewood are 3.52% and 3.33%. In Maharashtra, 55.53% of people consume LPG, 31.51% of people consume firewood, 4.77% of people consume kerosene and their respective share in total India's consumption of LPG, firewood and kerosene is 10.28%, 5.65% and 15.2%. Among these, the share of LPG and Kerosene is highest in India's total consumption of LPG and kerosene. 55.58% of people of Andhra Pradesh consume LPG, 39.61% of people consume firewood and their shares in India's total consumption of LPG and firewood are 8.82% and 6.09%, respectively. 48.41% of people of Karnataka consume firewood, 43.04% of people consume LPG, which are the two major fuels consumed in the state. But the share of Karnataka in total firewood and LPG consumption of India is 4.42% and 4.06%. In Goa, 85.46% of people consume LPG and 8.05% of people consume firewood, but their share in India's total consumption of LPG and firewood is 0.88% and 0.08% respectively. 61.72% of people of Daman & Diu consume LPG, 20.31% of people consume firewood, 14.06% of people consume kerosene and their respective shares in total India's consumption of LPG and firewood and kerosene are 0.18%, 0.06% and 0.71%. In Dadra & Nagar Haveli, 46.32% of people consume LPG, 41.58% of people consume firewood, 8.95% of people consume kerosene and their respective shares in total

India's consumption of LPG, firewood and kerosene are 0.2%, 0.18% and 0.67%. In Lakshadweep, 46.6% of people consume firewood and 29.32% of people consume LPG, but their share in India's total consumption of firewood and LPG is 0.2% and 0.13%, respectively. In Kerala, 50.96% of people consume firewood and 45.38% of people consume LPG, but their respective share in India's total consumption of firewood and LPG is 5.07% and 4.66%. In Tamil Nadu, 59.61% people consume LPG and 31.57% people consume firewood, but their shares in India's total LPG and firewood consumption is 9.11% and 4.67%, respectively. 73.96% of Pondicherry people consume LPG and 14.41% of people consume firewood, but their shares in India's total consumption of LPG and firewood is 0.98% and 0.18%, respectively. In Andaman & Nicobar Islands, 56.71% of people consume LPG and 20.32% of people consume firewood, but their shares in India's total consumption of LPG and firewood are 0.74% and 0.26%, respectively.

Table 12.4 suggests that the state having highest share of LPG consumption in India is Maharashtra (10.28%) followed by Tamil Nadu (9.11%) and Andhra Pradesh (8.82%). It is seen that Uttar Pradesh has the highest share (8.79%) followed by Orissa (6.49%) and Andhra Pradesh (6.09%) in India's total consumption of firewood. West Bengal (41.76%) has the highest share in India's coke or coal consumption followed by Jharkhand (37.94%) and Bihar (4.77%). Maharashtra (17.69%), Punjab (14.97%) and Karnataka (14.29%) have the highest shares in India's gobar gas consumption and Uttar Pradesh (47.24%), Bihar (17.7%) and Punjab (10.5%) in India's total dung cake consumption. Manipur, Orissa and Meghalaya lead in charcoal consumption and Maharashtra, Tamil Nadu and Gujarat lead in kerosene consumption. It is also observed that Jammu & Kashmir, Meghalaya and Orissa have the highest share in India's electricity consumption.

From Table 12.4, we have also analysed the fuel preference patterns of each state of India. We have ranked the fuel preferences in terms of most preferred fuel, second most preferred fuel and third most favoured fuel. From this analysis of most preferred fuels in Table 12.4, it is observed that LPG and firewood are the most favoured fuels in almost all states of India. It is also observed that 20 states (14 states and 6 union territories) consider LPG as the most preferred fuel and 15 states (14 states and 1 union territory) consider firewood as the most preferred fuel. The results are displayed using a Bar diagram. Figure 12.1 shows the most preferred fuels in India.

It is also reflected from Table 12.4 that firewood is the second most preferred fuel of 18

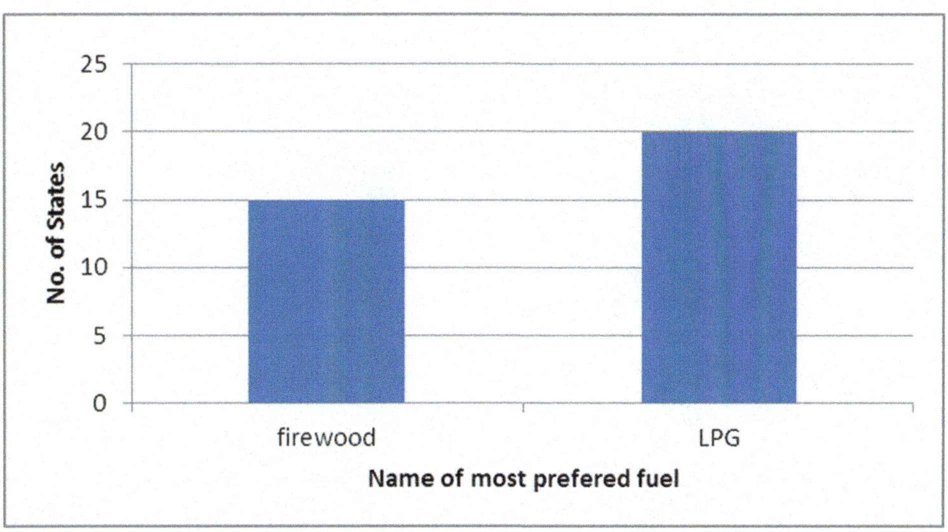

Fig. 12.1 Most preferred fuels in India

major states of India; LPG is the second most preferred fuel for 14 major states of India (see Fig. 12.2). It should be noted that coke and coal, Kerosene and no cooking arrangement also turned to be the second most preferred fuel for one major state of India for each fuels.

Table 12.4 also shows the third most preferred fuel of different states of India. The results are reflected from Fig. 12.3, which shows that kerosene is the third most preferred fuel of 16 major states of India followed by dung cake and no cooking arrangements in 5 major states and remaining fuels in the others states.

Now, this part of the study is mainly focused on fuel selection and their consumption in rural and urban India. Similar way, we also observe the consumption pattern of different fuels in rural as well as in urban India; however, the results are different.

Considering the fuel consumption pattern in rural India, it is observed that firewood is the most preferred fuel in 25 (23 states and 2 union

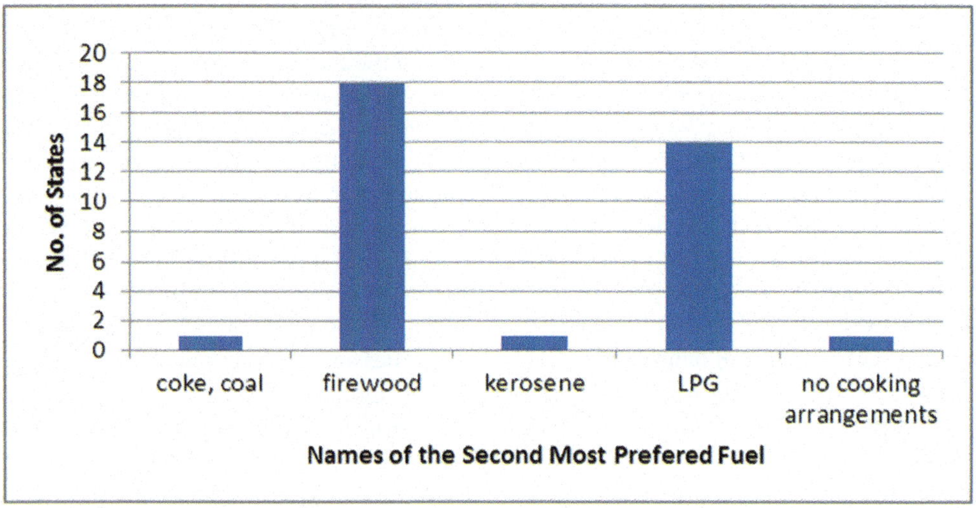

Fig. 12.2 Second most preferred fuels in India

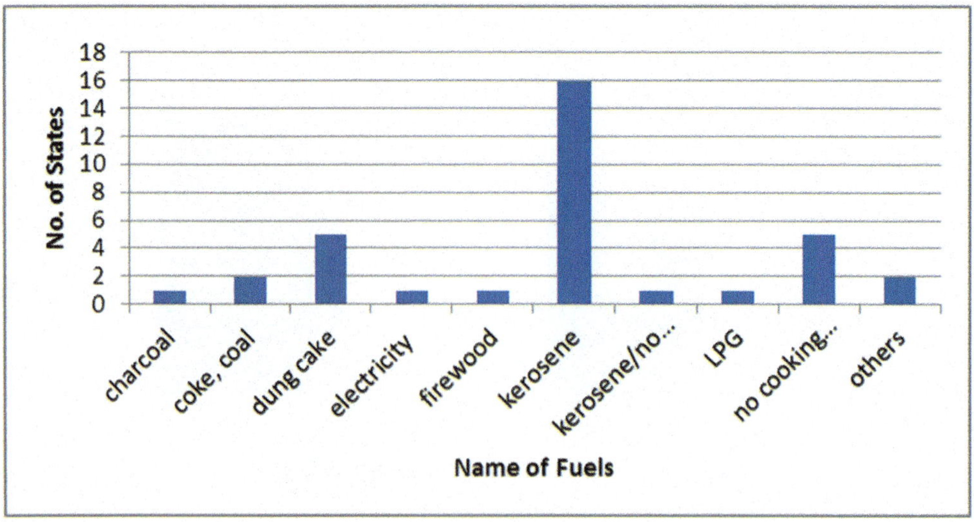

Fig. 12.3 Third most preferred fuels in India

territories) states, while LPG is the most preferred fuel in 10 states (5 states and 5 union territories). However, considering the fuel consumption pattern in urban India, it is noted that LPG is the most preferred fuel in all states and union territories while firewood is the most preferred fuel in no state. These results are reflected in Fig. 12.4.

So, clean fuel LPG is the most preferred fuel in urban India, while dirty fuel like firewood dominates in rural India. Hence, it can be inferred that firewood is the most preferred fuel of rural India and LPG in urban India among available sources. It is also observed that Uttar Pradesh, Orissa and Andhra Pradesh have the highest shares in firewood consumption in rural India. Shares of each state in rural India's firewood consumption are represented in Table 12.5.

Table 12.5 reflects that Uttar Pradesh has the highest share in rural India's firewood consumption followed by Orissa and Andhra Pradesh. The results are reflected using Fig. 12.5:

The top 18 states out of 35 (28 states and 7 territories) which have the highest shares in rural India's total firewood consumption are considered high firewood-consuming states in India and the remaining are considered as low firewood-consuming states in our preliminary analysis. Comparing the poor and non-poor states with high firewood and low firewood-consuming states, we make a comparative analysis of state-wise fuel consumption patterns in rural India. The results are represented in Table 12.6.

Table 12.6 reflects that red-coloured states are rural poor states i.e. those lying below the national average MPCE of rural India, whereas yellow-coloured states are rural rich states i.e. those lying above the national average MPCE. Comparing these results, we can say that mostly rural rich states consume less firewood, whereas rural poor states consume more firewood. Hence, Government of India should design policy for improving their capabilities and adopt it through proper trainings. From our analysis, this study suggests to reduce fuel consumption-related subsidies in urban area. To achieve sustainable consumption and production pattern, material consumption should be reduced drastically following circular economy model. In this context, we tangentially touch the material consumption scenario.

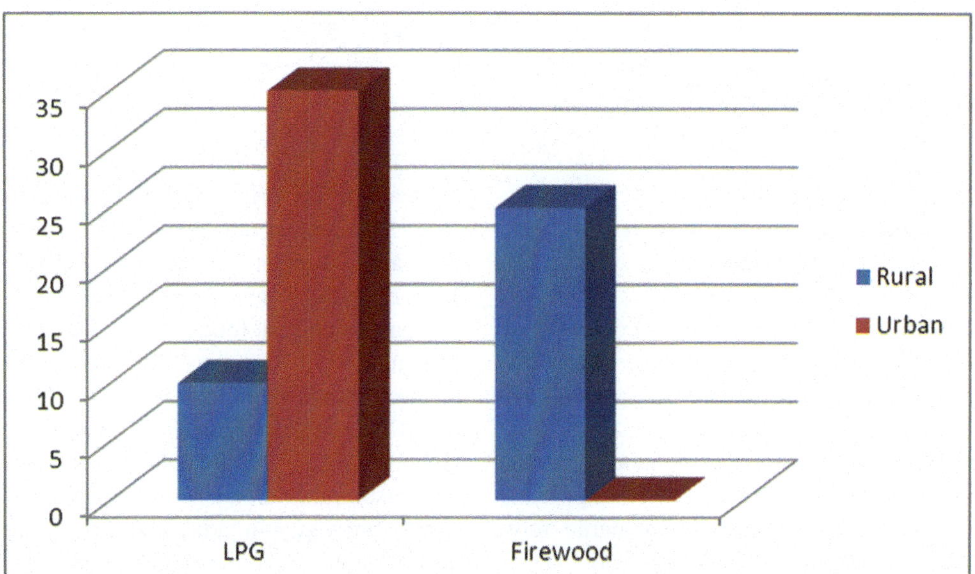

Fig. 12.4 State-wise most preferred fuel of rural and urban India in 2011–2012

Table 12.5 Shares of firewood consumption in rural area across of states in India

State	Firewood	State	Firewood	State	Firewood
Uttar Pradesh	8.43	Jammu & Kashmir	3.46	Mizoram	0.75
Orissa	6.67	Jharkhand	3.45	Nagaland	0.59
Andhra Pradesh	6.16	Chhattisgarh	3.45	Sikkim	0.44
Rajasthan	5.93	Gujarat	3.33	A&N Islands	0.29
Maharashtra	5.81	Himachal Pradesh	3.14	D&N Haveli	0.2
Madhya Pradesh	5.8	Tripura	3.09	Lakshadweep	0.14
West Bengal	5.74	Manipur	2.12	Pondicherry	0.08
Assam	4.93	Meghalaya	2.03	Goa	0.07
Bihar	4.49	Uttaranchal	1.96	Daman & Diu	0.05
Tamil Nadu	4.33	Arunachal Pradesh	1.94	Delhi	0.01
Kerala	4.16	Haryana	1.68	Chandigarh	0
Karnataka	4.1	Punjab	1.21		

Source: Author's calculation

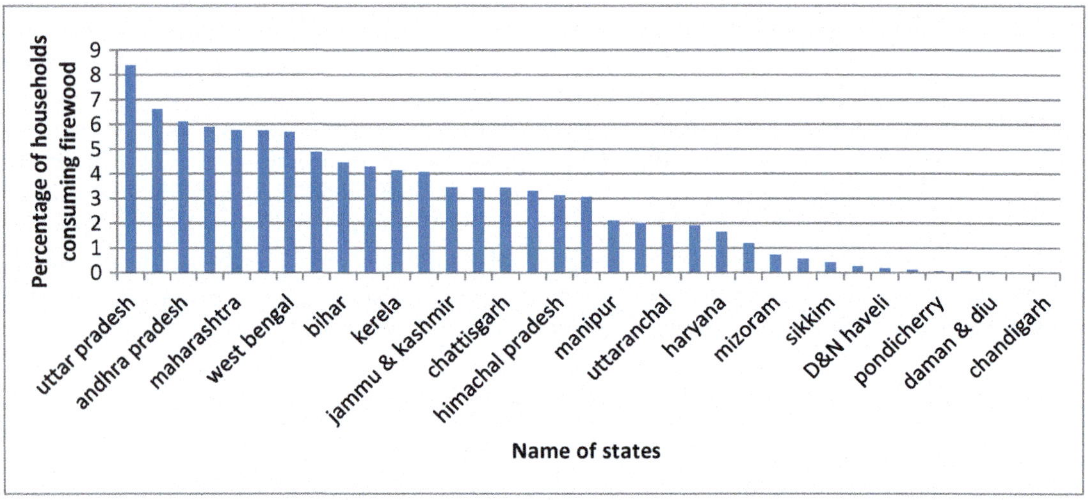

Fig. 12.5 State-wise share in rural India's firewood consumption

12.4 Material Consumption Footprint in India

From the global material consumption data, we extract relevant data of India for the period 2000–2010. Using these data, we estimate the trend line of material consumption per capita, material footprint per capita and we generate data for the period 2011–2030. Table 12.7 describes possible trends of material consumption, material footprint and its intensity during 2000–2030.

Considering the period 2000–2010 as the base for material consumption per capita (metric tons) and material footprint per capita (metric tons) in India, we forecast the trends of such consumptions and footprint for the period 2000–2030. Both material consumption per capita metric tons) and material footprint per capita (metric tons) increase over time (Table 12.7). However, material footprint per unit of GDP (Kilograms per US Dollar) or intensity of material footprint decreases during 2000–2030. Intensity of material consumption footprint ensures to achieve the sustainability goal in India. The existing trends of material consumption per capita (metric tons), material footprint per capita (metric tons) and material footprint per unit of GDP (Kilograms

Table 12.6 State-wise fuel consumption patterns in rural India

High firewood-consuming states	Low firewood-consuming states
Uttar Pradesh	**Manipur**
Orissa	**Meghalaya**
Andhra Pradesh	**Uttaranchal**
Rajasthan	**Arunachal Pradesh**
Maharashtra	**Haryana**
Madhya Pradesh	**Punjab**
West Bengal	**Mizoram**
Assam	**Nagaland**
Bihar	**Sikkim**
Tamil Nadu	**A&N Islands**
Kerala	**D&N Haveli**
Karnataka	**Lakshadweep**
Jammu & Kashmir	**Pondicherry**
Jharkhand	**Goa**
Chhattisgarh	**Daman & Diu**
Gujarat	**Delhi**
Himachal Pradesh	**Chandigarh**
Tripura	

Source: Author's calculation

per US Dollar) are clearly visible in Fig. 12.6. Table 12.8 describes the growth rate of domestic material consumption footprint in India for the period 1990–2010. Average annual growth rate of domestic material consumption per capita in India is 2.4% and that of domestic material consumption is 4.2%, and fossil fuel per capita growth rate is 5.9%, which are directly associated with India's GDP growth rate per annum over the said period. This raises the question of sustainable consumption in emerging economy like India. We need more focused study on it for identifying factors responsible such high domestic material consumption growth, and policy should guide de-growth material consumption for achieving SDG 12.

12.5 Conclusion

This chapter focuses on sustainable consumption and production in India highlighting sustainable development goals 12 (SDG 12). The chapter presents initially the basic development strategy with focusing targeted goals, and describes the development goals and related indicators through which one can assess the achievement of goals. SDG 12 ensures sustainable consumption and production pattern and related issues, which are discussed in earlier studies. The next part of the chapter provides India's evidence on energy consumption, highlighting the state-wise distribution of consumption pattern, which may indicate viability of sustainable consumption. Connecting with this lastly, it analyses the trends of domestic material consumption and traces out its footprints. On the basis of this discussion and limited analysis of consumption pattern, we try to identify the shortcomings of SDG 12 and provide some suggestion for incorporating certain indicators for assessing SDG 12 in a better way in future.

Mostly rural poor states of India consume more firewood, while rural rich states reduce firewood consumption. Hence, Government of India should design non-fossil fuel policy for improving the capabilities of rural poor states and adopt it through proper trainings. From our analysis, this study suggests to reduce fuel consumption-related subsidies in urban area. To achieve sustainable consumption and production pattern,

Table 12.7 Possible trends of material footprint in India during 2000–2030

Year	Domestic material consumption per capita (metric tons)	Material footprint per capita (metric tons)	Material footprint per unit of GDP (kilograms per US$)
2000	3.01	2.47	4.39
2001	3.06	2.55	4.37
2002	3.06	2.49	4.19
2003	3.17	2.55	4.02
2004	3.21	2.64	3.9
2005	3.33	2.79	3.83
2006	3.54	2.95	3.76
2007	3.81	3.26	3.83
2008	3.97	3.29	3.78
2009	4.08	3.45	3.7
2010	4.17	3.56	3.51
2011	4.267455	3.628	3.452909
2012	4.396727	3.747818	3.372636
2013	4.526	3.867636	3.292364
2014	4.655273	3.987455	3.212091
2015	4.784545	4.107273	3.131818
2016	4.913818	4.227091	3.051545
2017	5.043091	4.346909	2.971273
2018	5.172364	4.466727	2.891
2019	5.301636	4.586545	2.810727
2020	5.430909	4.706364	2.730455
2021	5.560182	4.826182	2.650182
2022	5.689455	4.946	2.569909
2023	5.818727	5.065818	2.489636
2024	5.948	5.185636	2.409364
2025	6.077273	5.305455	2.329091
2026	6.206545	5.425273	2.248818
2027	6.335818	5.545091	2.168545
2028	6.465091	5.664909	2.088273
2029	6.594364	5.784727	2.008
2030	6.723636	5.904545	1.927727

Note: (1) 12.2.1—Material footprint per capita—EN_MAT_FTPRPC—Metric Tons (metric tons), (2) 12.2.2—Domestic material consumption per capita—EN_MAT_DOMCMPC—Metric Tons (metric tons per capita), (3) 12.2.1—Material footprint per unit of GDP—EN_MAT_FTPRPG—Kilograms (kilograms per US$)

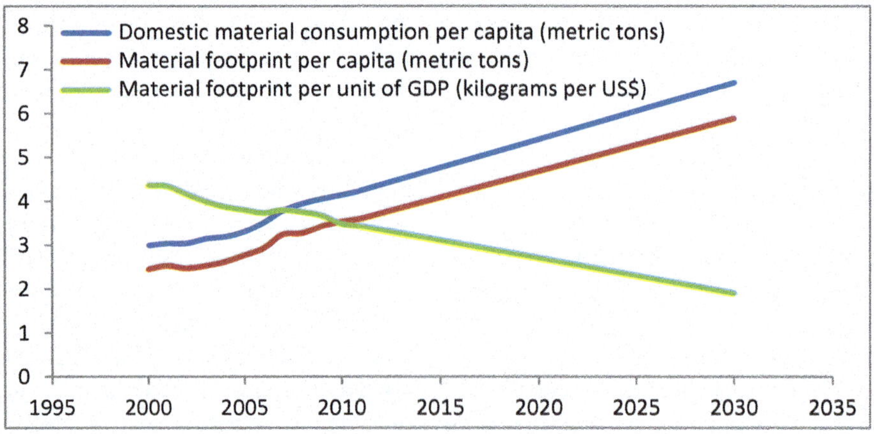

Fig. 12.6 Existing trends of material consumption and its footprint in India

Table 12.8 Growth rate of domestic material consumption in India during 1990–2010

Year	1990	1995	2000	2005	2007	2008	2009	2010	1990–2010
Domestic material consumption	3.3	5.9	0.5	5.3	9.1	5.5	4.2	3.5	4.2
Domestic material consumption per capita	1.2	3.9	−1.3	3.7	7.4	3.9	2.7	2.0	2.4
Domestic material consumption intensity	6.3	5.8	5.2	4.5	4.4	4.5	4.3	4.0	5.3
Fossil fuel per capita growth	10.1	−4.4	−11.5	−36.0	27.5	1.6	−12.0	−25.7	5.9

Source: UNESCAP (downloaded on 29.03.2019)

material consumption should be reduced drastically following circular economy model. Reuse is limited and recycle rate is lacking initiative due to the absence of awareness or feudal mindsets in India. New government may change this mindsets through new initiatives and hopes for better and green India.

References

Adrianzen MA (2013) Improved cooking stoves and firewood consumption: quasi-experimental evidence from the Northern Peruvian Andes. Ecol Econ 89:135–143

Akenji L, Bengtsson M (2014) Making sustainable consumption and production the core of the sustainable development goals. Sustainability 6(2):513–529

Arnold JEM, Kohlin G, Persson R (2006) Woodfuels, livelihoods and policy interventions: changing perspectives. World Dev 34(3):596–611

Banerjee M, Prasad R, Rehman IH, Gill B (2016) Induction stoves as an option for clean cooking in Rural India. Energy Policy 88:159–167

Barnes D, Khandker S, Samad HA (2010) Energy access, efficiency and poverty: how many households are energy poor in Bangladesh? Policy research working paper, WPS5332

Brouwer R, Falcao MP (2004) Wood fuel consumption in Maputo, Mozambique. Biomass Bioenergy 27(3):233–245

Campbell BM, Vermeulen SJ, Mangono JJ, Mabugu R (2003) The energy transition in action: urban domestic fuel choices in a changing Zimbabwe. Energy Policy 31(6):553–562

Couture S, Garcia S, Reynaud A (2012) Household energy choices and fuelwood consumption: an econometric approach using French data. Energy Econ 34:1972–1981

Dinda S (2014) A theoretical basis for green growth. International Journal of Green Economics 8(2): 177–189

Dinda S (2016) Creating natural resource base for green growth in India. International Journal of Green Economics 10(1):1–14

Dinda S (2017) A Circular Economy Approach for Sustainable Economic Growth, MPRA paper 89587

Davis M (1998) Rural household energy consumption: the effects of access to electricity—evidence from South Africa. Energy Policy 26(3):207–217

Gasper D, Shah A, Tankha S (2019) The framing of sustainable consumption and production in SDG 12. Global Policy 10(Suppl 1):83–95

Gebreegziabher Z, Mekonnen A, Kassie M, Kohlin G (2012) Urban energy transition and technology adoption: the case of Tigrai, Northern Ethiopia. Energy Econ 34:410–418

Goswami A, Bandyopadhyay KR, Kumar A (2017) Exploring the nature of rural energy transition in India: insights from case studies of eight villages in Bihar. Int J Energy Sect Manag 11:463

Gundimeda H, Kohlin G (2008) Fuel demand elasticities for energy and environmental policies: Indian sample survey evidence. Ecol Econ 30:517–546

Gupta G, Kohlin G (2006) Preference for domestic fuel: analysis with socio-economic factors and rankings in Kolkata, India. Ecol Econ 57:107–121

Hanna R, Oliva P (2015) Moving up the energy ladder: the effect of an increase in economic well-being on the fuel consumption choices of the poor in India. Am Econ Rev 105:242–246

Heltberg R (2005) Factors determining household fuel choice in Guatemala. Environ Dev Econ 10:337–361

Hosier RH, Dowd J (1987) Household fuel choice in Zimbabwe: an empirical test of the energy ladder hypothesis. Resour Energy 9:347–361

Jingchao Z, Kotani K (2012) The determinants of household energy demand in rural Beijing: can environmentally friendly technologies be effective? Energy Econ 34:381–388

Jumbe CBL, Angelsen A (2011) Modelling choice of fuelwood source among rural households in Malawi: a multinomial probit analysis. Energy Econ 33:732–738

Kebede E, Kagochi J, Jolly CM (2010) Energy consumption and economic development in Sub-Sahara Africa. Energy Econ 32:532–537

van der Kroon B, Brouwer R, van Beukering PJH (2014) The impact of the household decision environment on fuel choice behaviour. Energy Econ 4(1):236–247

Leach G (1992) The energy transition. Energy Policy 20(2):116–123

Ma Y, Rong K, Luo Y, Wang Y, Mangalagiu D, Thornton TF (2019) Value co-creation for sustainable consumption and production in the sharing economy in China. J Clean Prod 208:1148–1158

Masera OR, Saatkamp BD, Kammen DM (2000) From linear fuel switching to multiple cooking strategies: a critique and alternative to the energy ladder model. World Dev 28(12):2083–2103

Nayak BP, Werthmann C, Aggarwal V (2015) Trust and cooperation among urban poor for transition to cleaner and modern cooking fuel. Environ Innov Soc Trans 14:116–127

Ouedraogo B (2006) Household energy preferences for cooking in urban Ouagadougou, Burkin Faso. Energy Policy 34(18):3787–3795

Pachauri S, Jiang L (2008) The household energy transition in India and China. Energy Policy 36(11):4022–4035

Raman P, Ram NK, Murali J (2014) Improved test method for evaluation of bio-mass cookstoves. Energy 71:479–495

Scherer L, Behrens P, de Koning A, Heijungs R, Sprecher B, Tukker A (2018) Trade-offs between social and environmental sustainable development goals. Environ Sci Policy 90(1):65–72

Takama T, Tsephel S, Johnson FX (2012) Evaluating the relative strength of product-specific factors in fuel switching and stove choice decisions in Ethiopia. A discrete choice model of household preference for clean cooking alternatives. Energy Econ 34:1763–1773

Xu S, Chu C, Zhang Y, Ye D, Wang Y, Ju M (2018) Entangled stakeholder roles and perceptions of sustainable consumption: an evaluation of sustainable consumption practices in Tianjin, China. J Environ Manag 223:841–848

13 Climate Change Responses and Sustainable Development: Integration of Mitigation and Adaptation

Vaibhav Bangar, Rajat Goyal, and Rajiv Pandey

13.1 Introduction

The past century has witnessed extraordinary advances in technology, science and medicine. This advancement was achieved through the utilization of resources of intellectual imagination to provide solutions to various unsolved problems. However, there is no comparable worldwide advancement in issues related to morality such as discrimination based on race, caste, gender, ethnicity, religion or economic or other status (Woolcock 1998). Lately, these discriminations lead to conflicts and undermine the growth and development of humanity. In due course of time with advancement of knowledge and understanding about the humanity, the concept of well-being, dignity, prosperity, equality, justice, peace and spirituality has changed people's perception of development. It is now established that development is not only a mere growth in economy but a combination of various aspirations, which are hard to achieve even if a region is thriving economically (Schwab 2016). This recognition has led to the emergence of modern world development concept known as sustainable development. The notion of sustainable development ways its roots back in the fourth century BC as evident in Chanakya's (an ancient Indian philosopher and teacher) belief that the stability of an empire depends on the stability of its environment (Bhattacharya 2014). In this way, the concept of sustainable development is being accentuated as judicial utilization of available natural resources for continuous development and to satisfy human needs. Therefore, sustainability includes all along ecology, economy and equity (Pannell and Schilizzi 1999). With this in debate, in 2015, the United Nations General Assembly led to the establishment of sustainable development goals (SDGs). SDGs are comprehensive, far-reaching and people-centred set of universal and transformative goals (Simon et al. 2016). They are integrated and indivisible and balance the three dimensions of sustainable development: the economic, social and environmental (Fig. 13.1).

However, the success for achieving the SDGs is threatened by the negative impacts that climate change is posing on the ecology, economic and social objectives of the SDGs. Climate change among other things worsens poverty, elevates food insecurity, increases health risks and therefore increases the overall vulnerability of exposed people (UN Fact sheet 2010). Consequently, climate change will complicate the achievement of each and every goal of the sustainable development either directly or indirectly. Moreover, these complications will vary among nations depending upon their respective fragility, resources, climate exposure and sensitivity, along with differentiated responsibility for the climate

V. Bangar · R. Goyal · R. Pandey (✉)
Forest Research Institute, Dehradun, India

S. Hazra, A. Bhukta (eds.), *Sustainable Development Goals*, Sustainable Development Goals Series,
https://doi.org/10.1007/978-3-030-42488-6_13

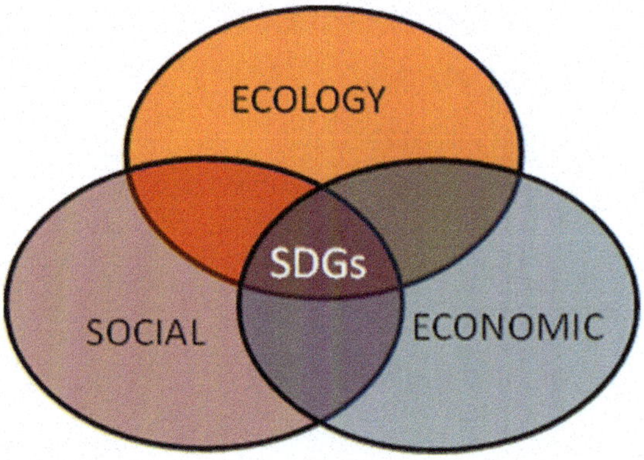

Fig. 13.1 Dimensions of sustainable development

change mitigation. Before establishing linkages between climate change response and sustainable development, it is imperative to apprehend how each sustainable development goal gets challenged by climate change impact. These linkages would facilitate achieving the human welfare through the common but differentiated responsibility of nations to mitigate the climate change. The following section would illustrate the challenges posed by climate change on each SDG as figured in various literatures and intellectual discourses.

13.2 Climate Change and SDGs

Climate change as a significant factor directly or indirectly affects the achievement of sustainable development goals (SDGs), which define sustainable development across the world. The sustainable development goals are the blueprint to achieve a better and more sustainable future for all. The goals interconnect and in order to leave no one behind, it is important that we achieve each goal and target by 2030. While climate change worsens poverty, elevates food insecurity, increases health risks, it can also drag societies towards social tipping points (forced migration, chronic poverty and war-like situation). Before establishing linkages between climate change response and sustainable development, it is imperative to apprehend how each sustainable development goal gets challenged by climate change impact. The following section would illustrate the challenges posed by climate change for each sustainable development goal and possible measures to account for the challenges in a brief and succinct manner on each SDG (Fig. 13.2).

13.2.1 SDG 1 (End Poverty in All Its Form Everywhere)

In the past decades, significant progress has been made to reduce global poverty from 35% of the world population in 1991 to 10% of the world population living under poverty in 2015. However, the World Bank categorically mentioned that ending poverty is threatened by climate change, and an additional 100 million people will be pushed into poverty by 2030 in the absence of stringent actions to reduce climate change impacts (World Bank 2015). Natural disasters such as tsunami, hurricane, flood and heatwaves destroy assets available to poor affecting their livelihood disproportionately and thus trapping these poor into the poverty net. A case study in Honduras reported that the poor people lost three times more assets than the non-poor when hit with hurricane Mitch in 1998 (Morris et al. 2002). Moreover, poor people often receive limited or no support from their family, do not have social safeties and no support from the financial system, which make it harder for them to avoid, cope, or adapt to climate change impacts (World Bank 2015).

With these impacts, development initiatives need to be coupled with climatic adaptation measures, such as constructing protective infrastructures like dikes, improved drainage systems and mangrove plantations to deal with flooding, installing disaster preparedness equipment for immediate response, and introducing climate-resistant breeds of crops and livestock (Sovacool et al. 2012). In the longer term, only immediate mitigation actions to reduce global emissions will save poor from climate change impacts and therefore mitigation policies should be designed by governments to protect and even benefit poor people. However, it is also important that the mit-

Fig. 13.2 Climate action (SDG13) links with other sustainable development goals

igative actions should be taken to reduce the burden of policies that often impose new costs (Laukkonen et al. 2009).

13.2.2 SDG 2 (End Hunger, Achieve Food Security and Improved Nutrition and Promote Sustainable Agriculture)

The implications of climate change on agricultural productivity can seriously impact the food security. Increased surface temperature and rainfall variability can elevate the vulnerability of people to crop failure and higher pest influx (Gromko and Abdurasulova 2019). African countries, least developed countries, landlocked developing countries and Small Island developing states are most vulnerable to food insecurity due to climate change (Sovacool et al. 2012). Moreover, due to population rise, the global food demand is projected to increase nearly by 60% in 2050 above 2006 levels, which makes it crucial to implement climate-resilient development to ensure sustainable food production (Godfray et al. 2010).

The adaptation measures to prevent food insecurity include transition to systems that are more productive, using efficient inputs, stabilizing outputs and technological shift (Gromko and Abdurasulova 2019). 'Climate smart agriculture' proposed by Food and Agriculture Organisation (FAO) advocates agriculture, which sustainably increases productivity, resilience (adaptation), reduces GHGs (mitigation) and leads towards food security (McCarthy et al. 2011).

13.2.3 SDG 3 (Ensure Healthy Lives and Promote Well-Being for All at All Ages)

Climate change is expected to cause around 250,000 additional deaths annually due to malaria, diarrhoea, malnutrition and heat stress between 2030 and 2050 with the direct damage cost to health due to climate change impacts expected to be USD 2–4 billion annually by 2030 (WHO 2018).

Several steps can be taken by governments to increase the adaptive capacity of people such as installation of early epidemic warning systems, which are informed by climate scenarios to intimate people of the possible threat at the earliest so that their responding ability may be increased (Sutherst 2004). Mitigation policies like cleaner energy systems, promoting use of public transport and promoting active moments such as walking or cycling as alternatives to using private vehicles have the potential to reduce carbon emission and sustain healthy lives in a longer term (WHO 2018).

13.2.4 SDG 4 (Ensure Inclusive and Equitable Quality Education and Promote Lifelong Learning Opportunities for All)

Climate change has a direct impact on education primarily due to the effects of extreme weather events, such as strong winds, hail storms and heavy rains accompanied by flash floods having both short- and long-term consequences leading to negative impacts upon educational attainment (UNDP 2017). The secondary impacts of climate change on education arise from the ways in which households adapt themselves to climate change. In most of the cases, children get involved in income-generating activities with household members and even some households choose to migrate (Hallegatte et al. 2015). Early child marriage is also adopted as a coping strategy at the time of food shortages because the bride price is a welcome income and it also means one less person to feed, clothe and educate. Thus, climate change undermines the right to education of every child, especially the most disadvantaged (UNDP 2017).

Introducing progressive policies, strategies, programmes and activities to increase adaptation and creating environment for the attainment of education for all is a crucial step in responding to changing climate. However, there are few or no policies within the education sector that addresses the challenges posed by climate change specifically (UNDP 2017). Policies should be implemented that enhance adaptation and social protection of orphan and vulnerable children such as school feeding, education assistance (UNDP 2017). Mitigation measures across all sectors to halt climate change can sustain the educational rights of all children for a longer term.

13.2.5 SDG 5 (Achieve Gender Equality and Empower All Women and Girls)

According to UN women watch, the adverse impact of climate change in areas such as agriculture, food security; water resources; human health; energy, transport, and human settlements affects the women disproportionately increasing inequality based on gender (UN Fact sheet 2010). Women are more vulnerable to the effects of climate change than men primarily because of two reasons, firstly, because they constitute the majority of the world's poor and secondly, women are more dependent to sustain their livelihood on natural resources that are threatened by climate change (Arora-Jonsson 2011). Furthermore, women face economic, social and political barriers that limit their coping capacity. They have limited mobility and unequal access to resources and decision-making, which make them highly vulnerable to climatic shocks such as food insecurity (Arora-Jonsson 2011).

For achieving gender equality and women empowerment in rural areas of developing countries, it is vital that climate change responses be integrated with development actions. It is crucial that the rights of rural women be ensured in regard to food security, equitable participation and non-discriminatory access to resources, while framing policies in response to climate change (Arora-Jonsson 2011). Mitigation and adaptation efforts should address the gender-specific impacts of climate change systematically and effectively.

13.2.6 SDG 6 (Ensure Availability and Sustainable Management of Water and Sanitation for All)

Water is a vital resource to both society and ecosystems. It is required in multiple sectors such as agriculture, energy production, recreation and manufacturing. A regular supply of clean drinking water is also crucial for sustaining health (WHO 2018). These multiple uses of water in various sectors are putting pressure on water resources that are likely to be exacerbated by climate change. The water cycle is a delicate balance of precipitation, evaporation and all of the steps in between. The rise in temperature as a result of climate change increases the rate of evaporation of water into the atmosphere, also increasing the capacity of atmosphere to 'hold' water. This leads to increased evaporation dry out in some areas as well as excess precipitation on other areas terribly disturbing the ecology of such areas (Titus and Anderson 2009).

Adaptation measures that can prove particularly effective include conservation tillage, rainwater harvesting, planting trees in steeply-sloping fields, maintaining vegetation cover, desalination, improved pasture management, reuse of water and more efficient soil and irrigation water management (Charlton et al. 2010). Protecting and restoring freshwater habitats and managing natural floodplains are some of the additional adaptive measures that are not a part of conventional management practices but can substantially reduce the impact of climate change on water resources (Lawler 2009).

13.2.7 SDG 7 (Ensure Access to Affordable, Reliable, Sustainable and Modern Energy for All)

Climate change presents increasing challenges for production of energy and its transmission. While the full impact on energy business is not entirely certain, the climate-related impacts may result in some new engineering challenges and increased costs for developing and accessing energy resources (Finley and Schuchard 2011). A progressive rise in temperature, changing precipitation patterns and an increase in number and severity of climate extreme events will negatively affect energy production and delivery. The supply of fossil fuels will also decline, along with decline in hydropower and thermal generation and transmission (Finley and Schuchard 2011).

Investment should be made to carry out extensive physical risk assessments to better prepare for extreme weather events or disasters, which can lead to equipment malfunctions (Finley and Schuchard 2011). Behavioural change as a response to climate change includes use of renewable energy sources, utilizing minimum energy, purchasing energy-efficient gadgets and use of biofuels (Chapman 2007).

13.2.8 SDG 8 (Promote Sustained, Inclusive and Sustainable Economic Growth, Full and Productive Employment and Decent Work for All)

The primary influence of global warming on economic growth is due to the damages caused to property and infrastructure, mass migration and security threats (Masson-Delmotte et al. 2018). The adverse impacts of climate change in areas such as biodiversity and ecosystems; water resources; agriculture and food security; human health; human settlements and energy, transport and industry can influence the economy of a country terribly (Handmer et al. 2012). Damage to the capital stock and labour supply due to extreme weather events will weaken the economic growth and the labour productivity will reduce as the world economy adjusts to increased temperatures (Harvey et al. 2014).

The economic growth of a country will depend upon the overall success of responses to climate change in other sectors such as agriculture, health, biodiversity, infrastructure, transport and industry. Intergovernmental agreements encompassing the major economies of the world will be the most effective in tackling climate change. Without a collective policy response, the efforts of a few countries to reduce greenhouse gas emissions will fall short of what is required to make a significant impact on a global level (Grafakos et al. 2018).

13.2.9 SDG 9 (Build Resilient Infrastructure, Promote Inclusive and Sustainable Industrialization and Foster Innovation)

Increased frequency of natural disasters due to climate change is creating a major threat to infrastructure across the world (NCA 2014). Sea-level rise, storm surge, floods, hurricanes, and heavy downpours in combination with continued development in coastal areas are increasing damage to infrastructure including industrial facilities, roads and buildings, and are also increasing risks to ports and coastal military installations worldwide (NCA 2014). Extreme heat and elevated temperature are also damaging transportation infrastructure such as roads, airport runways and railway lines (Mills and Andrey 2002).

Providing support to industries to develop low carbon growth and climate-resilient infrastructure through vulnerability assessments, policy advice and implementation of innovative technologies that compliment mitigation can be promising to achieve sustainable industrial development (NCA 2014). Adaptation measures include building climate shock proof transportation infrastructure and buildings and by modifying existing protection infrastructure, while taking into account possible future climatic impacts (Mills and Andrey 2002).

13.2.10 SDG 10 (Reduce Inequality Within and Among Countries)

Understanding the impact of climate change on the causes of global inequality is crucial for achieving equitable economic development across the world. According to a study, although inequality between countries has decreased over the past century, there is a 90% likelihood that climate change and global warming have slowed down that decrease (Burke et al. 2018). The primary driver of these growing inequalities is the relationship between temperature and economic growth, with warming decreasing growth in warm countries while increasing the growth in cooler countries (due to longer growing season) (Burke et al. 2018). Over the years, the impacts of climate change accumulated and distinctly declined the economic output in hotter, and poorer countries. Thus, the rising temperature due to greenhouse gases caused by fossil fuel use has substantially exacerbated the economic inequality associated with historical disparities in the consumption of energy (Stern and Stern 2007).

Given the magnitude of the warming-induced growth penalties that poor countries have already suffered, expansion of low-carbon energy sources can be expected to provide a substantial secondary development benefit (by curbing future warming-induced growth penalties), in addition to the primary benefits of increased energy access (Burke et al. 2018). Providing monetary and technological assistance to poorer countries for uplifting their economic growth can help them adapt to further risks of climate change. Promoting the development, dissemination, transfer, and diffusion of environmentally sound technologies to poorer countries on favourable terms can help reduce the difference in their coping strategies (Burke et al. 2018).

13.2.11 SDG 11 (Make Cities and Human Settlements Inclusive, Safe, Resilient and Sustainable)

Although cities cover less than 2% of the world's land surface, they are major contributors to climate change and account for 71–76% of the world's carbon dioxide from global final energy use (UN 2019). At the same time, cities (especially coastal urban areas) and towns are heavily vulnerable to climate change. Globally, hundreds of millions of people residing in urban areas will be affected by impacts of climate change such as rising sea levels, inland floods, increased precipitation and more frequent and stronger cyclones and storms (UN 2019). Climate change may also negatively impact urban infrastructure and worsen access to basic services and quality of life in cities (UN 2019). The most affected populations due to climate change are the urban poor i.e. slum dwellers, who tend to live along river banks, on slopes and hillsides prone to landslides, on decertified land, near polluted grounds and in unstable structures vulnerable to earthquakes and also along waterfronts in coastal areas. With more than one billion people living in slums and informal settlements, it is crucial to take immediate steps to prevent climate atrocities (Huq et al. 2007).

Despite these impacts, most of the cities have not yet addressed climate change; however, if properly planned, managed and capacitated through appropriate government structures and policies, cities can be places of efficiency and innovation (UN 2019). Adaptation measures needed to cope with climate change vary considerably depending upon the political, historical, cultural and climatic conditions (UN 2019). Such measures can range from climate proofing of infrastructure such as storm-drainage system, protection and relocation of energy or waste management facilities to 'working with nature' e.g. protecting mangrove and natural reef ecosystems and placing a greater emphasis on coastal resource management (NCA 2014). Mitigation strategies including a better planned city with reduced urban sprawl, better public transport and greener buildings and promoting use of alternative source of energy other than fossil fuel can reduce a city's carbon footprint while providing a better environment to its citizens (Tanner et al. 2009).

13.2.12 SDG 12 (Ensure Sustainable Consumption and Production Patterns)

Impact of climate change on agriculture and food security forces the affected community towards an inefficient utilization of natural resources, which can further increase their vulnerability (Locatelli et al. 2015). Increasing temperature means increased energy consumption for cooling purpose, which further leads to increase in emissions.

Capacity building for the industrial sector to improve the overall efficiency by reusing and recycling resources and preventing waste generation due to climate change impacts is crucial for the sustainable production (Shen 1995). Rationalizing inefficient fossil fuel subsidies and phasing out harmful subsidies that encourage wasteful consumption is important to reflect their impacts on environment (Whitley and Van Der Burg 2015).

13.2.13 SDG 14 (Conserve and Sustainably Use the Oceans, Seas and Marine Resources for Sustainable Development)

Climate change affects marine species including planktons, which form the basis of marine food chains. Coral bleaching shifts in ecological niches and stormy weather are other significant impacts of climate change on the marine environment (Doney et al. 2012). For mangrove ecosystems, rapid sea-level rise will be the greatest climate change challenge because they require stable sea levels for long-term survival. Oceans absorb a considerable proportion of carbon dioxide released as a result of anthropogenic activities; this leads to increased acidity of oceans which degrades marine biodiversity as well as decreases the potential of oceans to further absorb carbon dioxide (Doney et al. 2012).

Designing policies to protect and restore coastal ecosystem along with prevention of further conversion of these ecosystems to other land uses is vital to ensure the sustainable use of marine resources (Laffoley and Baxter 2016). Support for scientific research is also crucial to ensure the continued analyses and monitoring of the impacts of climate change on marine ecosystem, and the knowledge gained can be used to design and implement adequate and appropriate adaptation and mitigation.

13.2.14 SDG 15 (Protect, Restore and Promote Sustainable Use of Terrestrial Ecosystems, Sustainably Manage Forests, Combat Desertification and Halt and Reverse Land Degradation and Halt Biodiversity Loss)

Emissions of pollutants and greenhouse gases from power plants, motor vehicles, deforestation, and other anthropogenic sources are warming the Earth and damaging ecosystems and human well-being (Gonzalez 2010). Significant changes in terrestrial ecosystems have occurred due to changing climate rather than other factors that could alter global biogeochemical cycles, convert extensive land areas from one biome to another, and isolate or drive numerous species to extinction (Gonzalez 2010).

Use of efficient land-use policies that integrate disaster risk reduction, climate change mitigation and adaptation, and sustainable development, while providing biodiversity benefits, is important to combat challenges posed by climate change to terrestrial ecosystem along with management options to avoid or reverse greenhouse gas emissions from ecosystems, leading to benefits for biodiversity as well as assistance in the continued delivery of important ecosystem services (Gonzalez 2010).

13.2.15 SDG16 (Promote Peaceful and Inclusive Societies for Sustainable Development, Provide Access to Justice for All and Build Effective, Accountable and Inclusive Institutions at All Levels)

Climate change has generated issues related to violence, political instability and disparities between nations. In fragile states, climate change will complicate the achievement of SDG 16. It is increasingly accepted that climate change can be a continuous contributor—at times significant, at times subtle—to the causal network that generates conflict and threatens human security (Scheffran and Battaglini 2011). Regions such as the Sahel, Horn of Africa and Middle East are highly fragile states in terms of peace where the worst climate impacts are anticipated (Brown et al. 2007).

The climate change responses, specifically mitigation in any sector, can create opportunities for such fragile states to prevent them from impacts of climate change on peace and human life.

13.2.16 SDG 17 (Strengthen the Means of Implementation and Revitalize the Global Partnership for Sustainable Development)

Climate change is a global threat and most of the climatic actions are only globally effective to reduce emissions. The countries have to be in partnership for the social, ecological and economic development through technology transfer, managing impacts, innovation sharing and financial aid required in case of a crisis, conflicts or natural disaster. The aim is to combat the changing climate, taking humanity as one and providing aid to the country in need.

13.3 Linking Integrated Climate Change Responses (Mitigation and Adaptation) with SDGs

The previous section addressed the challenges that sustainable development faces due to the ill-impacts of climate change. It is a major roadblock for the success of sustainable development goals. Response options to avoid these challenges were also addressed for every sustainable development goal in previous section. It is clear that climate change is seriously threatening every aspect of sustainable development that is social, ecological and economic. Therefore, to succeed in meeting carbon emission reduction targets, to limit projected climate change impacts, to improve the livelihood standards of people living in climate-sensitive remote indigenous areas and for development of a sustainable climate-resilient future, it is imperative that improved synergies be developed between climate change responses (mitigation and adaptation) and sustainable development agendas.

Recognizing the complementarity between adaptation strategies and sustainable development objectives is crucial to form policies that are climate-resilient as well as fulfil sustainable development agenda. Since, many of the policies proposed for adaptation (such as income diversification or capacity building) have long been implemented as part of development projects, there exists a continuum between them (Locatelli et al. 2015). Adaptation activities, which attempt to reduce future climate change impacts, cannot achieve success unless appropriate development measures to deal with current problems are effectively undertaken (Ngum et al. 2019).

The more progress towards adaptation means the more it will be necessary for policies to consider the uncertainties of future climate so that adequate modifications can be made in sustainable development strategies as the change occurs (Ngum et al. 2019). Failure of this can significantly increase the vulnerability of population, making it necessary to systematically integrate adaptation into development processes. Mitigation measures to combat changing climate are equally essential to increase the sustainability of development actions for a longer term. Despite these promising links between climate change response and sustainable development, knowledge of how the approach should be implemented 'on the ground' is lacking.

Thus, to create a 'triple wins' situation, i.e. development with emission reduction, decision-making committee should encourage the development of such synergies in every sector by generating enabling factors such as designing efficient policies, laws and strategies; reducing knowledge gaps by intensifying scientific research; by facilitating projects geared towards synergy and by capacity building of institutions involved in framing policies. Thus, if effectively implemented, integrated climate change responses can provide the best opportunity to make an intense progress in terms of sustainable development.

13.4 Building Blocks of Climate-Resilient Sustainable Future

Integrating means of achieving sustainable development goals such as technology transfer and financing with climate change response would promote the development of a climate-resilient sustainable future in both developed and developing countries. Inadequate financial resources and insufficient technology are opined as the most important challenge faced by developing countries to adopt an integrated approach for policy-making (Ngum et al. 2019). Figure 13.3 shows the building blocks of a climate-resilient sustainable future, especially in developing countries.

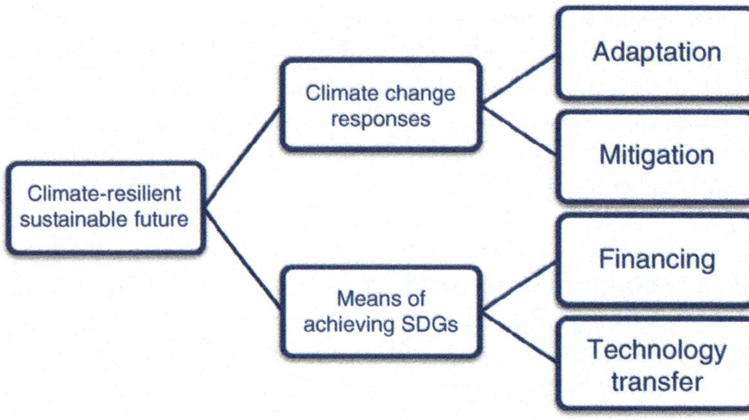

Fig. 13.3 Building blocks of climate-resilient sustainable future

13.5 Conclusion

Climate change challenges the success of each and every sustainable development goal in one way or other. It is clear that the impacts of climate change are threatening every aspect of sustainable development i.e. social, ecological and economic aspects. However, most of the adaptation policies generally are implemented already as part of development projects rather than as climate change response. This continuum between climate change adaptation and sustainable development needs to be deeply analysed to frame policies that are sustainable as well as climate-resilient. Moreover, most of the policies, strategies and programs do not explicitly mention 'climate change', but propose some actions and activities that indirectly address climate change. Other challenges that hamper the effectiveness of integrated and holistic development strategy include inadequate coordination, insufficient capacity building, ineffective implementation, lack of proper transparency and inadequate public participation (Ngum et al. 2019).

Further, four areas have been identified as the critical building blocks for creating a climate-resilient sustainable future: adaptation, mitigation, technology transfer and financing. Mitigation and adaptation are climate change responses and technology transfer and financing are means for achieving sustainable development goals. Together, these four building blocks can help create a 'win–win' situation. This is especially important in areas where the development policies are overlooking the future impacts of climate change as well as in remote indigenous areas where basic life-sustaining processes can be adversely affected by changing climate. Therefore, to succeed in meeting carbon emission reduction targets, to limit projected climate change impacts, to improve the livelihood standards of people living in climate sensitive remote indigenous areas and for the development of a sustainable climate-resilient future, it is imperative that improved synergies be developed between climate change responses (mitigation and adaptation) and sustainable development agendas.

References

Arora-Jonsson S (2011) Virtue and vulnerability: discourses on women, gender and climate change. Glob Environ Chang 21(2):744–751

Bhattacharya S (2014) Forest and biodiversity conservation in ancient Indian culture: a review based on old texts and archaeological evidences. Int Letters Soc Humanist Sci 30:35–46

Brown O, Hammill A, McLeman R (2007) Climate change as the 'new' security threat: implications for Africa. Int Aff 83(6):1141–1154

Burke M, Davis WM, Diffenbaugh NS (2018) Large potential reduction in economic damages under UN mitigation targets. Nature 557(7706):549

Chapman L (2007) Transport and climate change: a review. J Transp Geogr 15(5):354–367

Charlton M, Bailey A, Arnell N (2010) Water for agriculture—implications for future policy and practice:

reviewing and modelling the impacts of climate change in future food production. Royal Agricultural Society of England, Warwickshire

Doney SC, Ruckelshaus M, Emmett Duffy J, Barry JP, Chan F, English CA, Galindo HM, Grebmeier JM, Hollowed AB, Knowlton N, Polovina J (2012) Climate change impacts on marine ecosystems. Ann Rev Mar Sci 4:11–37

Finley T, Schuchard R (2011) Adapting to climate change guide for the energy and utility industry. Business School of Responsibility, San Francisco

Godfray HCJ, Beddington JR, Crute IR, Haddad L, Lawrence D, Muir JF, Pretty J, Robinson S, Thomas SM, Toulmin C (2010) Food security: the challenge of feeding 9 billion people. Science 327(5967):812–818

Gonzalez P (2010) Impacts of climate change on terrestrial ecosystems and adaptation measures for natural resource management. In: Dodson J (ed) Changing climates, earth systems and society. Springer, New York, pp 5–20

Grafakos S, Pacteau C, Delgado M, Landauer M, Lucon O, Driscoll P (2018) Integration of climate mitigation and adaptation: opportunities and challenges. In: Rosenzweig C, Solecki W, Romero-Lankao P, Mehrotra S, Dhakal S, Ibrahim SA (eds) Climate change and cities: second assessment report of the urban climate change research network. Cambridge University Press, Cambridge

Gromko D, Abdurasulova G (2019) Climate change mitigation and food loss and waste reduction: exploring the business case

Hallegatte S, Bangalore M, Bonzanigo L, Fay M, Kane T, Narloch U, Rozenberg J, Treguer D, Vogt-Schilb A (2015) Shock waves: managing the impacts of climate change on poverty. The World Bank, Washington, DC

Handmer J, Honda Y, Kundzewicz ZW, Arnell N, Benito G, Hatfield J, Mohamed IF, Peduzzi P, Wu S, Sherstyukov B, Takahashi K (2012) Changes in impacts of climate extremes: human systems and ecosystems. In: Managing the risks of extreme events and disasters to advance climate change adaptation special report of the intergovernmental panel on climate change. Intergovernmental Panel on Climate Change, Geneva, pp 231–290

Harvey CA, Rakotobe ZL, Rao NS, Dave R, Razafimahatratra H, Rabarijohn RH, Rajaofara H, MacKinnon JL (2014) Extreme vulnerability of smallholder farmers to agricultural risks and climate change in Madagascar. Philos Trans Roy Soc B Biol Sci 369(1639):20130089

Huq S, Kovats S, Reid H, Satterthwaite D (2007) Reducing risks to cities from disasters and climate change. Environ Urban 19(1):39–64

Laffoley DDA, Baxter JM (eds) (2016) Explaining ocean warming: causes, scale, effects and consequences. IUCN, Gland

Laukkonen J, Blanco PK, Lenhart J, Keiner M, Cavric B, Kinuthia-Njenga C (2009) Combining climate change adaptation and mitigation measures at the local level. Habitat Int 33(3):287–292

Lawler JJ (2009) Climate change adaptation strategies for resource management and conservation planning. Ann N Y Acad Sci 1162(1):79–98

Locatelli B, Pavageau C, Pramova E, Di Gregorio M (2015) Integrating climate change mitigation and adaptation in agriculture and forestry: opportunities and trade-offs. Wiley Interdiscip Rev Clim Change 6(6):585–598

Masson-Delmotte V, Zhai P, Pörtner HO, Roberts D, Skea J, Shukla P, Waterfield T (2018) Global warming of 1.5 °C. An IPCC special report on the impacts of global warming of 1.5 °C above pre-industrial levels and related global greenhouse gas emission pathways, in the context of strengthening the global response to the threat of climate change, sustainable development, and efforts to eradicate poverty. IPCC. Sustainable development, and efforts to eradicate poverty

McCarthy N, Lipper L, Branca G (2011) Climate-smart agriculture: smallholder adoption and implications for climate change adaptation and mitigation. Mitigation of climate change in agriculture series 4, vol 3. Food and Agriculture Organization of the United Nations, Rome, pp 1–37

Mills B, Andrey J (2002) Climate change and transportation: potential interactions and impacts. In: The potential impacts of climate change on transportation. The National Academies Press, Washington, DC, p 77

Morris SS, Neidecker-Gonzales O, Carletto C, Munguía M, Medina JM, Wodon Q (2002) Hurricane Mitch and the livelihoods of the rural poor in Honduras. World Dev 30(1):49–60

NCA (2014) National climate assessment. Retrieved from the Internet, https://nca2014.globalchange.gov/highlights/report-findings/infrastructure, Accessed 30 May 2019

Ngum F, Alemagi D, Duguma L, Minang PA, Kehbila A, Tchoundjeu Z (2019) Synergizing climate change mitigation and adaptation in Cameroon: an overview of multi-stakeholder efforts. Int J Clim Change Strat Manag 11(1):118–136

Pannell DJ, Schilizzi S (1999) Sustainable agriculture: a matter of ecology, equity, economic efficiency or expedience? J Sustain Agric 13(4):57–66

Scheffran J, Battaglini A (2011) Climate and conflicts: the security risks of global warming. Reg Environ Change 11(1):27–39

Schwab K (2016) The fourth industrial revolution. World Economic Forum, Geneva, pp 51–59

Shen TT (1995) Industrial pollution prevention. Springer, Berlin, pp 15–35

Simon D, Arfvidsson H, Anand G, Bazaz A, Fenna G, Foster K, Jain G, Hansson S, Evans LM, Moodley N, Nyambuga C (2016) Developing and testing the Urban Sustainable Development Goal's targets and indicators—a five-city study. Environ Urban 28(1):49–63

Sovacool BK, D'Agostino AL, Rawlani A, Meenawat H (2012) Improving climate change adaptation in least developed Asia. Environ Sci Policy 21:112–125

Stern N, Stern NH (2007) The economics of climate change: the Stern review. Cambridge University Press, Cambridge

Sutherst RW (2004) Global change and human vulnerability to vector-borne diseases. Clin Microbiol Rev 17(1):136–173

Tanner T, Mitchell T, Polack E, Guenther B (2009) Urban governance for adaptation: assessing climate change resilience in ten Asian cities. IDS Working Papers 315:1–47

Titus JG, Anderson KE (2009) Coastal sensitivity to sea-level rise: a focus on the mid-Atlantic region, vol 4. Government Printing Office, Washington, DC

UN (2019) United Nations habitat. Retrieved from the Internet, https://unhabitat.org/urban-themes/climate-change/, Accessed 25 May 2019

UN Fact sheet (2010) Women, gender inequality and climate change. Retrieved from the Internet, http://www.un.org/womenwatch/. Accessed 15 Apr 2019

UNDP (2017) Climate change and education, Zimbabwe Human Development Report 2017. Retrieved from the Internet, https://www.undp.org/content/dam/zimbabwe/docs/key%20documents/UNDP_ZW_2017ZHDR_Briefs%20-%20Climate%20Change%20and%20Education.pdf. Accessed: 7 May 2019

Whitley S, Van der Burg L (2015) Fossil fuel subsidy reform: from rhetoric to reality. New Clim Econ 10:3

WHO (2018) Climate change and health. Retrieved from the Internet, https://www.who.int/news-room/fact-sheets/detail/climate-change-and-health. Accessed 15 May 2019

Woolcock M (1998) Social capital and economic development: toward a theoretical synthesis and policy framework. Theory Soc 27(2):151–208

World Bank (2015) Rapid, climate-informed development needed to keep climate change from pushing more than 100 million people into poverty by 2030. Retrieved form the Internet, http://www.worldbank.org/en/news/feature/2015/11/08/rapid-climate-informed-development-needed-to-keep-climate-change-from-pushing-more-than-100-million-people-into-poverty-by-2030. Accessed 12 May 2019

Ocean and Sustainable Development: A Preliminary Assessment for India

14

Tuhin Ghosh and Somnath Hazra

14.1 Introduction

In 1987, World Commission on Environment and Development (WCED) first interpreted the concept of sustainability and they have developed a definition of sustainable development as 'Development that meets the needs of the present without compromising the ability of future generations to meet their own needs'. According to the experts and researchers, the understanding on the concept of sustainable development in the academic community has also emerged from the year 1987 (Tisdell 1988; Lele 1991; Barbier and Burgess 2017). In the discipline of economics, the concept of sustainable development is defined through a systemic approach based on the three pillars—environmental, economic and social systems (Barbier 1987; Barbier and Markandya 2012; Elliott 2006) with an objective of maximization of goals of all these systems through a process of adaptation.

In the year 2015, after completion of the Millennium development Goal deadline, the United Nations has adopted a new, fresh agenda for sustainable development termed as 'The 2030 Agenda for Sustainable Development', which is basically a framework for 'peace and prosperity for people and the planet, now and into the future' (UN 2015). The main objective of this agenda is to measure the progress of Sustainable Development Goals (SDGs). The agenda has 17 goals comprising 169 targets and 232 indicators. The aim of the SDGs is to measure the progress of combination of economic development, environmental sustainability and social inclusion through some of the scheduled indicators and targets. Each goal is interlinked and integrated in nature. The SDG goals are well suited with the systems approach of sustainable development.

Till now, the assessment of these goals mainly focused on the formulation of targets, indicators and collection of baseline data, but this is also an important step to articulate the actions in three domains (environmental, economic and social policy) and to evaluate the indicators in future. In India, after this step, there is no quantitative assessment on the viability of targets, indicators and getting reliable data. This chapter is focused on Goal 14: 'Conserve and sustainably use the oceans, seas and marine resources for sustainable development'. Till now, the Indian government has developed few indicators based on the scheduled targets. SDG 14 covers economic, social and environmental impacts on marine resources and coastal communities. More specifically, coastal communities are basically impacted by the economic pressure and in terms of socio-economic activity, they are majorly dependent on the oceans and marine resources.

T. Ghosh · S. Hazra (✉)
School of Oceanographic Studies, Jadavpur University, Kolkata, India

S. Hazra, A. Bhukta (eds.), *Sustainable Development Goals*, Sustainable Development Goals Series,
https://doi.org/10.1007/978-3-030-42488-6_14

Apart from the introduction, In Sect. 14.2, the chapter describes the rationality of SDG 14 in the Agenda and this section also tries to analyse the framework of SDG 14 with different targets and indicators. Specifically, this section will address the conceptual framework of indicators of SDG 14 and multidimensional nature of SDG 14. Section 14.3 will discuss the effectiveness of targets of SDG 14 in terms of monitoring and evaluation in the context of India. Some preliminary assessment of the different targets will be discussed in Sect. 14.4. On the basis of the analysis of Sect. 14.4, some future research framework will be explored in Sect. 14.5, and Sect. 14.6 is for the conclusion.

14.2 The Rationality of SDG 14 in the Agenda

The SDG is the first global agenda for an integrated sustainable development with all-round sustainability (Economic, Social and Environmental). The schedule targets and indicators are intended to guide the next 14 years to facilitate the scheduled actions with all levels of the society. This comprehensive framework will enhance the capacity of the State actors to achieve the anticipated results. In order to monitor the progress, each and every country has to inform or rearrange the policy to ensure the responsibility of the specific departments for a specific goal. According to the UN Report,[1]the world monitoring process must be based on the available comparable national data of countries that must be reported to the international statistical system. According to the UN Statistical Commission, to track the success at those scales, the member states also have to develop more comprehensive indicators at the regional and national levels.

There are quite a few noble reasons to add ocean in the sustainable development goals. The dependency of GDP and the local economy ensures the importance of ocean in the regional agenda and now it has been seen that the ocean economy is gaining platform in the Indian Ocean region. It has also been seen that the economy of the coastal states is mainly linked with the health of the coast and oceans. Economists are also agreed that the outcomes of their GDP are exhaustively linked with the health of the ocean. In this respect, SDG 14 is highly important for India like other ocean-dependent economy because India is highly dependent on the ocean for some critical issues like food security, livelihood security, minerals, energy security, industrial activities and various key ecological services activities.

Sea is a very economical and less carbon-emitting mode of transportation for world trade activities. It has been estimated that 90% of the global trade is occurring through sea routes. Advances in technology improved the efficiency of shipping and contributed to the sustained rise in the world seaborne trade. It is also seen that near about 50,000 merchant ships with container ships, bulk carriers, ferries, and cruise ships are involved in global trade activity.[2] According to UNCTAD reported data (2013), the seaborne trade size is roughly 9600 million ton; out of that, more than 70% comes from dry cargo. It is interesting to note that more than two-thirds of world seaborne trade activities are conducted through the Indian Ocean.[3]

The ocean is giving us some tangible and intangible benefits, which have some economic importance. These are provisioning services (fisheries, aqua fauna, offshore oil and gas), cultural and supporting services (tourism and offshore wind energy) and some economic activities like ocean trade. The Indian Ocean is one of the major sea routes which connects other continents of the world. Its fishes are growing importance of domestic consumption to the Asian countries. Russia, Japan, South Korea and Taiwan depend

[1]UN, Report of the Inter-agency and Expert Group on Sustainable Development Goal Indicators. Document E/CN.3/2017/2, Annexe III, IV and V. New York: UN, 2017a.

[2]International Chamber of Shipping (http://www.ics-shipping.org/shipping-facts/shipping-and-world-trade).

[3]UNCTAD (2014), The Oceans Economy: Opportunities and Challenges for Small Island Developing States, New York and Geneva.

on the Indian Ocean for shrimp and tuna fish. It has been estimated that 40% of the oil offshore production comes from the Indian Ocean. Eastern Indian Ocean region is one of the important regions and the fourth largest fish-producing region in the world with 8%, or 6,387,659 MT, of the global catch in 2016; out of that, the contribution of India is 1,382,504 MT. On the other hand, the Western Indian Ocean region is the world's sixth largest fish-producing region with 6% or 4,931,124 MT of the global catch; out of that, India's contribution is 2,217,189 MT.[4] Fishing activity is not the only major economic activity, but the ocean also plays an important in terms of social well-being.

Ocean health is one of the most important factors to achieve sustainable development and the impacts of major stresses are reducing the ability of the ocean to provide economic, social, and environmental benefits. It has been seen that over-fishing, exploitation of marine resources, pollution, invasive species, damages to habitat and climate change are the major pressures of the ocean (OECD 2017; UNDESA 2014).

While ensuring healthy oceans is vital for achieving sustainable development, the impacts of key pressures are compromising the ability of the oceans to deliver economic, social and environmental benefits (UNCSD 2012).

According to FAO's State of World Fisheries and Aquaculture 2016, it has been seen that 68.6% of fish stocks were at sustainable levels and 31.4% of fish stocks were at a biologically unsustainable level and therefore overfished. The world fish trade grew from 15 million tonnes in 1991 to 45 million tonnes in 2014. In 2014, developing countries contributed 56% of total fish exports in the world.[5]The ocean is also an important source of income. Globally, near about 60 million people are working in the fisheries and aquaculture sector, and it is also assessed that nearly 200 million jobs are linked (directly or indirectly) with this sector.[6]

According to an estimate, it has been seen that India has the highest population of fishing communities around the world and spread over near about 3600 fishing villages. In India, near about 14.50 million people are purely dependent on the fisheries sector for their livelihood. To ensure sustainable development of the fisheries sector, a lot of activities have been taken by the government, with a major emphasis on livelihood creation and resource conservation. The other measures are the implementation of Potential Fishing Zone Advisory programme, fishing centres, as well as the application of the prohibition of mechanized fishing for sensitive areas[7]; 80% of marine pollution is coming from land-based sources. Major sources of pollution are ship pollution, atmospheric pollution and deep-sea mining, direct discharge, etc. (OECD 2017). It has also been seen that 80% of all litter in the oceans is plastic. It is also estimated that, by 2050, the ocean will carry more plastic than fish (UNEP and GRID-Arendal 2016). Overexploitation of fish, poor agricultural practices, irregular coastal development, mining, dredging, and tourism are the major causes of habitat destruction. Due to such destruction, the aquatic plants and animals are facing significant challenges to survive (Table 14.1).

14.3 Effectiveness of Targets and Indicators of SDG 14 in Terms of Monitoring and Evaluation: In the Context of India

The aim of Goal 14 is basically sustainable management of marine and coastal ecosystems from pollution, and address the influence of ocean acidification. Sustainable use of marine resources through international law is also helping to mitigate some of the issues of oceans towards sus-

[4]https://www.cia.gov/library/publications/the-world-factbook/geos/xo.html

[5]The State of World Fisheries and Aquaculture. Contributing to food security and nutrition for all. See http://www.fao.org/3/a-i5555e.pdf.

[6]https://unctad.org/en/PublicationsLibrary/ditcted2017d3_en.pdf

[7]https://sustainabledevelopment.un.org/content/documents/15836India.pdf

Table 14.1 Targets of SDG 14

SDG 14 targets	
14.1	By 2025, prevent and significantly reduce marine pollution of all kinds, in particular from land-based activities, including marine debris and nutrient pollution
14.2	By 2020, sustainably manage and protect marine and coastal ecosystems to avoid significant adverse impacts, including by strengthening their resilience, and take action for their restoration in order to achieve healthy and productive oceans
14.3	Minimize and address the impacts of ocean acidification, including through enhanced scientific cooperation at all Levels
14.4	By 2020, effectively regulate harvesting and end overfishing, illegal, unreported and unregulated fishing and destructive fishing practices and implement science-based management plans, in order to restore fish stocks in the shortest time feasible, at least to levels that can produce maximum sustainable yield as determined by their biological characteristics
14.5	By 2020, conserve at least 10% of coastal and marine areas, consistent with national and international law and based on the best available scientific information
14.6	By 2020, prohibit certain forms of fisheries subsidies which contribute to overcapacity and overfishing, eliminate subsidies that contribute to illegal, unreported and unregulated fishing and refrain from introducing new such subsidies, recognizing that appropriate and effective special and differential treatment for developing and least developed countries should be an integral part of the World Trade Organization fisheries subsidies negotiation
14.7	By 2030, increase the economic benefits to small island developing States and least developed countries from the sustainable use of marine resources, including through sustainable management of fisheries, aquaculture and tourism
14.a	Increase scientific knowledge, develop research capacity and transfer marine technology, taking into account the intergovernmental oceanographic commission criteria and guidelines on the transfer of marine technology, in order to improve ocean health and to enhance the contribution of marine biodiversity to the development of developing countries, in particular small island developing states and least developed countries
14.b	Provide access for small-scale artisanal fishers to marine resources and markets

(continued)

Table 14.1 (continued)

SDG 14 targets	
14.c	Enhance the conservation and sustainable use of oceans and their resources by implementing international law as reflected in the United Nations convention on the law of the sea, which provides the legal framework for the conservation and sustainable use of oceans and their resources, as recalled in paragraph 158 of "the future we want".

Source: UN (2015)

tainable development. Goal 14 has ten targets with respect to the prime objective of conservation and sustainable development of oceans. To achieve the objective of Goal 14, a total of 13 indicators are formulated. The basic distribution of indicators, the monitoring and data availability is as follows:

Sl no	Item	Number
1	Number of targets	10
2	Number of targets with no indicator	2
3	Total number of indicators	13
4	Number of indicators with data availability	3

Source: http://mospi.nic.in/sites/default/files/publication_reports/SDG_Baseline_report_4.3.2019_0.pdf

Human life is very much related to oceans, seas and marine resources. Basically, the economy of all countries highly depends on the ocean. There are totally 13 indicators with respect to the 10 targets, out of which the direct data are available for 3 indicators and for 10 indicators, the data are not available directly. For some indicators, the data need to be derived, and in some cases, the data are partially available. To develop some indicators, we need to be dependent on government and private agencies data or data from research organizations.

It has been seen that, for most of the indicators, data collection frequencies are not available or not applicable for India. In some cases, the data are collected annually. In some case cases, the national data are only available at the public portal and it has been seen that for the Indicator 14.5.1 (Coverage of protected areas in relation to marine areas), the state-level data are available (Table 14.2).

Table 14.2 SDG 14 targets with scheduled national indicators and corresponding data availability

Target	Indicator	Data availability
14.1	14.1.1: Health index of the area of coastal water (percentage change)	Not available
	14.1.2: Number of sewage treatment plants installed along the coast and construction of toilets under Swachh Bharat Mission	Not available
	14.1.3: Percentage change in the use of nitrogen fertilizers in the coastal states	This indicator measures the percentage change in the use of nitrogen fertilizers. The data source for this indicator is the Department of Agriculture Cooperation & farmers welfare (DAC&FW), Ministry of Agriculture and farmers welfare (MoAFW). The indicator is computed
14.2	14.2.1: Percentage change in area under mangroves	The indicator has to be computed; data are available for 2 years. The data source for this indicator is the Ministry of Environment, forests and climate change (MoEF&CC).
	14.2.2: Implementation of coastal zone regulation notification of 2011	Data are not available. The data source for this indicator is MoEF&CC.
	14.2.3: Percentage change in marine protected areas (MPA)	Data are not available. The data source for this indicator is MoEF&CC.
14.3	14.3.1: Coral health index of exclusive economic zone (EEZ)	Data are not available. The data source for this indicator is the Ministry of Earth Science.
14.4	14.4.1: Maximum sustainable yield (MSY) in fishing	Data are not available. The data source for this indicator is DAC&FW, MoAFW.
14.5	14.5.1: Coverage of protected areas in relation to marine areas	Data are not available. The data source for this indicator is MoEF&CC.
	14.5.2: Percentage change in area under mangroves	The indicator has to be computed
14.6	The national indicator is yet to develop for this target.	
14.7	The national indicator is yet to be developed for this target.	
14.a	14.a.1: Allocation of budget resources for research as per the EEZ or coastal line	Data are not available. The data source for this indicator is the Ministry of Earth Sciences (MoES).
14.b	14.b.1: Assistance to the traditional/ artisanal fishers for procurement of FRP boats and other associated fishing implements	Data are not available. The data source for this indicator is the Department of Agriculture & cooperation, MoAFW.
14.c	14.c.1: Percentage compliance of international laws	Data are not available. The data source is MoES

Source: http://mospi.nic.in/sites/default/files/publication_reports/SDG_Baseline_report_4.3.2019_0.pdf

14.4 Some Preliminary Assessment of Different Targets

To boost, protect and conserve the coastal and marine ecosystems, India has taken a number of steps. These are as follows:

14.4.1 Mangroves and Coral Reefs

Historically, India has a strategy for managing mangrove forest. The scientific management of mangrove forest has been started in the Sundarbans mangroves, which is located in the Bay of Bengal; this is the first scientific management of mangrove in the world. Government of India has given importance to develop special research and development activities on mangrove biodiversity. According to the forest survey of India, it has been seen that as per the current assessment, the mangrove cover of the country is 4921 sq. km, which is 15% of the total geographical area of the country. Out of the total mangrove cover of India, 30.10% is under dense mangrove cover, 30.07% is under moderately dense mangrove forest and 39.89% is under open forest[8] (FSI report 2017). As compared to the 2015 assessment, it has been seen that the net increase in the mangrove cover of the country is 181 square km. It is also seen that the major reason of increment of mangrove cover is plantation and regeneration and more than 80% of mangroves have been planted with the energetic participation of local communities under the ICZMP (Integrated Coastal Zone Management project). India is running an initiative 'Mangroves for the Future' with IUCN and UNDP. Under this initiative, India has identified four major coral reefs for rigorous conservation and management. Further, India has 25 Marine Protected Areas in the peninsular region and 106 in islands, collectively covering approximately 10,000 square km of the country's geographical areas.[9] A marine protected area (MPA) is basically an area in the ocean where human activities are strictly regulated compared to the surrounding waters. These areas are special in terms of protections for natural or historic marine resources by the government authorities.

14.4.2 Ensuring Sustainability of Fisheries

To ensure sustainable development of the sector, lots of measures have been taken by the Indian government, through special importance on livelihood generation and conservation of resources. Some of the measures are Potential Fishing Zone Advisory programme, modernization and upgradation of fishing centres and prohibition of mechanized fishing in certain areas. For sustainable and responsible fishing with nutritional and income security, Government of India has emphasized the requirement of a revolution named 'Blue Revolution'. The Blue Revolution is essentially a multi-dimensional activity that mainly focuses on improvement of fisheries production and on enhancing the productivity of aquaculture and fisheries resources, both marine and inland. The government has developed a vision plan and formulated an integrated development and management of fisheries. A national Fisheries Action Plan has also been formulated. It is predicted that near about 15 million beneficiaries will be getting livelihood opportunities from these interventions.[10] Further, the government has also given importance to ecological security of the marine and coastal environment with special importance on endangered marine species.

14.4.3 Coastal and Marine Ecosystems Protection

To protect and conserve the marine and coastal environment, Government of India has developed a lot of national and sub-national legislation. Regarding the use of oceans and their resources, Government of India has done the ratification with some of the international conventions including the United Nations Convention on the Law of the Sea. An online mechanism has been developed to predict the movement of oil spills; in 2015, an Oil Spill Advisory System has been launched. The government has also developed the National Oil Spill Disaster Contingency Plan, 2015, which basically reveals the important national and international regulations relating to marine and oceans. Regarding controlling of marine pollution, government has established some monitoring stations along the coastline. Further, levels of marine pollution are being monitored by the government at various locations along the country's coastline through the Coastal Ocean Monitoring and Prediction System. India is also setting up a Marine Observation System along the coast to gain a better understanding of coastal processes and monitor water quality.

Through COMAPS Programme, the Ministry of Earth Sciences, Government of India, has been

[8]http://fsi.nic.in/isfr2017/isfr-mangrove-cover-2017.pdf

[9]http://wiienvis.nic.in/Database/MPA_8098.aspx

[10]http://dadf.gov.in/fisheries-blue-revolution

monitoring the marine pollution level at about 80 locations along the entire coastline of the country. The prime objective of this programme is to do a long-term assessment and know the tendency of coastal and marine environmental quality. Nearly 25 environmental parameters (including physical, chemical, biological and microbiological characteristics of water and sediment) are being collected with the help of research and development institutions in 0–10 km sector of these locations.

14.4.4 All-Inclusive Islands and Coastal Area Development

To ensure the holistic development of the island and coastal areas, in 2016, Government of India has launched a flagship programme named as 'Sagarmala'. The basic objective of this programme is the enhancement of the port connectivity, modernization and establishment of a new port, coastal community development and port-linked industrialization. The vision of Sagarmala is to decrease the logistics cost (domestic and EXIM cargo) through optimization of infrastructural investment. The Sagarmala programme predicts that the overall cost savings will be around INR 35,000–40,000 cr. per annum.[11] The Sagarmala programme will continue up to 2025. Sustainable development of coastal communities is one of the key pillars of this programme. For better livelihood opportunities, Government of India is promoting coastal tourism under the Sagarmala programme.

According to the SDG 19 report, out of four estimated parameters, it has been seen that three (1) Mean area that is protected in marine sites important to biodiversity (%) 29.0; (2) Ocean Health Index Goal-Clean Waters (0–100) 22.7; and (3) Fish caught by trawling (%) 10.2 are in stagnating stage and only one is in improving stage (i.e. Percentage of Fish Stocks overexploited or collapsed by EEZ (%) 12.4).[12]

14.5 Framework for Future Research Requirement

In India, the scheduled SDG 14 indicators are in developing stage, so it may change over time. In the previous section, as we have seen, some of the national indicators are not in line with global indicators and as a consequence, we should include some proxy data (like 14.1.1) on the management of marine and coastal ecosystem. In some cases, the state-wise data are also not available; so, for these indicators, we should develop a state-wise data collection method.

According to data requirement of SDG 14, it has been found that, in some cases, the country has to provide some national data according to age group, sex, caste, income level, physical ability, etc., but in India, the Ministry of Statistics has not kept these according to the above categories. So, from now onwards, the department has to develop such a level of disaggregation, which has to be incorporated in the data collection process. This entire process will need a huge investment in data collection and data keeping.

The targets of the SDG 14 not only have the social and economic dimension, but also there are enough scientific dimensions. India has developed to monitor water pollution, the monitoring stations, but till now, India has very limited capacity to monitor human influences on the oceans. There are a lot of data gaps and knowledge gaps regarding plastic pollution, the influence of heavy metals and hazardous materials and how these are impacting ocean health and marine ecosystems. These data gaps can be minimized through scientific research on ocean ecology and ecosystems. Some research is also required in ocean acidification and how this impacts ocean biodiversity (UN 2017).

In case of conservation and protection of marine and coastal resources, we have very limited scientific understandings on effective conservation measures based on the local ecology and how marine ecology is helping nutrition, medicine and human well-being. The concept of Integrated Coastal Zone Management Project

[11] http://sagarmala.gov.in/about-sagarmala/vision

[12] https://github.com/sdsna/2019GlobalIndex/blob/master/country_profiles/India_SDR_2019.pdf

(ICZMP) was developed in 1992, but until now, some of the parameters are not very much clear (like ecosystem-based management).

To review the SDGs indicators, countries can follow the multi-layer review process in terms of governance as well as in terms of the state actors. In the case of a multi-layer system of governance, the country has to follow international, regional, national, state and local level governance system of the same parameters so that it can be compared with others. To maintain the regional review process, one can introduce a peer-review mechanism with common characteristics.

The perception of marine ecosystem services is also very much necessary to ensure the sustainable development of oceans. Scientists have to distinguish the long-term and short-term benefits and how these benefits can improve the human well-being of the coastal communities. It also needs to be examined how a small environmental change in the marine ecology can impact human welfare and what will be the actual carrying capacity of the biophysical systems. The proper quantification of the value of marine ecosystems can improve the conservation strategy, which leads to the sustainable development of oceans. It has been found that for the target 14.4, intergenerational equity is very much required, and for the target 14.5, equitable management is necessary. Steady coastal and marine ecosystem services are very much like the other SDGs like SDG1, SDG2, SDG 3 and SDG 6. So, a common approach and methodology have to be developed.

14.6 Conclusions

From the above analysis, it has been found that India has huge Knowledge gaps with respect to all the SDG 14 targets and also with other goal targets relating to it. It is also believed to be suffiecient ground for such gaps. For example, the data on conservation, protection of marine resources, poverty alleviation strategy for coastal communities, impact of climate change on marine ecosystems, age-wise / sex wise climate migration of coastal communities etc are either not updated regularly or very much fragmented in India. Moreover, sometimes the knowledge gap is not only for the lack of data but also for the limited accessibility of the data, lack of common data collection procedures, lack of political or sectoral boundaries, or lack of departmental convergence, lack of analytical capability etc. According to the assessment of SDG 14 targets and indicators, it can be said that an integrated research is very much required with proper capacity building of the government officials, data collectors and data management officials. To generate data related to SDG 14 targets India has to start a lot of scientific research relating to ocean and marine ecosystems. Marine open-access data platforms should be created. The economists are seen to be arguing regarding the appreciation of the joint interface of economic and environmental systems to determine the market and nonmarket outcomes which is very much crucial to develop a model that appraise the leadership of this "safe operating space" (Smith 2017). It is also found that the enriched welfare analysis among the sustainable development goals will be very much required for the policy analysis. Finally, from the above analysis, it is also clear that quantitative assessment and welfare analysis across interlinking goals are necessary, so that their outcomes can be applied for a policy decision.

References

Barbier EB (1987) The concept of sustainable economic development. Environ Conserv 14(2):101–110

Barbier EB, Burgess JC (2017) The sustainable development goals and the systems approach to sustainability. Economics 11(2017–28):1–22. https://doi.org/10.5018/economics-ejournal.ja.2017-28

Barbier EB, Markandya A (2012) A new blueprint for a green economy. Routledge/Taylor & Francis, London

Elliott JA (2006) An introduction to sustainable development, 3rd edn. Routledge, London

Forest Survey of India (FSI), 2017, India State of Forest Report – 2017: Mangrove cover, Ministry of Environment and Forests, New Delhi: Government of India. Available online at: http://fsi.nic.in/isfr2017/isfr-mangrove-cover-2017.pdf. (Accessed on 20 April 2019)

Lele SM (1991) Sustainable development: a critical review. World Dev 19(6):607–621

OECD (2017) Marine protected areas: economics, management and effective policy mixes. OECD, Paris

Smith VK (2017) Environmental economics and the anthropocene. Oxford Research Encyclopedia of Environmental Science. https://doi.org/10.1093/acrefore/9780199389414.013.386

Tisdell C (1988) Sustainable development: differing perspectives of ecologists and economists, and relevance to LDCs. World Dev 16(3):373–384

UN (2015) Draft outcome document of the United Nations summit for the adoption of the post-2015 development agenda. Draft resolution submitted by the President of the General Assembly, Sixty-ninth session, Agenda items 13 (a) and 115, A/69/L.85, August 12, 2015. UN, New York

UN (2016) The first global integrated marine assessment: world ocean assessment I, Group of Experts of the Regular Process (Innis, L. and Simcock, A., Joint Coordinators), United Nations Regular Process for Global Reporting and Assessment of the State of the Marine Environment, including Socioeconomic Aspects. United Nations, New York

UN (2017) Technical abstract of the First Global Integrated Marine Assessment on the Ocean and the United Nations' Sustainable Development Goals under the Agenda 2030. In: Regular process for global reporting and assessment of the state of the marine environment, including socioeconomic aspects. United Nations, New York

UNCSD (2012) Rio Ocean declaration, co-chairs' statement of the oceans day at Rio+20, United Nations conference on sustainable development. UN, New York

UNDESA (2014) How oceans- and seas-related measures contribute to the economic, social and environmental dimensions of sustainable development: local and regional experiences. UNDESA, New York

UNEP and GRID-Arendal (2016) Marine litter vital graphics. UNEP and GRID-Arendal, Nairobi

15 Biodiversity Conservation and the UN's Sustainable Development Goals: India's Responses Evaluated—Particularly in Relation to SDG15

Clement A. Tisdell

15.1 Introduction

Sustainable Development Goal 15 (as part of the UN's Global Agenda 2030 development) is focused on conserving life on land. Its stated aim is primarily 'to foster the adoption of policy measures to protect, restore and promote sustainable use of terrestrial ecosystems, sustainably manage forests, combat desertification and halt and reverse land degradation while also integrating ecosystems and biodiversity into national and local planning developmental processes, poverty reduction strategies and national accounts'. In addition, it 'seeks to promote a fair and equitable sharing of benefits arising from the utilization of genetic resources, promote appropriate access to such resources and prevent poaching and trafficking of protected species of flora and fauna'. It sets out 12 targets to be achieved by (or before) 2030 for this purpose, all of which relate to the conservation of natural biodiversity and natural ecosystems on land.

As discussed below, India has determined quantitative performance values for achieving 3 of the 12 targets listed in Sustainable Development Goal 15 (SDG15) by 2030.[1] It is also recognized by India that SDG15 is closely interrelated with the achievement of most of the other SDGs. This interdependence will be taken into account in this discussion which primarily concentrates on India's plans to satisfy SDG15.

The aim of this chapter is to discuss the adequacy of India's plans for meeting SDG15 and interrelated targets and to consider the likelihood of India meeting these targets. Because conserving biodiversity is central to SDG15 and raises issues about what constitutes biodiversity conservation (which includes ecosystem conservation), the multiple dimensions and nature of biodiversity are discussed first. This is followed by a general examination of the relevance of biodiversity conservation and methods for achieving sustainable development.

In turn, the following points are then critically analysed:

1. India's quantified biodiversity targets for achieving SDG15.
2. The relationship between India's targets for SDG15 and its other SDG targets.
3. India's prospects for meeting its biodiversity SDG targets.

C. A. Tisdell (✉)
School of Economics, The University of Queensland, Brisbane, QLD, Australia
e-mail: c.tisdell@uq.edu.au

[1]All the references to India's SD plans in this discussion are drawn from the United Nations and NITI Aayog (2018) Report.

S. Hazra, A. Bhukta (eds.), *Sustainable Development Goals*, Sustainable Development Goals Series,
https://doi.org/10.1007/978-3-030-42488-6_15

A discussion and concluding comments wrap up this chapter.

15.2 The Multiple Dimensions of Biodiversity

The stock of biodiversity is multi-faceted (Magurran 2003) and not easily quantified by a single measure (Juhász-Nagy 1993). Both the diversity of ecosystems and that of genetic material are components of the extent of biodiversity. The exact measurement of the extent of biodiversity is hampered by the fact that different ecosystems are not discrete and can be defined at different spatial scales (Tisdell 2015, Chap. 2; Tansley 1935). Judgement is required about the appropriate scale on which to classify different ecosystems and the categorization of differences can alter with each problem being considered. For example, in relation to the UN's SDGs (Sustainable Development Goals), many ecosystems on land (SDG15) are interdependent with those below the water (SDG14).

There is also the problem that not all genetic material is of equal value for achieving SD. Furthermore, differences exist in the extent of the disparity between types of genetic material and their rarity. This adds to the difficulty of devising a meaningful or single index of the stock of biological diversity. As yet, no satisfactory index of the extent of biodiversity has been devised. This is probably because biodiversity involves the diversity of diversities (Juhász-Nagy 1993). Nevertheless, it is clear that the extent of natural biodiversity has declined significantly due to human activity. This decline has been substantial in modern times, as for example evidenced by the global loss of wild vertebrate animal species (Whitmee et al. 2015; World Wide Fund for Nature 2014).

A further factor that needs to be taken into account in evaluating biodiversity conservation is the status of heritage human-developed genetic material and ecosystems, for example, those pertaining to agriculture and other forms of human-managed biological production (Tisdell 2015, Chaps. 2 and 8). Evolution of human selection and production of new genetic combinations for cultured organisms and human changes in ecosystems for cultivating organisms often threaten the conservation of pre-existing heritage biodiversity as well as the stock of natural biodiversity. Therefore, trade-off issues need to be considered when biodiversity conservation policies are devised to foster SD.

Whether or not adequate attention will be paid to conserving biodiversity in India (and elsewhere) is doubtful for at least two reasons.

1. Biodiversity conservation issues are extremely complex and bounded rationality limits the scope for the rational determination of the trade-offs.
2. Different government bodies have control over public decisions relating to different aspects of biodiversity and to a large extent act independently. They often have and do pursue different agendas. Public policy decisions are, therefore, commonly made on a partially uncoordinated basis.

Because different biodiversity targets are assigned to different SDGs (not all of which appear to be well aligned with one another), this may encourage different government bodies to try to fulfil several of the potentially conflicting targets independently, e.g. government agricultural bodies may act independently of forest and other conservation bodies.

15.3 The Relationship Between Sustainable Development and Biodiversity Conservation: Conflicting Perspectives

The overarching aim of the UN's SDGs and associated targets is to foster the achievement of sustainable development (SD). However, that raises the question of what exactly is meant by SD. The problem is that diverse definitions exist of what constitutes SD. Economic definitions of SD are invariably anthropocentric in nature. For example, one economic view of SD is that it requires common economic activity to be managed in a way that ensures that the income per capita of future generations is no less than that of current generations. Another is that the economic activities of present generations should be such that future generations should be left sufficient scope to enjoy no less income per capita than that of the present generation. To what extent (if any) must present generations forgo income in order to achieve these goals? Is it necessary for them to do this? How are these sustainability goals to be achieved? We are still struggling to provide definitive answers to these types of questions.

Not all citizens are willing to accept the types of anthropocentric criteria for the SD proposed by economists. For example, 'dark green ecologists' place a high weight on conserving natural ecosystems and natural biodiversity. They support policy measures to do this even if it involves some decrease in the incomes and economic welfare of mankind.

Individuals differ in their willingness to conserve nature. Consequently, because of conflicting values, it is probably impossible to devise SDGs and associated targets which will satisfy everyone. Therefore, for political acceptability reasons, SDG goals and targets tend to be expressed in general terms and are, to some degree, open-ended. Furthermore, they may not be always entirely consistent. In fact, the 2030 Sustainable Development Agenda provides individual nations with a huge amount of flexibility (leeway) in devising policies to satisfy the SDG goals and targets.

15.4 India's Quantified Biodiversity Targets for Achieving SDG15, That Is, for Conserving Life on Land

15.4.1 Constraints on India's Choice of Targets and Procedural Considerations

The UN's Sustainable Agenda places a heavy emphasis on the quantification of the targets which individual nations decide on in order to achieve their various global sustainable goals. Nevertheless, it is left to individual nations to determine which targets they will pursue, how they will quantify their aspirational targets and how they will monitor their progress in achieving these targets. The premium placed on the quantification of SD targets has had the effect that only a limited amount of targets are being selected to satisfy the SD agenda. One reason is that insufficient data are available to quantify all targets.

The policy planning procedure is to first determine baseline values for each of the selected targets which a nation hopes to achieve. Targets are then decided on. Only 3 of the 12 targets for satisfying SDG15 have been quantified by the Union Government of India. It is said that these performance indicators 'have been selected based on availability of data at the national level and to ensure comparability across States and Union Territories (UTs)' (United Nations and NITI Aayog 2018, p. 167). Because India has a federal structure of government and the government of its individual states and UTs has considerable control over their management and conservation of forests, the prospects of achieving targets for conserving nature (especially forests) depend heavily on the policies pursued by each of the states and UTs.

15.4.2 India's Aspirational Levels for SD Targets 15.1 and 15.2

India's Union government has decided to pursue three targets as its contribution to achieving SDG 15. The first two of these targets are as follows:

15.1 By 2030, ensure the conservation, restoration and sustainable use of terrestrial and inland freshwater ecosystems and the services, in particular forests, wetlands, mountains and drylands, in line with obligations under international agreements.

15.2 By 2020, promote the implementation of sustainable management of all types of forests, halt deforestation, restore degraded forests and substantially increase afforestation and reforestation globally.

In relation to SD target 15.1, India proposes to increase its forest cover by one-third (33%) by 2030 compared to its baseline coverage in 2015

(21.54%). This would result in its forest cover increasing to around 28–29% by 2030. It also seems that India wishes to maintain (for India as a whole) the same areal proportion of water bodies within its forested areas as in 2015. Consequently, as its expansion of forest cover occurs, it wants to ensure no decline in the proportionate area covered by water bodies in its forested areas.

Its plans for contributing to SD target 15.2 are rather vague. However, it appears that it may want to maintain the type of forest management policies which prevailed between 2015 and 2017 and which are claimed to have been associated with 0.21% increase in Indian forest cover.

15.4.3 Shortcomings in India's Choice of SD Targets 15.1 and 15.2

The SD targets 15.1 and 15.2 chosen by India only partially reflect the actual targets suggested for the Global Agenda. First, the nature and quality of forest cover is not addressed. Degraded forests are likely to have a negative effect on the conservation of nature biodiversity and the ecosystems services emanating from forested land. Second, plantation forests are given equal weight to those forests that are more natural in character. The former often lack diversity and can have a negative impact on the conservation of biodiversity. Third, there needs to be greater transparency in how the nature of forest cover is determined, that is how much an area must be forested to be considered to be forested? Fourth, plans for conserving drylands and high mountain areas that are naturally not forested have been left up in the air. Such areas are often unsuitable for conservation by means of tree planting (Tisdell and Xue 2013).

Maintaining the proportionate water area in its forests has also been selected by India as part of its contribution to meeting SD target 15.1 of the SD Global Agenda. However, such a strategy may not be very effective in conserving natural biodiversity. This is because the water area in forests would include dams constructed for meeting human demands, e.g. for irrigation, industrial and household water use. Usually, these structures result in reduced natural biodiversity, e.g. as a result of flooding forests and wetlands. The qualities of the water areas conserved need to be taken into account and this has not been done in this case. It also needs to be borne in mind that India's surface waters are highly polluted (Lélé et al. 2018). This is mainly as a result of the discharge of effluents from industry, agriculture and households into water bodies. Maintaining the proportionate areas of water bodies in forests may do little to ameliorate this pollution problem.

As pointed out above, India's stated aspirations for contributing to the realization of SD target 15.2 are unclear although it indicates that it will be contributing to the satisfaction of this target. Its plans provide no specific target for the sustainable management of all types of forests. Nevertheless, it might be claimed that this target is partly satisfied by India's plans to increase its forest cover by one-third by 2030. However, it can be concluded that target 15.2 is not effectively addressed. Consequently, India is only focusing on 2 not 3 of the 12 SD15 targets for promoting the SD Global Agenda, namely targets 15.1 and 15.7.

15.5 India's Plans to Contribute to the Fulfilment of SD 15.7

15.5.1 The Target Quantified by India

The stated UN Global Agenda 2030 SD 15.7 target is to 'take urgent action to end poaching and trafficking of protected species of flora and fauna and address both demand and supply of illegal wildlife products'. India's quantitative plan for contributing to this target is to ensure that its population of wild elephants is sustained at its level in 2017.

The reason given for this chosen target, however, is not directly related to SD target 15.7. India's agenda states 'Since elephants have high dietary requirements, their population can be supported only by forests that are under optimal conditions' (United Nations and NITI Aayog 2018, p. 173). The reason given for conserving elephants appears to be more relevant to SD target 15.2 than 15.7.

15.5.2 Discussion of India's Selected Component of SD 15.7

If India achieves its plans for maintaining the level of its population of wild elephants, a number of factors could contribute to that objective. These include avoiding a reduction in forest habitat suitable for the sustenance of wild elephants and reducing any trafficking in elephant products, such as ivory. However, the Asian elephant (unlike the African elephant) is not so valuable as a source of ivory. Possibly, a greater number of Asian elephants are killed as agricultural pests than for their ivory. Wild elephants can be significant agricultural pests (see, for example, Bandara and Tisdell 2002). No mention is made of how India plans to tackle the problem of reducing the number of elephants killed because of the damages they do to agriculture.

The role of wild elephants in promoting or reducing biodiversity in forests is not discussed in India's SD plans. Many ecologists classify elephants as umbrella species which by their activities help to promote natural biodiversity within their habitats. While this is true in some cases, elephant populations can increase to such an extent that they exceed the carrying capacity of forests and reduce biodiversity within them. Consequently, their population may need to be reduced by culling to preserve an ecological balance.

The question also needs to be asked of why concentrate only on the conservation of the Asian elephant when so many other wild species in India are threatened with extinction. Is it because the elephant has great cultural significance for most Indians (Tisdell and Bandara 2004)? As mentioned above, it is said that one of the reasons is that the population of elephants is a barometer of the health of the forest. If that is really true, the health of India's forests must have improved dramatically in the 5-year period 2012–2017 because it is estimated in the baseline report that the population of wild elephants in India in that period rose by nearly 20%. It is, however, pertinent to ask: how reliable were the population estimates? Also to what extent was the estimated increase in the elephant population due to more reliable estimates in 2017 than earlier? Because elephants are only capable of increasing the level slowly, (they are *k*-selected species) it is possible that the estimated increase in their populations of 20% in 5 years could well be on the high side. It needs to be compared with the biological possibility of this happening.

15.6 Other Aspects of India's Plans to Contribute to the Fulfilment of SD Goal 15

India is intending to use an index score to measure how well it performs in achieving SD goal 15 and to specify this for each of its states and UTS. One of the ideas behind this is that it will enable each of India's states and UTs to determine how well it has performed in achieving India's chosen targets for contributing to SD goal 15. It will enable each to compare its performance with other states and UTs. The index is a normalized one in which each target is equally weighted. In effect, this gives each equal importance. However, it is debatable whether all the chosen targets should be weighted equally. At least doing this ought to be justified.

The use of such an index may also tend to conceal the fact that many SD targets are not addressed in the sustainable development plans of India. For example, no specific attention is given to addressing the following:

1. Combating desertification (a part of global SD target 15.3);
2. Conserving mountain ecosystems (a part of global SD target 15.4);
3. Overall action to protect and prevent the extinction of threatened species (Target 15.5); and
4. No mention is made of 'measures to prevent the introduction and significantly reduce the impact of invasive alien species on land and

water systems and control or eradicate the priority species' (Target 15.8). Alien pests can be a serious threat to natural ecosystems and biodiversity as well as to agroecosystems and agricultural biodiversity.

15.7 The Links of SD Goal 15 with Other SD Goals

India's baseline report on its plans for contributing to the Global Agenda for SD states that its plans for achieving SD Goal 15 are closely linked with the fulfilment of several other SD goals (United Nations and NITI Aayog 2018, p. 166), but it fails to spell out the links. It claims that meeting its stated targets for Goal 15 will make a positive contribution to the following:

1. Eliminating poverty (Goal 1);
2. Ending hunger (Goal 2);
3. Ensuring good health and well-being (Goal 3);
4. Bringing about greater gender equality (Goal 5); and
5. Creating clean water and improved sanitation. Furthermore, it is claimed that its targets for Goal 15.
6. Reducing economic inequality (Goal 10);
7. Promoting sustainable consumption and production (Goal 12); and
8. Strengthening resilience and adaptive capacity to react to climate change (Goal 13).

A shortcoming of the plan is that it does not provide a clear guide to most of these interconnections and there is no indication of the size of the synergies involved. For example, India's support for the objective of promoting a fair and equitable sharing of genetic resources (especially natural genetic material in the wild in India) is likely to do little to improve income inequality in India, particularly after the administrative costs of distributing any fees obtained for the use of this material are taken into account (Tisdell 2015, Chap. 10). Furthermore, no mention is made of the real possibility that increasing forest cover could contribute to life below the water (Goal 14).

More attention should be paid to the fact that increasing forest cover in India is unlikely to be sufficient to significantly increase the quality of its freshwater resources. India's surface waters are highly polluted (Lélé et al. 2018). Pollutants enter these waters from mining, agriculture and households e.g. sewage and other sources (Fig. 15.1). Polluted water is a threat to human health; can lower agricultural production e.g. because the water may be toxic to livestock and unsuitable for irrigating crops; and can reduce biodiversity, e.g. by diminishing the diversity of water-based biota. A reduction in the extent of forest cover and natural vegetation adds to these problems. Increasing and improving the quality of this cover can help ameliorate water pollution, but it must be complemented by policy measures to substantially reduce the other sources of water pollution.

The lack of treatment of urban sewage and the discharge of sewage into water bodies are major sources of water pollution in India. In around 2015, the urban sewage treated before discharge was only 37.58%. As part of its contribution to the attainment of Goal 6, India plans to increase this to 68.79% by 2030, that is virtually double the percentage of sewage treated before its release into water bodies. Even if this target is achieved, the discharge of urban sewage into water bodies is still likely to cause considerable water pollution. Also problems of water pollution associated with mining, manufacturing, agriculture and other economic activities will need to be addressed, if India's water pollution is to be substantially reduced. India's target for raising agricultural productivity as part of its goal to achieve zero hunger by 2030 is, in part, likely to result in increased water pollution.

India aims to double its yield of rice, wheat and coarse grains from its 2015 baseline figure of 2509.22 to 5018.44 kg/ha by 2030, as part of its contribution to achieving zero hunger by 2030 (SD Goal 2). This requires India to double the yield of these crops. One wonders whether this is a reasonable objective and how this target is going to be achieved. The fact that yields in the Punjab have almost reached this target is mentioned by the United Nations and NITI Aayog (2018) as an indicator that this target could be met. However, the scope for attaining similar yields in many other parts of India is constrained by poorer environmental conditions. In order to

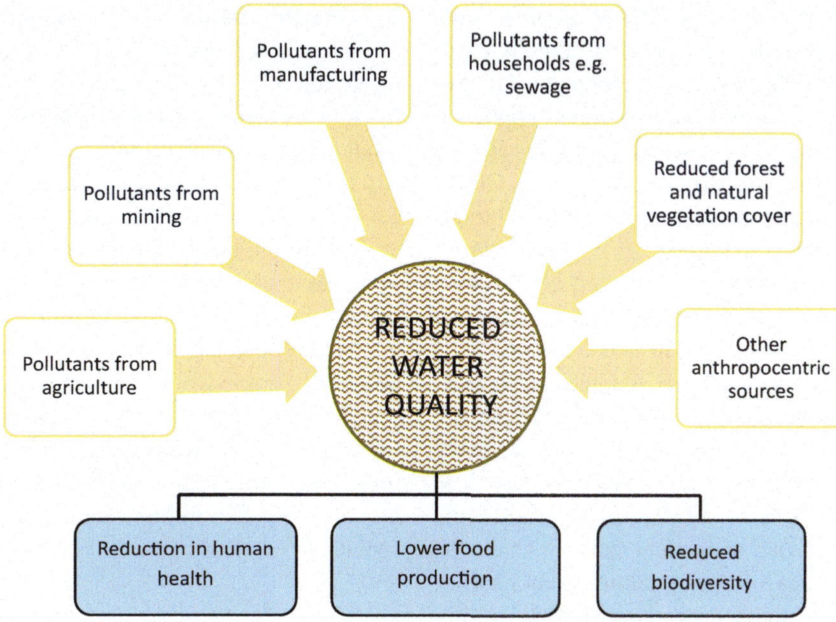

Fig. 15.1 Multiple contributors to reduced water quality and consequences of water pollution. Note also that a reduction of water flows can add to the degree to which water is polluted

achieve the increase in agricultural yields aspired to, it is highly likely that more water, larger amounts of chemical fertilizers and greater quantities of pesticides will be used in agricultural production. Consequently, greater pollution of water bodies as a result of more intense agricultural activity can be expected.

Increased yields may partially come about as a result of improved varieties of agricultural genomes. However, this type of development is a double-edged sword because it is usually associated with a loss of heritage varieties of crops (Tisdell 2015, Chap. 5). In addition, the agricultural intensification strategy can have adverse consequences for the conservation of biodiversity generally, especially if the area allocated to agricultural production rises at the same time as intensification occurs.

It should also be noted that while target 2.5 of the SD Agenda 2030 states that efforts should be made to 'maintain genetic diversity of seeds, cultural plants and domesticated animals and their related wild species', India has not proposed any specific targets for doing this. It is likely that heritage biodiversity will continue to decline in India.

There is an urgent need to undertake research to prioritize the types of genetic material which needs to be conserved and for greater account to be taken of the economics of this conservation (see for example, Tisdell 2016). Although India is very supportive of policies to ensure a fair and equitable sharing of benefits from the utilization of genetic resources and associated traditional knowledge, there is no guarantee that this policy will be very effective in conserving genetic resources and traditional knowledge or that it will result in the most desirable selection of those resources and knowledge.

15.8 Discussion

Devising targets to contribute to Global Agenda 2030 is a formidable task, especially since the UN has a preference for individual nations to quantify their targets. The targets selected by India have been partly determined by data availability, the political inputs of its Central Ministries and the possibility of their relevance to at least half of its states and UTs. This has resulted in many targets not being addressed or only being partially addressed. There are undoubtedly cases where relevant data do not exist for at least half of India's states and UTs, but associated conservation targets (and others) would be highly relevant for contributing to the fulfilment of Global Agenda 2030. Procedures for selecting targets are clearly subject to biases.

Although the United Nations and NITI Aayog baseline report (2018) claims that a holistic approach has been adopted in selecting India's SD targets, as pointed out above, the interconnections between the targets are not well articulated. Furthermore, in this chapter, possible conflicts between the selected targets have been noted. In addition, it is likely to be left to individual Central Ministries to foster individual targets which fall within their area of public administration. They may tend to do this independently and jeopardize India's performance in achieving all its SD targets.

Again, India's policy interventions for achieving its SD targets appear to be weak, because no targets are available yet for its individual states and UTs. It appears that consultation and persuasion by NITI Aayog with pubic authorities in India's states and UTs is going to be the main means for getting them to contribute to India's SD goals. How well this will work remains to be seen.

India's targets are not set in stone. Politicians are likely to have a major influence on whether serious steps are taken to achieve them. India is a democratic country and politicians compete for votes. The pro-economic growth sentiment still remains very strong in India, as in other parts of the world. Environmental conservation may still be sacrificed in India for some time yet in favour of economic growth. It is generally believed that the BJP (led by Narendra Modi) strongly favours economic growth policies. In 2019, this party was re-elected to govern India with Narendra Modi as Prime Minister. Therefore, it could be an uphill battle for India to achieve its major environmental conservation targets by 2030.

It is clear that the Indian baseline report on biodiversity conservation, especially SDG 15 does not pay enough attention to the varied biogeographic zones of India and the biodiversity hotspots of India (see PMF IAS 2019, https://www.pmfias.com/biodiversity-hotspots-india/). This is partly because the focus of this report is on the political divisions of India. Although its prime focus on conserving wild elephants is useful, it is too restricted as a measure for conserving biodiversity in the wild. Most of India's biodiversity hotspots are under increasing pressure due to economic growth. This is ignored.

In addition, there should have been more attention paid to policy measures to ensure more effective enforcement of nature conservation in India. Poaching of protected animals and the removal of protected flora from conservation areas still remain a problem as does the use of these areas for illegal grazing of livestock. However, enforcing conservation regulations and laws in LDCs is much harder than in higher income countries for social and economic reasons.

15.9 Conclusions

As was shown, shortcomings in India's plans for contributing to the fulfilment of the SD targets, set out in the UN's Global Agenda 2030, reflect deficiencies in the specification of the SD targets in this agenda and in the type of SD that ought to be pursued by nations. Furthermore, the agenda does not prioritize the desirability of conserving different types of biodiversity and ecosystems. The imprecision of Global Agenda 2030 has probably been dictated by the need to obtain the political support of most sovereign nations for it. In addition, it has been left to individual nations to determine which SD targets they will pursue. However, a preference has been expressed by the UN for the selection of targets that can be quantified. Again leaving individual nations to select the specific SD targets to be pursued by them probably reflects political realities, that is, the need to take account of their national sovereignty. The proposed global agenda possibly had to be broad enough (and sufficiently vague) for it to be adopted by most nations. India's responses to Global Agenda 2030 reflect the 'fuzzy' nature of the agenda itself.

As was shown above, India has only selected a small number of targets for contributing to the fulfilment of SD Goal 15 (the conservation of life on land) and other SD goals. SD Goal 15 mainly focuses on the conservation of natural biodiversity and the preservation of natural ecosystems on land and within these selected targets, India has only chosen very restricted items as targets to be achieved. Consequently, India's response to the Global Agenda 2030 is very patchy.

Furthermore, what to do about conflicts between different targets has not been resolved by India. For example, the possible conflict between raising agricultural yields and conserving biodiversity and ecosystems has yet to be tackled. It has also been pointed out that India's Central Ministries might independently pursue SD targets which come within their ambit of administration. Thus, no coordinated attempts may be made to resolve conflicts between different Indian SD targets.

India has both a democratic and a federal political system. Meeting its SD targets will depend heavily on the cooperation of the governments of its states and UTs and the agendas of political parties. Consultation and persuasion seem to be the main chosen vehicle for getting India's states and UTs on board with the Union's SD targets. These may not be strong policy instruments. In addition, elected political parties may still find it opportunistic to pursue pro-economic growth policies at the expense of environmental conservation if these are vote-winners.

While Agenda 2030 is likely to maintain awareness about the desirability of achieving SD and the important contribution that conserving biodiversity and ecosystems can play in the SD process, its practical effects might be quite limited in India and elsewhere. There is a risk that the main impact of Global Agenda 2030 could be to provide extra-employment for national and international public administrators. It may also be that some of its SD targets will be met independently of the agenda. That is not to deny that Global Agenda 2030 is trying to address issues that warrant urgent attention in a difficult global political environment.[2]

[2]Some additional problems associated with the UN's Global Agenda 2030 are raised in Svizzero and Tisdell (2016).

References

Bandara R, Tisdell CA (2002) Asian elephants as agricultural pests: economics of control and compensation in Sri Lanka. Nat Resour J 42(3):491–519

Juhász-Nagy P (1993) Notes on compositional diversity. Hydrobiologia 249(1):173–182. https://doi.org/10.1007/bf00008852

Lélé S, Srinivasan V, Thomas BK, Jamwal P (2018) Adapting to climate change in rapidly urbanizing river basins: insights from a multiple-concerns, multiple-stressors, and multi-level approach. Water Int 43(2):281–304. https://doi.org/10.1080/02508060.2017.1416442

Magurran AE (2003) Measuring biological diversity. Blackwell Publishing, Oxford

PMF IAS (2019) Biodiversity hotspots of India. Available at https://www.pmfias.com/biodiversity-hotspots-india/. Accessed 13 June 2019

Svizzero S, Tisdell CA (2016) The post-2015 development agenda: a critical analysis. J Self-Govern Manag Econ 4(1):72–94. https://doi.org/10.22381/JSME4120163

Tansley AG (1935) The use and abuse of vegetational concepts and terms. Ecology 16(3):284–307. https://doi.org/10.2307/1930070

Tisdell CA (2015) Sustaining biodiversity and ecosystem functions: economic issues. Edward Elgar, Cheltenham

Tisdell CA (2016) Genetic loss in food crops in the Pacific: socio-economic causes and policy issues. J Pac Stud 36(2):23–40

Tisdell CA, Bandara R (2004) Human societies, economics and the fate of the Asian elephant. Sci India 7(9/10):67–72

Tisdell CA, Xue D (2013) Managing ecosystem services for human benefit: economic and environmental policy challenges. In: Crighton E, Davovich P (eds) Environmental policy: management, legal issues and health aspects. Nova Science Publishers, New York, pp 87–106

United Nations and NITI Aayog (2018) SDG India index: baseline report 2018. United Nations and NITI Aayog, New Delhi. Available at https://www.niti.gov.in/content/sdg-india-index-baseline-report-2018

Whitmee S, Haines A, Beyrer C, Boltz F, Capon AG, de Souza Dias BF, Ezeh A, Frumkin H, Gong P, Head P, Horton R, Mace GM, Marten R, Myers SS, Nishtar S, Osofsky SA, Pattanayak SK, Pongsiri MJ, Romanelli C, Soucat A, Vega J, Yach D (2015) Safeguarding human health in the Anthropocene epoch: report of the Rockefeller Foundation—*Lancet* Commission on planetary health. Lancet 386(10007):1973–2028. https://doi.org/10.1016/S0140-6736(15)60901-1

World Wide Fund for Nature (2014) Living planet report 2014: species and space, people and places. WWF, Gland. Available at https://www.wwf.or.jp/activities/data/WWF_LPR_2014.pdf

16 Sustainable Development Goals: Implementation of Goal 16 by India

Sanjay Upadhyay and Avani Dubey

16.1 Sustainable Development Goals (SDGs): An Overview

16.1.1 (People, Planet, Prosperity, Peace and Partnership)

One of the universally accepted definition of sustainable development is by the Brundtland Report, which states that it is a kind of development that meets the needs of the present without compromising the ability of future generations to meet their own needs.[1] It calls for concerted efforts towards building an inclusive, sustainable and resilient future for the people and the planet. To achieve the same, it is crucial to harmonize three core elements: economic growth, social inclusion and environmental protection.[2] These elements are interconnected and are crucial for the well-being of the countries and their people, and the planet at large. The SDGs embody these three core elements.

As is well known, the Sustainable Development Goals are a collection of 17 global goals set by the United Nations General Assembly in 2015 for the year 2030. The SDGs are part of Resolution 70/1 of the United Nations General Assembly, the 2030 Agenda.

The 2030 Agenda for Sustainable Development, adopted by all United Nations Member States in 2015, provides a blueprint for peace and prosperity for people and the planet. It has at its core the 17 Sustainable Development Goals (SDGs), which are an urgent call for action by all countries, including developing countries in a global partnership. They recognize that ending poverty and other deprivations must be taken together with strategies that improve health and education, reduce inequality and spur economic growth—all while tackling climate change and working to preserve oceans and forests.[3]

The SDGs build on the Millennium Development Goals in the Millennium Declaration adopted by the Millennium Summit in 2000, which consisted of eight goals which were supposed to decrease poverty by 2015.[4]

[1]World Commission on Environment & Development, *Our Common Future: Report of the World Commission on Environment and Development* (1987), https://sustainabledevelopment.un.org/content/documents/5987our-common-future.pdf accessed 15th May 2019.

[2]Niti Aayog, Government of India, https://niti.gov.in/ accessed on 14th May 2019.

[3]Sustainable Development Goals Knowledge Platform https://sustainabledevelopment.un.org/?menu=1300 accessed on 13th May 2019.

[4]*Id.*

S. Upadhyay (✉) · A. Dubey
Managing Partner, Enviro Legal Defence Firm,
New Delhi, India
e-mail: sanjay@eldfindia.com; vish@eldfindia.com

S. Hazra, A. Bhukta (eds.), *Sustainable Development Goals*, Sustainable Development Goals Series,
https://doi.org/10.1007/978-3-030-42488-6_16

At the United Nations Conference on Sustainable Development (Rio+20) in Rio de Janeiro, Brazil, in June 2012, Member States adopted the outcome document '*The Future We Want*' in which they decided, *inter alia*, to launch a process to develop a set of SDGs to build upon the MDGs and to establish the UN High-level Political Forum on Sustainable Development. The Rio+20 outcome also contained other measures for implementing sustainable development, including mandates for future programmes of work in development financing, small island developing states, and more.

In 2013, the General Assembly set up a 30-member Open Working Group to develop a proposal on the SDGs. In January 2015, the General Assembly began the negotiation process on the post-2015 development agenda. The process culminated in the subsequent adoption of the 2030 Agenda for Sustainable Development, with 17 SDGs at its core, at the UN Sustainable Development Summit in September 2015.

The SDGs are to be monitored at the global level according to the global indicators framework developed by the Inter-Agency Expert Group on SDGs, agreed by the UN Statistical Commission (UNSD) and adopted by the UN General Assembly in July 2017.[5]

The Addis Ababa Action Agenda governs the implementation of the SDGs, the outcome document of the Third International Conference on Financing for Development.[6]

16.2 Implementation Support and Review

The HLPF (High Level Political Forum) is the follow-up body to the SDGs which hosts Voluntary National Reviews (VNR), aimed at facilitating the sharing of experiences, including successes and challenges, to accelerate the implementation of the 2030 Agenda. The reviews are state-led, voluntary and 'include developed and developing countries as well as relevant United Nations entities and other stakeholders, including the civil society and the private sector'.[7]

The Division for Sustainable Development Goals (DSDG) in the United Nations Department of Economic and Social Affairs provides substantive support and capacity-building for the SDGs and certain related issues, including water, energy, climate, oceans, urbanization, the Global Sustainable Development Report, partnerships and Small Island Developing States. DSDG plays a key role in the evaluation of UN system-wide implementation of the 2030 Agenda and on advocacy and outreach activities relating to the SDGs.[8]

16.3 SDG 16

SDG 16 calls upon governments to 'Promote peaceful and inclusive societies for sustainable development, provide access to justice for all and build effective, accountable and inclusive institutions at all levels'.

SDG 16 can be divided into four sub-groupings:

[5] Global indicator framework for the Sustainable Development Goals and targets of the 2030 Agenda for Sustainable Development, https://unstats.un.org/sdgs/indicators/Global%20Indicator%20Framework%20after%202019%20refinement_Eng.pdf accessed on 15th May 2019.

[6] Niti Aayog, Government of India, https://niti.gov.in/ accessed on 14th May 2019.

[7] https://sustainabledevelopment.un.org/hlpf accessed on 13th May 2019.

[8] Sustainable Development Goals Knowledge Platform https://sustainabledevelopment.un.org/?menu=1300 accessed on 13th May 2019.

- Peaceful societies
- Rule of law and access to justice
- Effective, accountable and inclusive institutions
- Human rights and fundamental freedoms[9]

Commitment to implementing Goal 16 can be said to rest on five prongs:

1. Fostering bottom-up empowerment
2. Supporting home-grown institutions and reformers
3. Facilitating selective decentralization
4. Harnessing the power of technology and
5. Following through on universality.[10]

SDG 16 Has 12 Sub-Goals

- Significantly reduce all forms of violence and related death rates everywhere.
- End abuse, exploitation, trafficking and all forms of violence against and torture of children.
- Promote the rule of law at the national and international levels and ensure equal access to justice for all.
- By 2030, significantly reduce the illicit financial and arms flows, strengthen the recovery and return of stolen assets and combat all forms of organized crime.
- Substantially reduce corruption and bribery in all their forms.
- Develop effective, accountable and transparent institutions at all levels.
- Ensure responsive, inclusive, participatory and representative decision-making at all levels.
- Broaden and strengthen the participation of developing countries in the institutions of global governance.
- By 2030, provide legal identity for all, including birth registration.
- Ensure public access to information and protect fundamental freedoms, in accordance with national legislation and international agreements.
- Strengthen relevant national institutions, including through international co-operation, for building capacity at all levels, in particular in developing countries, to prevent violence and combat terrorism and crime.
- Promote and enforce non-discriminatory laws and policies for sustainable development.[11]

The unanimously agreed upon objectives provide the overarching framework for international development for the next 15 years. The goals embody a strong commitment to a rights-based approach towards development, bound by the aspiration to 'leave no one behind'.

SDG 16 imbues the whole development agenda with good governance. The SDGs recognize that good governance is key to sustainable development. [12]

[9]Trinanjan Radhakrishnan Gavin Davies, SDG 16 India: Mapping The Datascape, Commonwealth Human Rights Initiative http://www.humanrightsinitiative.org/download/SDG%2016%20India%20Mapping%20the%20Datascape.pdf accessed on 13th May 2019.

[10]Terra-Lawson Remer, *How can we implement Sustainable Development Goal 16 on institutions?* The Brookings Institution (Oct. 1, 2015) https://www.brookings.edu/blog/future-development/2015/10/01/how-can-we-implement-sustainable-development-goal-16-on-institutions/accessed on 14th May 2019.

[11]NITI Aayog, SDG India Index: Baseline Report 2018, https://niti.gov.in/writereaddata/files/SDX_Index_India_21.12.2018.pdf accessed on 14th May 2019.

[12]Trinanjan Radhakrishnan Gavin Davies, SDG 16 India: Mapping The Datascape, Commonwealth Human Rights Initiative http://www.humanrightsinitiative.org/download/SDG%2016%20India%20Mapping%20the%20Datascape.pdf accessed on 13th May 2019.

16.4 Implementation by India

According to the SDG Index and Dashboard, **India is ranked 110 out of 149 in the amount of progress it has already made in relation to the 17 goals.**[13]

At the HLPF 2017, the Government of India made its first Voluntary National Review (VNR) Report, but included only seven (SDGs 1, 2, 3, 5, 9, 14 and 17) out of the 17 goals in its VNR,[14] which were the prime focus areas of the HLPF for 2017.[15] In 2018, the NITI Aayog released the First Baseline Index Report on India's progress on SDGs after formulation of a National Indicator Framework for monitoring national progress on SDGs.[16] A High Level Steering Committee has been constituted for a yearly review of the framework.[17]

The second edition of the Sustainable Development Goals (SDG) India Index, which comprehensively documents the progress made by India's States and Union Territories towards achieving the 2030 SDG targets was released as late as December 2019. It was very aptly put by the UN Resident Coordinator that In 2020, the world enters the final decade for achieving the SDGs—the 'Decade for Action'. And the Intergovernmental Panel on Climate Change tells us that we have 12 years left to save the planet from the worst effects of climate change. So, the time to act is now. The SDG India Index 2.0 and the dashboard enables India to both track and encourage accelerated progress to meet the SDGs across all its States and Union Territories. Implementation in India for all SDGs, including SDG 16, is done by several governmental agencies. The lastest press release states, India's composite score has improved from 57 in 2018 to 60 in 2019, thereby showing noticeable progress. According to Niti Ayog, the SDG India Index 2019 is more robust than the first edition on account of wider coverage of goals, targets, and indicators with greater alignment with the NIF. The Index spans 16 out of 17 SDGs with a qualitative assessment on Goal 17. This marks an improvement over the 2018 Index, which covered only 13 goals. While the NITI Aayog is responsible for overall coordination of the SDGs in India,[18] each SDG has a specific nodal ministry and other collateral ministries for its actual implementation.[19] For SDG 16, the nodal ministry for implementation, as for 2016, is the Ministry of Home Affairs. The Ministry of Law and Justice and the Ministry of External Affairs are also engaged in advancement of several key targets.[20]

[13] Sustainable Development Solutions Network (SDSN), and BartelsmannStiftung., (2016), Index Map, http://www.sdgindex.org/data/index/.

[14] Davies, *supra* note 10, at 3.

[15] Voluntary National Review Report, India, *On the Implementation of Sustainable Development Goals* https://niti.gov.in/writereaddata/files/Final_VNR_report.pdf accessed on 14th May 2019.

[16] Ministry of Statistics and Programme Implementation, http://www.mospi.gov.in/national-indicator-framework accessed on 15th May 2019.

[17] Ministry of Statistics and Programme Implementation, http://mospi.nic.in/sites/default/files/announcements/SDG_07122018.pdf?download=1 accessed on 15th May 2019.

[18] Ministry of Health and Family Welfare, Press Information Bureau (Feb. 26, 2016) Press Release, http://pib.nic.in/newsite/PrintRelease.aspx?relid=136880 accessed on 15th May 2019; Davies, *supra* note 7. See Also https://pib.gov.in/PressReleaseIframePage.aspx?PRID=1597981https://pib.gov.in/PressReleaseIframePage.aspx?PRID=1597981

[19] *Sustainable Development Goals (SDGs), Targets, CSS, Interventions, Nodal and other Ministries* (As on 04.04.2016) https://niti.gov.in/writereaddata/files/Mapping-SDGs%20V19-Ministries%20Feedback%20060416_0.pdf accessed on 14th May 2019.

[20] *Id.*

IMPLEMENTING AGENCIES OF SDG 16

- **The Ministry of Home Affairs (MHA)** at the Centre is the nodal agency for SDG 16 in India.
- **NITI Aayog**, the GoI's premier policy think-tank, is tasked with the overall responsibility of **coordinating the SDGs**. It **collates data** from various government agencies, **monitors implementation** and **reports** at the national, regional and global levels.

- **Dicennial census by the Registrar General and Census Commissioner** under the MHA is the largest single source of statistical information in India.
- The Ministry of Statistics and Programme Implementation **(MoSPI)** is responsible for development of the statistics system, coordination b/w agencies for stats, **nation-wide statistical exercises**. Its Central Statistics Office **prepares national accounts**
- **NSSO**
- **Directorate of Economics and Statistics (DES)** is the apex body responsible for the coordination of statistical activities at the state level

- Under MHA, NCRB
- Bureau of Police Research and Development, National Judicial Data Grid, Ministry of Law and Justice, NALSA, Ministry of Social Justice and Empowerment, NHRC, SIC/CIC, MWCD.

16.4.1 Government Schemes/ Initiatives for Various SDG 16 Targets

Since the SDGs represent the three pillars of sustainable development: social, economic, and environmental, they are all interlinked in nature. This necessarily means that fulfilment of one might lead to the fulfilment of another. SDG 16 has crucial links with many other SDGs namely: 1, 4, 5, 8, 10, 11 and 17. It can also be found related to several others, such as SDG 3 and 13.

16.4.1.1 Developing Inclusive Societies

Financial inclusion is necessary for the achievement of sustainable development. The vision of the Narendra Modi government, i.e. '*Collective Effort, Inclusive Growth*' itself signifies the inclination of each government initiative to achieve the same. The government has also released a draft of 'Three Years Action Agenda' covering years 2017–2018, 2018–2019 and 2019–2020 for the achievement of the same. The Agenda contains action plans for financial inclusion by reducing severe underemployment through creation of highly productive, high-wage jobs and boosting agricultural productivity through advanced technology, shifting to high-value commodities, improved access to irrigation, introduction of a modern land-leasing law, and the Fasal Bima Yojana.[21] The Make in India initiative is a significant step in achieving financial inclusion.

The Pradhan Mantri Jan Dhan Yojana was launched in 2014 to provide universal access to banking facilities for all households across the country through a bank branch or a fixed point Business Correspondent within a reasonable distance. It seeks to cover all households with at least one basic banking account with RuPay

[21] India: Three Year Action Agenda 2017–18 and 2019–20, https://niti.gov.in/writereaddata/files/coop/IndiaActionPlan.pdf accessed on 17th May 2019.

Debit Card with an inbuilt Rs. 1 lakh accident insurance cover. It also comprises a financial literacy programme at village level, which was undertaken during Phase I (2014–2015). It provides life insurance of Rs. 30,000 and direct benefit transfers of government subsidies through Aadhaar. It also covers pension schemes such as Swavlamban.[22] The scheme has so far been successful acccording to news reports.[23]

BharatNet, or Bharat Broadband Network Limited, is the world's largest rural broadband project, meant to provide broadband connectivity to all 2.5 lakh Gram Panchayats (Village Council) in India. It is part of the scheme Digital India, discussed below. In December 2018, it crossed its halfway mark, covering 1.2 lakh Panchayats. When complete, the BharatNet initiative will connect 250,000 village blocks or more than 600,000 villages. It is 100% funded by the Centre and had a completion deadline of 31 March 2019.[24] The Department of Telecommunications (DoT) is the nodal department for this project.[25] The project has been the subject of a lot of debate due to weak implementation[26] and has been weakened by several glitches such as the entry of Reliance Jio, which, while having increased urban tele-density, has severely decreased the inflow of funds to the Universal Service Obligation Fund (USOF) which is utilized to build rural connectivity infrastructure (such as BharatNet).[27] This has increased the urban–rural divide which has watered down the attempts at inclusivity through BharatNet.

16.4.1.2 Target 16.1: Significantly Reduce All Forms of Violence and Related Death Rates Everywhere

Target 16.1 holistically speaks of all forms of violence.

As to homicide among children, mortality rate (deaths per 100,000) due to homicide among adolescents aged 10–19 years, in 2015 is less than 5 in India. Eighty to ninety percent women aged 15–19 have experienced forced sex by intimate partners. The percentage of adolescent girls who have sought professional help is almost zero.[28] In 2018, India promulgated an ordinance which significantly increased the punishment for rape of women and girls. The Criminal Law (Amendment) Ordinance, 2018 laid down a minimum punishment of 10 years for rape and a minimum punishment of 20 years rigorous imprisonment/maximum punishment of life imprisonment or death penalty for rape of girls aged less than 12. It also disallows anticipatory bail for persons accused of rape of girls aged less than 16.

With regard to violence against children, in India, there is no ban on corporal punishment in

[22]Pradhan Mantri Jan Dhan Yojana, https://pmjdy.gov.in/ accessed on 17th May 2019.

[23]The Jan Dhan Yojana is gathering steam (14 May 2019) Business Line https://www.thehindubusinessline.com/opinion/editorial/the-pm-jan-dhan-yojana-is-delivering-on-inclusion-but-awareness-among-account-holders-must-rise/article27129894.ece accessed on 17 May 2019.

[24]*BharatNet crosses halfway mark, links 1.2 lakh panchayats* (Dec. 13, 2018) https://economictimes.indiatimes.com/tech/internet/bharatnet-crosses-halfway-mark-links-1-2-lakh-panchayats/articleshow/67066989.cms accessed on 16 May 2019.

[25]Ministry of Electronics & Information Technology, https://digitalindia.gov.in/content/broadband-highways accessed on 16 May 2019.

[26]Manoj Gairola, *In 'Digital India', Not Even 2.5% Panchayats Have Commercial Broadband,* The Wire (Nov. 19, 2018) https://thewire.in/government/narendra-modi-government-digital-india-village-broadband-connections accessed on 16 May 2019; Manoj Gairola & Anuj Srinivas, *PMO Anger Over BharatNet Project Reveals Shoddy State of Broadband Initiative,* The Wire (8 Jan. 2019) https://thewire.in/government/pmo-bharat-net-project-shoddy-state-of-broadband-initiative accessed on 16 May 2019.

[27]Manoj Gairola, *Has Reliance Jio's Entry Delayed Digital India's Goals?*, The Wire (28 Apr. 2017) https://thewire.in/business/reliance-jio-digital-india accessed on16 May 2019.

[28]United Nations Children's Fund, A Familiar Face: Violence in the lives of children and adolescents, https://www.unicef.org/publications/files/Violence_in_the_lives_of_children_and_adolescents.pdf accessed on 17 May 2019.

schools; Section 89 of the Indian Penal Code (IPC) allows the same. Data collected in 2009 found that in India's Andhra Pradesh and Telangana states, 78% of children aged 8 years and 34% of adolescents aged 15 years reported that they had been physically punished at school.[29]

16.4.1.3 Target 16.2: End Abuse, Exploitation, Trafficking and All Forms of Violence Against and Torture of Children.

The Protection of Children from Sexual Offences Act, 2012 was legislated in order to curb sexual violence against children, but has been lukewarm in its performance.[30] Children also suffer due to corporal punishment allowed under Section 89 of the IPC.

The Child Labour (Prohibition & Regulation) Act, 1986 which was updated in 2015 also provides for protection for children against employment in hazardous activities.

The Integrated Child Development Scheme is a core scheme for the fulfilment of SDG 16 through working towards target 16.2 which seeks to cease exploitation of children. It provides supplementary nutrition (SNP), immunization, and pre-school education to children, and was launched in 1975. It aims at pregnant, lactating mothers, and children below 6 years. Under SNP, beneficiaries are given hot meals along with take-home rations. The scheme has, however, not quite lived up to expectations due to inadequate infrastructure, low payments to Anganwadi workers such as Auxiliary Nurse Midwife (ANMs), lack of awareness and proper utilization by the local people, poor monitoring and corruption, etc.[31]

[29] *Id.*

[30] Abhishek Angad, *POCSO: Why cases of child sexual abuse mostly end in acquittal.*

[31] Jyotiranjan Sahoo et al., *Operational Assessment of ICDS Scheme at Grass Root Level in a Rural Area of Eastern India: Time to Introspect* 10 J. Clinical & Diagnostic Research (2016).

The POSHAN Abhiyaan launched in 2018 is also aimed at the same goal, through focused action towards reducing malnutrition amongst children aged 0–6 from 38.4% to 25% by 2022.[32] India has inked a $200 million loan agreement with World Bank for financing the scheme.[33]

The practice of bonded labour still thrives in India according to reports by NGOs, and there have been no studies conducted by the government, despite the fact that India has an obligation to do so under ILO 29 and 105.[34]

16.4.1.4 Target 16.4: Decreasing Illicit Financial and Arms Flows

The Drivers and Dynamics of Illicit Financial Flows from India: 1948–2008, released by Washington-based Global Financial Integrity (GFI) has found that the faster rates of economic growth since economic reform started in 1991 had led to a deterioration of income distribution which has led to more illicit flows from the country.

In 2018, the government demonetized currency notes in India, which had three broad objectives to fight black money, corruption and terror funding. It has been criticized as according to some experts on the basis of RBI's annual report that it has neither brought about a dip in counterfeit currency, nor brought about a substantial decrease in high-denomination bank notes.[35] MSMEs were especially affected due to

[32] POSHAN Abhiyaan, https://www.india.gov.in/spotlight/poshan-abhiyaan-pms-overarching-scheme-holistic-nourishment accessed on 16 May 2019.

[33] World Bank, Press Release, Project Signing : Government of India and World Bank Sign Agreement for Additional Financing to India's National Nutrition Mission (POSHAN Abhiyaan) to 315 Districts across all States and UTs (7 May 2018).

[34] Trinanjan Radhakrishnan Gavin Davies, SDG 16 India: Mapping The Datascape, Commonwealth Human Rights Initiative http://www.humanrightsinitiative.org/download/SDG%2016%20India%20Mapping%20the%20Datascape.pdf accessed on 13 May 2019.

[35] *How Successful was Demonetisation? Four Takeaways From the RBI's Annual Report*, (30 Aug 2018) https://thewire.in/economy/how-successful-was-demonetisa-

it, as per the report by RBI.[36] It also did not succeed in ushering in a cashless economy in India, as even though a short spike was seen after demonetization in digital transactions, the same ended when cash availability normalized.[37] Black money worth Rs. 4900 crores was deposited in the PM Garib Kalyan Yojana.

The government has enacted the Fugitive Economic Offenders Act, 2018 which bars a person from putting forward or defending any civil claim once an individual has been declared a fugitive economic offender by the virtue of Section 12 of the Act by the Special Court. This bar includes any claim which may or may not be related to the offence. This enables ex-parte proceedings against the economic offender. This provision has been questioned by critics as a violation of a person's right to access to justice, which might run counter to target 16.2.

To curb illicit arms flows through India, the government has set up the Inter-Ministerial Task Force in 2002 whose members include representatives from the Ministries of External Affairs, Defence and Home Affairs and National Security Council Secretariat, which regularly provides inputs related to policy guidance, research and monitoring of measures taken to control the same. The Arms Act 1959 and the Arms Rules 1962 govern prevention and control of illicit arms trade in India.

India is also a signatory to all 13 United Nations Conventions and Protocols dealing with international terrorism.[38]

16.4.1.5 Target 16.5 Significantly Reduce Bribery and Corruption in All Their Forms

India's ranking on the Corruption Perception Index, 2018 developed by Transparency International is 78 out of 180 countries, with a score of 41 out of 100. India has shown little progress over its score in 2017, which was 40 out of 100.[39]

The Prevention of Corruption (Amendment) Act, 2018 was passed which made bribe-giving a specific offence and has introduced corporate criminal liability for the acts of bribery. It has expanded the definition of a bribe to include any 'undue advantage', i.e. any pecuniary or non-pecuniary advantage taken by a public servant. It has also included provisions for the liability of the person advancing such undue advantage.[40] However, Section 17 A of the same makes it mandatory for police officers to obtain prior sanction by the concerned government if any investigation is to be undertaken against a public servant. Moreover, the definition of criminal misconduct has been significantly truncated. It has been challenged in the Supreme Court of India by CPIL as unconstitutional.[41] In 2016, the Supreme Court expanded the definition of 'public servants' to

---

tion-four-takeaways-from-the-rbis-annual-report accessed on 18 May 2019.

[36] Prosenjit Datta, *Let's admit it! Demonetisation was a failure*, (31 Aug. 2018) https://www.businesstoday.in/opinion/prosaic-view/lets-admit-it-demonetisation-was-a-failure/story/281860.html accessed on 18 May 2019.

[37] Nupur Anand, *India's central bank just proved demonetisation was for nothing*, (29 Aug 2018) https://qz.com/india/1373030/modis-demonetisation-did-nothing-for-india-shows-rbi-report/ accessed on 18 May 2019.

[38] India: Integrated National Report on The Implementation of International Instrument to Enable States to Identify and Trace, In a Timely and Reliable Manner, Illicit Small Arms and Light Weapons (31 Jan. 2010) http://www.poa-iss.org/CASACountryProfile/PoANationalReports/2010@90@India's%20Integrated%20National%20Report.pdf accessed on 21 may 2019.

[39] Transparency International, Corruption Perceptions Index, 2018 https://www.transparency.org/files/content/pages/2018_CPI_Executive_Summary.pdf accessed on 17 May 2019.

[40] Abhimanyu Pal, Anshul Prakash, Kruthi N. Murthy, *India: Prevention Of Corruption (Amendment) Act 2018 – Booster For The Honest Or The Corrupt?* (2 Aug. 2018) http://www.mondaq.com/india/x/724856/White+Collar+Crime+Fraud/Prevention+Of+Corruption+Amendment+Act+2018+Booster+For+The+Honest+Or+The+Corrupt accessed on 20 May 2019.

[41] Apoorva Mandhani, *Prior Sanction For Probing Public Servants: SC Issues Notice On CPIL's Plea Challenging Amendments To Prevention Of Corruption Act*, (26 Nov. 2018)https://www.livelaw.in/prior-sanction-for-probing-public-servants-sc-issues-notice-on-cpils-plea-challenging-amendments-to-prevention-of-corruption-act-read-petition/ accessed on 20 May 2019.

include officers of private banks, as their duties are public in nature.[42]

Demonetization was also a major step towards reducing corruption, which however could not quite achieve its objectives fully.[43] The following enactments were also passed in the recent years:

1. The Lokpal and Lokayukta (Amendment) Act, 2016 (This requires public servants to declare their assets and liabilities but not those of their spouses and dependent children, criticized as being regressive).
2. Black Money (Undisclosed Foreign Income and Assets) and Imposition of Tax Act, 2015 (This penalizes the concealment of foreign income and assets and any related tax evasion).
3. Benami Transactions (Prohibition) Amendment Act, 2016 (empowers the specified authorities to provisionally attach benami properties which can eventually be confiscated).
4. Companies (Amendment) Act, 2017 (regarding corporate fraud penalty for amount not exceeding Rupees one million and not affecting public interest).
5. Whistleblowers Protection (Amendment) Bill 2015 (has not been enacted yet due to protests from RTI activists alleging dilution of the law).

A former Supreme Court Judge has been appointed as the first Lokpal of India in 2019. His appointment process has, however, come to be criticized, as it did not contain any opposition party members.[44]

16.4.1.6 Target 16.6: Develop Effective, Accountable and Transparent Institutions at All Levels (+Target 16.3: Enhancing Rule of Law)

The government of India has taken several steps to ensure that there can be inclusive, efficient, accountable, transparent institutions. Many of these have in fact been implemented even before the SDGs materialized.

1. **The Right to Information Act, 2005**

   The RTI Act, for instance, is an example of a governmental venture which has significantly bolstered public access to information for the citizens, and hence is a step in fulfilling target 16.9 for SDG 16. The RTI system under the Act, however, has been observed to be languishing due to corrupt information officers, a dismal number of penalties imposed and a long winding procedure of appeals, a plethora of which are pending currently.[45] A 6% fall in 2015–2016 and 2016–2017 has been reported in RTI applications by the Central Information Commission.[46] Attempts by the government to weaken the law have further deteriorated its status.[47]Moreover, it has been observed that the RTI Act has failed to show inclusivity contemplated by SDG 16 due to having failed to reach the poor.[48] Nevertheless, it has resulted

[42] *Central Bureau of Investigation, Bank Securities and Fraud Cell v Ramesh Gelli* Criminal Appeal Nos. 1077-1081 of 2013 and W.P. (Crl.) No. 167 of 2015.

[43] *Supra* note 34.

[44] Lok Sabha Congress leader Mallikarjun Kharge was called as a special invitee, with say or voting power, and hence did not attend, see *Former Supreme Court Judge Pinaki Chandra Ghose Appointed India's First Lokpal* (Mar. 19, 2019) https://www.news18.com/news/india/former-supreme-court-judge-justice-pc-ghose-appointed-indias-first-lokpal-2072235.html accessed on 20 May 2019.

[45] Chetan Chauhan, *How RTI Act is dying a slow death in India*, (3 May 2018) https://www.hindustantimes.com/india-news/how-rti-is-dying-a-slow-death-in-india/story-sTpdC63K7s42vxgV1bxwTI.html accessed on 15 My 2019.

[46] *Id.*

[47] *Id.* The government's proposal to reduce the status of CICs and ICs from that equal to Supreme Court Judges to a government secretary had raised concerns about affecting their independence and efficiency; Centre had also proposed new RTI rules allowing an applicant to withdraw his appeal and abatement of an appeal if the applicant dies.

[48] Nita Bhalla, *India's "revolutionary" RTI Act fails to reach the poor* (2 July 2010) https://in.reuters.com/article/

in exposures of major scams.[49] The most recent amendments in the RTI law, the Right to Information (Amendment) Bill, 2019, was passed by the Rajya Sabha (Upper House) on Thursday passed by a voice vote amid a walk-out from the Congress, Rashtriya Janata Dal (RJD), Trinamool Congress (TMC) and the other opposition parties, clearly suggesting that the recent attempts are more towards diluting the Act.

2. **National Data Sharing and Accessibility Policy (NDSAP)**

In 2012, Government of India (GoI) adopted the National Data Sharing and Accessibility Policy (NDSAP) which enables the public to access government shareable data to promote data sharing for national planning and development. NDSAP differs from RTI in that it is a proactive rather than a reactive form of data sharing by the government. Shareable data are those which are 'not sensitive' and do not fall under the negative list. The NDSAP applies to all data and information created, generated, collected and achieved using public funds provided by GoI directly or through agencies by various ministries, departments, organizations, agencies and autonomous bodies. The nodal agency is the Department of Science and Technology on policy matters, while implementation is overseen by the Ministry of Electronics and Information Technology, along with the National Informatics Centre.[50]The National Informatics Centre is the custodian of e-Government and e-Governance applications as well as a promoter of digital opportunities for sustainable development, and, through its software NICNET, has institutional linkages with all the ministries and departments of the central government as well as locally with many state and district administrations. It has initiatives such as National Prison Information Portal, CCTIS Tamil Nadu (Criminal Tracking InfoSys.) and online civil registration systems.[51]

However, the collection, processing and dissemination of data need to be streamlined, as different circumstances of states (availability of server systems, electricity problems, language uniformity, etc.) need to be taken into account.[52]

3. Another such scheme aimed at advancing good governance through accountable, effective, inclusive institutions is PRAGATI or Pro-Active Governance and Timely Implementation, launched in 2015, which utilizes the grievance database Centralized Public Grievance Redress and Monitoring System (CPGRAMS), the Project Monitoring Group and MoSPI. The scheme consists of a monthly interaction between the PMO, Union Government Secretaries and Chief Secretaries of States redressing grievances and monitoring the implementation of public projects, etc. through discussions and directions. It takes into consideration various correspondences to PM's office by the common people or from high dignitaries of States and/or developers of public projects, etc.[53]

idINIndia-49806820100702 accessed on 15th May 2019.

[49]Betwa Sharma, *5 Scams The RTI Act Helped Bust In Its First 10 Years*, (15 July 2016) https://www.huffingtonpost.in/2015/10/12/5-most-critical-scams-exp_n_8263302.htmlaccessed on 15 May 2019.

[50]Ministry of Electronics and Information Technology, https://www.meity.gov.in/open-data accessed on 15 May 2019.

[51]Trinanjan Radhakrishnan Gavin Davies, *SDG 16 India: Mapping The Datascape*, Commonwealth Human Rights Initiative http://www.humanrightsinitiative.org/download/SDG%2016%20India%20Mapping%20the%20Datascape.pdf accessed on 13 May 2019.

[52]Dinsa Sachan, *Time to reveal*, Down to Earth (7 Jun. 2015) https://www.downtoearth.org.in/coverage/time-to-reveal-37769 accessed on 15 May 2019.

[53]*PM Narendra Modi launches PRAGATI platform for redressal of grievances*, (25 Mar. 2015) https://economictimes.indiatimes.com/news/politics-and-nation/pm-nar-

The scheme brings an essential character of co-operative federalism to the table as it allows Union government Secretaries and the Secretaries of States to share one stage and discuss issues put forth to them through the databases. It hence advances the objectives of SDG 16 through boosting the quality of institutions and promoting access to justice.

4. The above scheme forms part of the Digital India campaign which was launched in 2015 and has rapidly taken shape in the recent years through various schemes. It consists of three core components: the development of secure and stable digital infrastructure, delivering government services digitally and universal digital literacy. It has nine key deliverables: broadband highways, Universal Access to Mobile Connectivity, Public Internet Access Programme, e-governance, e-Kranti (electronic delivery of services), information for all, IT for jobs, electronics manufacturing, early harvest programmes.[54] It also seeks to make services such as e-courts, e-police, etc. available online.[55] Digital India hence conceptually fulfils three objectives of providing access to justice, building inclusive societies and effective, inclusive and accountable institutions under SDG 16.

Some schemes which have been launched under the same are BharatNet, Make in India, Startup India, Standup India, UMANG and MyGov.[56]

For boosting access to justice, India has had the Legal Services Authority Act, 1987 in place for decades, which has been functioning through Lok Adalats (or People's Courts), which are a form of alternate dispute resolution in India.[57] Lok Adalats have been reported to take up as many as 12 lakh cases per day.[58] Lok Adalats do not charge the parties and any court fees paid before referral by the court is also refunded to the parties. However, it has been noted by scholars that people have grown increasingly dissatisfied with them, as the system has acquired an adversarial character over the years. Lack of adequate infrastructure has also contributed to the same.[59]

The government had also launched the Gram Nyayalayas Act, 2008 which established rural mobile courts in villages, which were supposed to hold weekly/monthly court sessions. Its preamble states that it is an Act to provide Gram Nyayalayas (Village Courts) at the grassroot level to ensure that citizens, regardless of their social, economic or other disabilities, have access to justice at their doorsteps. However, the actual establishment of the Gram Nyayalayas has been lagging behind the number which initially had to be established. Most of the Gram Nyayalayas have either not been established or are not functional.[60]Several factors have contributed to this: inadequate funds, lack of staff, the lack of political will and the lack of coordination between the High Courts and the state governments[61]. It has been noted that High Courts have been averse to the idea of mobile courts, noting that they can erode the sanctity of a for-

endra-modi-launches-pragati-platform-for-redressal-of-grievances/articleshow/46693007.cms?from=mdr accessed on 16 May 2019.

[54] Ministry of Electronics & Information Technology, https://www.digitalindia.gov.in/ accessed on 16 May 2019.

[55] *Id.*

[56] *Id.*

[57] National Legal Services Authority, https://nalsa.gov.in/lok-adalat accessed on 16 May 2019.

[58] Soibam Rocky Singh, Lok Adalats clear over 12 lakh cases in one day (20 Feb. 2018) https://www.thehindu.com/news/cities/Delhi/lok-adalats-clear-over-12-lakh-cases-in-one-day/article22794628.ece accessed on 16 May 2019.

[59] Tameem Zainulbhai, *Justice for All: Improving the Lok Adalat System in India* 35 Fordham Int'l L. J. 248, 250 (2016).

[60] Jitendra, *Where are rural courts?* Down To Earth (17 Aug. 2015) https://www.downtoearth.org.in/coverage/where-are-rural-courts-44754accessed on 16 May 2019.

[61] *Id.*

mal justice system.[62] Some scholars have dismissed the idea of Gram Nyayalayas as one creating new litigation instead of dispensing with the existing, and also compromising with the quality of decision-making, at the cost of lowering litigation expenses and time.[63]

16.4.1.7 Target 16.7: Ensure Responsive, Inclusive, Participatory and Representative Decision-Making at All Levels

There continues to be under-representation of women, minorities, marginalized populations in the Legislature, Judiciary and the Executive. Women continue to be under-represented as the list by Inter-Parliamentary Union[64] shows that representation of women in India's parliament is only about 12.6%. India ranked 149 with that score in the same.[65]

The representation of youngsters is also low in the Parliament and state legislatures. The outgoing Parliament has 53% of the members aged 55 or above.[66] Minority representation also lags behind. The representation of women in public services is meagre[67] and so is that of the Other Backward Classes (OBCs).[68] Women's representation is also sparse in the judiciary and so is the representation of minorities.[69] This is especially disadvantageous as sensitive cases such as the Babri Masjid case did not have a Muslim judge on the Bench.[70]

The Rights of Persons with Disability Act, 2016 has been enacted to increase the participation of persons with disabilities by reserving 4% government jobs for such persons.

The Women's Reservation Bill [The Constitution (108th Amendment) Bill] is one of the longest pending legislations in the Indian Parliament which seeks to reserve 33.3% seats for women in the Parliament, state assemblies and local bodies according to the 73rd and 74th Amendments to the Constitution. The Bill has never been successfully passed due to lack of political consensus.[71]

To promote inclusive governance, the government has also introduced the MyGov app which allows users to participate in formulation and implementation of government projects through discussions, through uploaded documents, etc.

[62] *Id.*

[63] Sudhir Krishnaswamy, *Gram nyayalayas and the hazards of informal justice*, SUNDAY GUARDIAN http://www.sunday-guardian.com/analysis/gram-nyayalayas-and-the-hazards-of-informal-justice accessed on 16 May 2019.

[64] http://archive.ipu.org/english/issues/wmndocs/world.htm.

[65] Shreehari Paliath, *Data check: India lags behind its neighbours on women's representation in Parliament,* (16 Mar 2019).

[66] Trinanjan Radhakrishnan Gavin Davies, *SDG 16 India: Mapping The Datascape*, Commonwealth Human Rights Initiative http://www.humanrightsinitiative.org/download/SDG%2016%20India%20Mapping%20the%20Datascape.pdf accessed on 13 May 2019.

[67] *Id.*

[68] *Id.*

[69] *India has token representation of women in higher levels of judiciary: Hawaii Supreme Court judge* (7 Apr. 2019) https://economictimes.indiatimes.com/news/politics-and-nation/india-has-token-representation-of-women-in-higher-levels-of-judiciary-hawaii-supreme-court-judge/articleshow/68761699.cms accessed on 20 May 2019.

[70] John Sebastian & Faiza Ahmed, *Babri Masjid Case: Why Representation of Minority Judges Matters* (18 Jan. 2019) https://thewire.in/law/babri-masjid-case-why-representation-of-minority-judges-matters accessed on 20 March 2019.

[71] Shreehari Paliath, *Data check: India lags behind its neighbours on women's representation in Parliament,* (16 Mar. 2019).

Women Members Of Parliament In Lok Sabha, 1952 To 2014

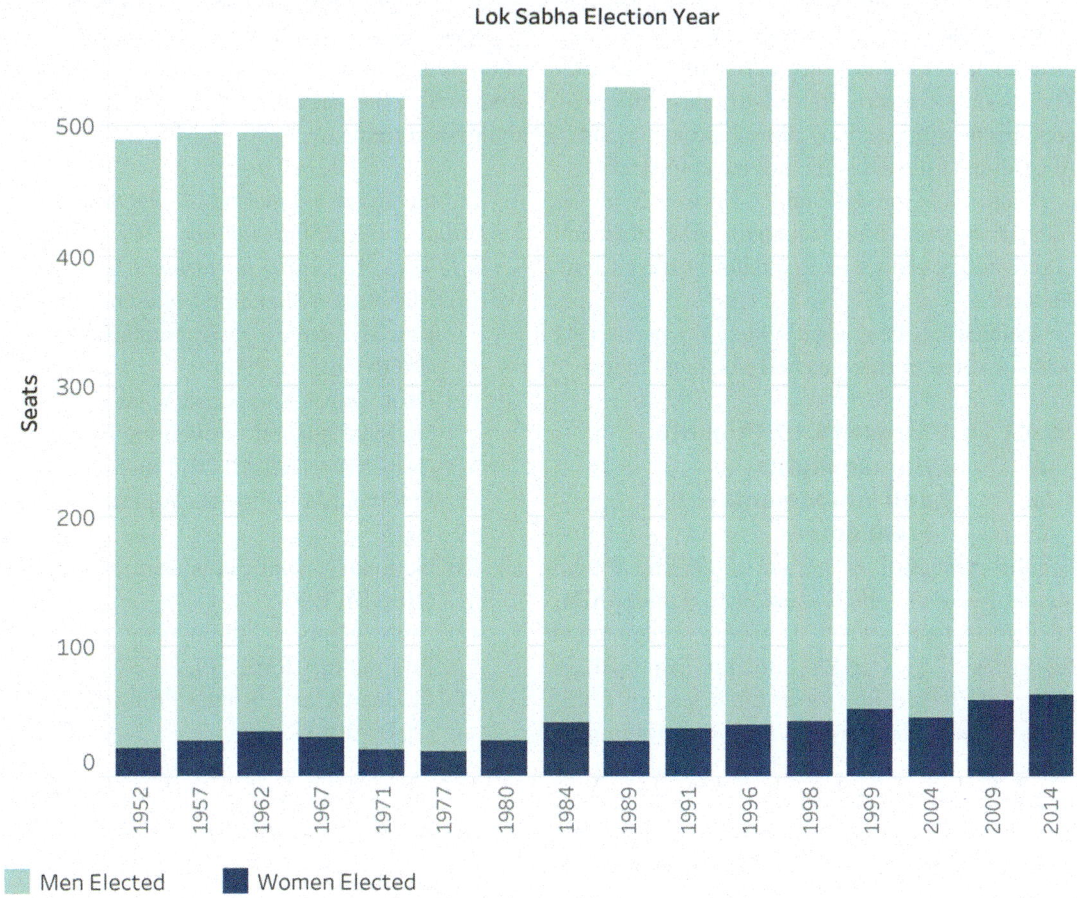

16.4.1.8 Target 16.8: Broaden and Strengthen the Participation of Developing Countries in the Institutions of Global Governance

India is a member of four out of the five World Bank Group Institutions; it has also been a voting member of the Human Rights Council for 8 years. India is also a part of the General Assembly of the UN, and was a member of the UNSC from 2011 to 2013 and WTO.

16.4.1.9 Target 16.9: By 2030, Provide Legal Identity for All, Including Birth Registration

It has been estimated that approximately 42% of the births each year go unregistered in India, despite the Registration of Births & Deaths Act, 1969.[72] The lowest performing states in India are UP, Bihar, AP, Madhya Pradesh and Rajasthan.

[72] UNICEF India, http://unicef.in/Story/1133/Birth-Registration-the-picture-in-India accessed on 20 May 2019.

For facilitating the target for providing legal identity for all and birth registration under Target 16.9, the government has introduced "Aadhaar". The Government of India has disbursed a cumulative amount of $25 Billion to 329 Million beneficiaries through Direct Benefit Transfer.[73] The privacy concerns raised by the same have now been addressed by the Supreme Court.[74] According to the decision, private companies can no longer ask for data under Section 5 of the Aadhaar Act, and Bank Accounts, SIM cards, and examinations cannot be mandatorily linked with Aadhaar.

Aadhaar has been used widely for registering births, such as in Haryana and Madhya Pradesh.[75]

16.4.1.10 Target 16.10: Promote Human Rights and Fundamental Freedoms

For the promotion of human rights and fundamental freedom, which is target 16.9 under SDG 16, many schemes which have been implemented for SDGs 1, 2, 3 and 4 fulfil the same. SDGs being interlinked in nature, often lead to fulfilment of each other through their own fulfilment. Therefore, action for SDG 17 in revitalizing the global partnership for sustainable development also suffices as action to boost global inclusiveness by enabling all countries to participate effectively in the quest for sustainable development under SDG 16.

Hence, the Anganwadi scheme for the fulfilment of SDGs 1 and 2 will also contribute to fulfilling the human right to health and hence fulfil target 16.9 for SDG 16 as well. Similarly, schemes which boost adaptability and resilience to climate change and other natural calamities also protect and fulfil human rights such as the right to shelter, the right to a healthy environment, the right to health, etc. It is imperative that citizens of a country develop resilience and adaptability to climate change and other natural disasters so that they do not suffer deprivation of their basic rights.

1. *Ending extreme Poverty ($1.25 per day) & Reduce by Half the Number of People Living in Poverty According to National Definitions*
   - (a) Mahatma Gandhi National Rural Employment Guarantee Act (MGNREGA), 2006
   - (b) Deen Dayal Antyodaya Yojana (DAY)—National Rural Livelihood Mission (NRLM) + NULM, 2011 and 2013
   - (c) Pradhan Mantri Awaas Yojana—Gramin, 2015
   - (d) National Social Assistance Programme (NSAP), 1995
   - (e) Prime Minister Employment Generation Programme (PMEGP), 2008
   - (f) Pradhan Mantri Rojgar Protsahan Yojana (2016–2017)
   - (g) Special Central Assistance to Tribal SubScheme (SCA to TSS) (1977–1978)
   - (h) Assistance to Disabled Persons for purchase of Fitting Devices (ADIP), 2005
   - (i) Pradhan Mantri Jan Dhan Yojana, 2014
   - (j) Pradhan Mantri Jeevan Jyoti Bima Yojana, 2015
   - (k) Pradhan Mantri Suraksha Bima Yojana (PMSBY), 2015
   - (l) Atal Pension Yojana (APY), 2015
   - (m) Pradhan Mantri Vaya Vandana Yojana (PMVVY), 2017
   - (n) PDS (1947)
2. *Implementation of Social Protection Systems* (schemes the same as above + the following)
   - (a) Vanbandhu Kalyan Yojana, 2014
   - (b) Relief and Rehabilitation for Migrants and Repatriates
   - (c) Freedom Fighters (Pension and Other Benefits)
3. *Equal Access to Economic Resources, Natural Resources, Inheritance, Access to Basic*

[73] Kumar Deepak, *Challenges Before India in Achieving Sustainable Development Goals*, (31 Jul. 2018) https://medium.com/@kumardeepak_4023/challenges-before-india-in-achieving-sustainable-development-goals-61a9fcffafde accessed on 17 May 2019.

[74] Justice K. Puttaswamy v. Union of India.

[75] Menaka Rao, *Haryana is making babies enrol in Aadhaar before it will issue them birth certificates*, (9 May 2017) https://scroll.in/pulse/835970/babies-in-haryana-are-being-enrolled-in-aadhaar-before-they-are-issued-birth-certificates accessed on 20 May 2019.

Services; Appropriate New Technology; Ownership and Control over Land and Other Forms of Property (same schemes as list 1 + the following)

(a) Pradhan Mantri Gram Sadak Yojana (PMGSY), 2000
(b) LPG Connection to poor households—UJJWALA, 2016
(c) Swachh Bharat Mission, 2014
(d) National Rural Drinking Water Programme, 2014
(e) AMRUT (Atal Mission for Rejuvenation and Urban Transformation), 2015
(f) PM Mudra Yojana, 2015
(g) Umbrella Programme for Development of Scheduled Castes, OBCs and Other Vulnerable Groups.
(h) Pradhan Mantri Jan Vikas Karyakram, 2008–2009
(i) Shyama Prasad Mukherjee RURBAN Mission, 2016[76]
(j) Pradhan Mantri Sahaj Bijli Har Ghar Yojana—'Saubhagya' (Rural and Urban) (2017)
(k) Rashtriya Gram Swaraj Abhiyan (RGSA), 2018.
(l) Deendayal Upadhyaya Gram Jyoti Yojana (DDUGJY), 2015.
(m) Equity Support to the National/State Scheduled Tribes Finance and Development Corporations for Financial Support to Tribal Entrepreneurs.

4. *Building Adaptability and Resilience towards Climate Change and Other Environmental Disasters*
   (a) National Cyclone Risk Mitigation Project (NCRMP) (e.g. Early Warning Dissemination System, Odisha) (2015)
   (b) Flood Management & Border Areas Programme (2018–2019)
   (c) Development of Water Resources Information System
   (d) Atmosphere and Climate Research—Modelling, Observing Systems and Services (ACROSS) (2018–2019).
   (e) Ocean Services, Technology, Observations, Resources, Modelling and Science (OSTORMS)
   (f) Seismology and Geosciences (SAGE)
5. *Ensure Mobilization of Resources*
   (a) Sarva Shiksha Abhiyan (2002)
   (b) National Programme of Mid-Day Meal in Schools (MDM) (1995)
   (c) Rashtriya Madhyamik Shiksha Abhiyan (2009)
   (d) Pandit Madan Mohan Malaviya National Mission on Teachers and Teaching (2015).
   (e) National Scheme for Incentive to Girl Child for Secondary Education (NSIGSE), 2008.
   (f) Interest subsidy and contribution for Guarantee Funds
   (g) Scholarship for College & University Students.
   (h) Pre-matric Scholarship to students with disabilities.
   (i) Pre- and post-matric scholarships for SC, OBC and other vulnerable groups.

- **SDG 2 (End Hunger, Promote Food Security) and SDG 16**[77]

  – Anganwadi Services, National Nutrition Mission, Scheme for Adolescent Girls, Child Protection Services and, National Creche Mission (Umbrella ICDS), 1975.
  – Pradhan Mantri Matru Vandana Yojana (PMMVY), 2016.
  – National Health Mission, 2005
  – National AYUSH Mission, 2014
- **SDG 3 (Health)**
- **SDG 4 (Inclusiveness through Education)**:

[76]NITI Aayog, SDG India Index: Baseline Report 2018, https://niti.gov.in/writereaddata/files/SDX_Index_India_21.12.2018.pdf accessed on 14 May 2019.

[77]NITI Aayog, SDG India Index: Baseline Report 2018, https://niti.gov.in/writereaddata/files/SDX_Index_India_21.12.2018.pdf accessed on 14 May 2019.

- Sarva Shiksha Abhiyan, 2002
- National Programme of Mid-Day Meal in Schools (MDM), 1995
- Rashtriya Madhyamik Shiksha Abhiyan, 2009
- National Means-cum-Merit Scholarship Scheme (NMMSS), 2008
- Pandit Madan Mohan Malaviya National Mission on Teachers and Teaching, 2015
- National Scheme for Incentive to Girl Child for Secondary Education (NSIGSE), 2008
- Pre-matric scholarships for SC, OBC and vulnerable groups; Pre-matric scholarships for children of those engaged in unclean occupations and prone to health hazards; Pre-matric Scholarship to students with disabilities; Other scholarships & fellowships for SC, OBC Students; Free coaching for SC students; Hostels for SC & OBC students
- National Child Labour Project, 1988
- Rashtriya Uchhatar Shiksha Abhiyan, 2013
- Technical Education Quality Improvement Programme (EAP), 2002
- Umbrella ICDS, 1975
- Multi-Sectoral Development Programme for Minorities (MsDP):
  Seekho aur Kamao—Skill Development Initiatives for minorities, 2013
  Upgrading Skills and Training in Arts/Crafts for Development (USTAAD) for minorities, 2014–2015
  Nai Manzil—The Integrated educational and Livelihood initiatives, 2015
- Pradhan Mantri Kaushal Vikas Yojana, 2015
- Vocational Training Centres in Tribal Areas, 2009
- National Scheme for Incentive to Girl Child for Secondary Education (NSIGSE), 2008.
- Pradhan Mantri Kaushal Vikas Yojana, 2015
- Saakshar Bharat, 2009 (It is important to emphasise that most programs after 2014 is in vogue now due to the incumbent Government and previous programs have been integrated or assimilated in these new schemes.

16.4.1.11 Target 16.10 A and 16.10 B

Target 16.10 A requires that every country has independent National Human Rights Institutions. India has an independent statutory body mandated by the Human Rights Protection Act, 1993, called the National Human Rights Commission. It is entrusted with the protection and promotion of human rights, defined by the Act as 'rights relating to life, liberty, equality and dignity of the individual' guaranteed by the Constitution or embodied in the ICCPR and ICESCR of 1966. From 1996 to 2006 it was accredited with an 'A' grade by GANHRI as it complies with the Paris Principles, which are a set of non-binding principles adopted by the UNGA in 1993, provide a set of international standards which frame and guide the work of NHRIs across the world.

The NHRC has, however, not adequately performed according to the standards set out by the UNHRC in its communication to the Government of India[78] which consisted of the creation of a civil society and independent expert advisory panel; establishment of three regional offices to increase public access; a national helpline to field urgent calls of severe human rights abuse; and empowering the NHRC to cover all relevant cases and inquire into allegations of human rights violations and abuses by paramilitary and armed forces of India, including in Jammu and Kashmir. A 2017 report by International Federation for Human Rights has reported that the recommendations have not been followed.[79]

Target 16.10 B requires countries to take anti-discriminatory measures for minorities, backward classes, women and the LGBTQ community.

---

[78] Trinanjan Radhakrishnan Gavin Davies, *SDG 16 India: Mapping The Datascape*, Commonwealth Human Rights Initiative http://www.humanrightsinitiative.org/download/SDG%2016%20India%20Mapping%20the%20Datascape.pdf accessed on 13 May 2019.

[79] Dimitris Christopolous, *Re: Downgrading of the National Human Rights Commission of India, FIDH, Paris, France*, 8 November 2017.

India has taken several measures such as ratifications of the ILO Convention 111, and above all Article 15, 16, 17, 18 of the Constitution. The Employment of Manual Scavenging and Construction of Dry Latrines (Prohibition) Act, 1993 was enacted for the upliftment of manual scavengers who majorly consist of backward castes, and the National Safai Karamcharis Finance & Development Corporation (NSKFDC), which was set up in 1997, following Prohibition of Employment as Manual Scavengers and Their Rehabilitation (Amendment) Act, 2013.

The Anti-Discrimination and Equality Bill, 2016 was introduced in the Lok Sabha but has not been enacted yet.

16.5 Conclusion: Challenges in Implementing SDGs

- One major factor which keeps India from fulfilling the goals laid out under SDG 16 is the fact that despite a proliferation of schemes and legislation, it has not been able to implement them effectively. Examples include the RTI Act and the Gram Nyayalayas Act, which, though enacted, have been weakened with ineffective implementation.
- Certain areas require more initiative by the government such as boosting representation of women in legislatures at the national and local levels. Enactment of the aforementioned Bill can help the cause to a great extent.
- **Measurement of Progress**: NDSAP is an important venture in implementing the SDGs. However, data need to be specific, measureable, accessible, reliable and timely. A majority of the government departments and their data collection agencies are not yet computerized and the majority of digital records are not machine-readable. Lack of inter-operability of data and low levels of proactive disclosure of information by the government hampers access to information.
- Data are limited in scope and sporadic in certain cases, such as sex trafficking and bonded labour; data are untimely such as in case of NCRB (data is published 8 months after it is reported); data lack specificity; reliability is lacking (e.g. discrepancies in figures exist within government ministries and departments— in 2015, the MHA mentioned three cases of communal violence in Haryana, while the NCRB recorded 201.[80] This lack of data is reflected in the Baseline Index Report 2019 as well: the indicators do not include SDGs 12, 13 and 14 as there is no comparative data across states. SDG 17 has not been included as indicators have not been identified for this goal.
- **Financial Constraints**: It has been observed by several institutional reports that India faces a dire financial gap for implementing the SDGs. The first level of estimates indicates a financial shortfall of INR 533 lakh crores (USD 8.5 trillion) over 15 years for achieving SDGs. This equals approximately INR 36 lakh crores or USD 565 billion per year.[81] The estimates for financial requirements are even higher as calculated by the UNCTAD.[82] Such a deficit is true for many developing countries.[83] India has also iterated this financial incapacity in its Voluntary National Review Report, 2017, and, on the basis of the principle of common but differentiated responsibilities, has called upon developed countries to financially assist developing countries and LDCs in implementing the SDGs.[84]

---

[80] Trinanjan Radhakrishnan Gavin Davies, *SDG 16 India: Mapping The Datascape*, Commonwealth Human Rights Initiative http://www.humanrightsinitiative.org/download/SDG%2016%20India%20Mapping%20the%20Datascape.pdf accessed on 13 May 2019.

[81] Technology and Action for Rural Advancement, *Achieving the Sustainable Development Goals in India: A Study of Financial Requirements and Gaps*, https://www.devalt.org/images/L3_ProjectPdfs/AchievingSDGsinIndia_DA_21Sept.pdf accessed on 15th May 2019.

[82] *Id.* at 12.

[83] *Id.* at 12.

[84] Prabhakar Mohandas, *Sustainable Development Goals (SDGs)-Challenges for India* 9 IND. J. PUB. HEALTH & RESEARCH 1,5 (2018).

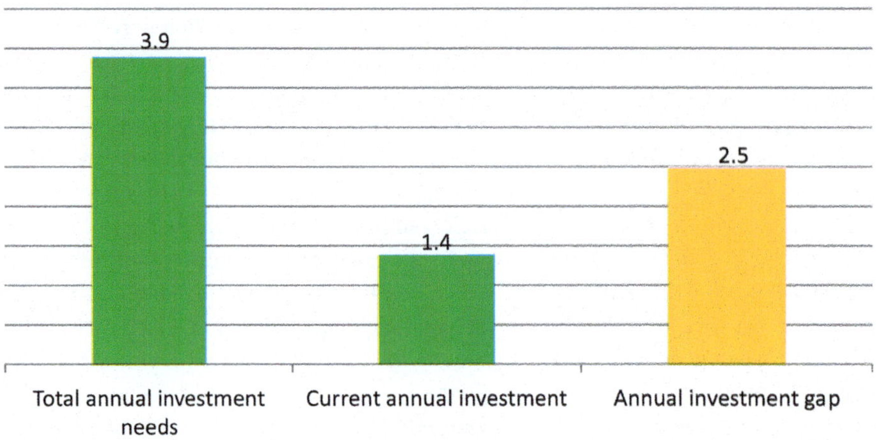

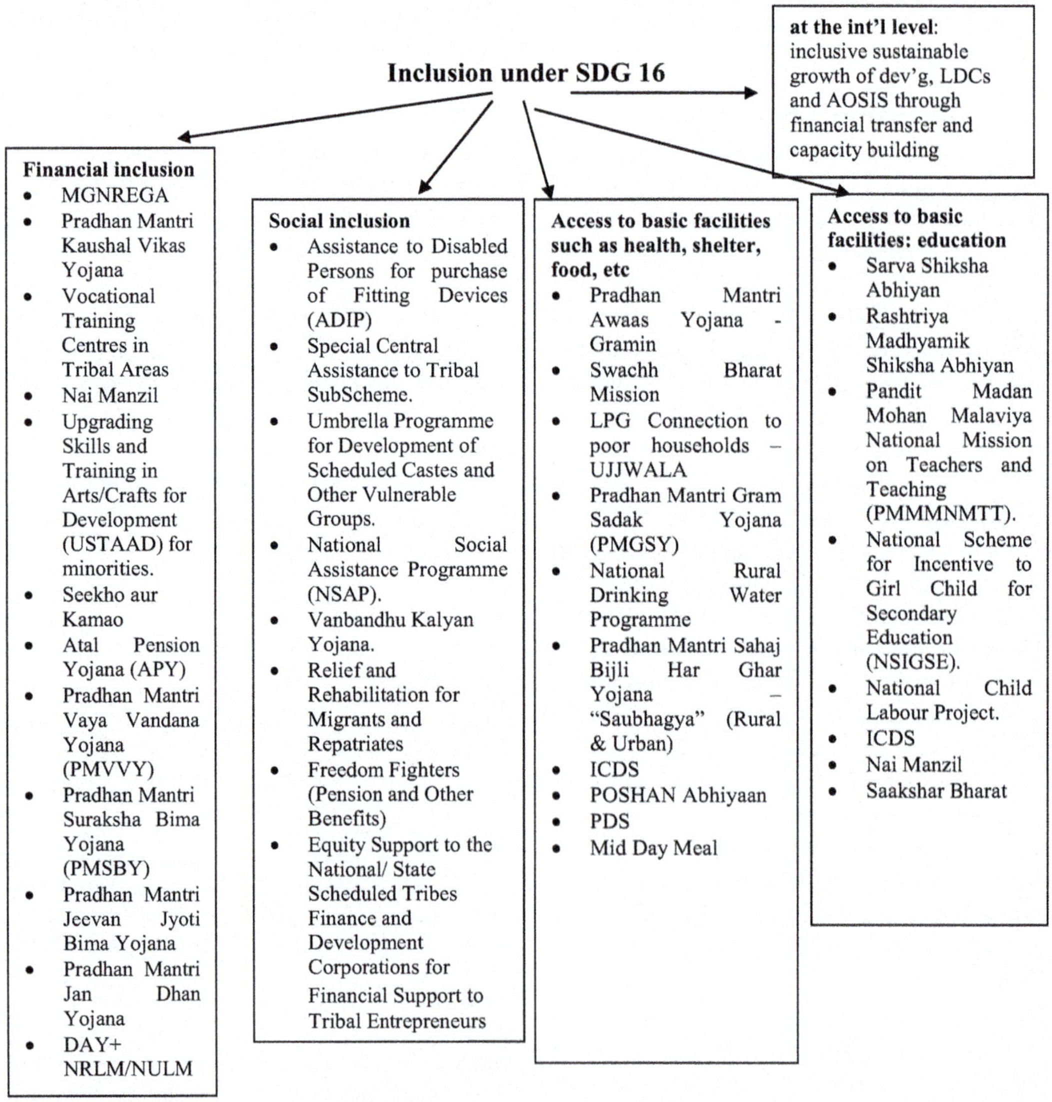

Avani Dubey is a Researcher of Enviro Legal Defence Firm. She has worked on several issues in environmental law such as reparation in international environmental law, wetland conservation and the Sustainable Development Goals.

Trade and Sustainable Development Goals: A Multiscenario Analysis for India and Other Asia-Pacific Countries

17

Badri Narayanan Gopalakrishnan
and Shounkie Nawani

17.1 Introduction

In concurrence with the idea of ensuring sustainable development, now and in future, the 2030 Agenda for sustainable development was adopted at the United Nations Sustainable Development Summit on September 25, 2015. Although the current agenda represents a transformative shift from the Millennium Development Goals (MGDs), yet they reflect a convergence of the MDGs and the global discussions on sustainable development while ensuring that economic growth, environmental protection, and social inclusion are considered as part of an integrated and indivisible development agenda (WTO 2018). Adopted in the year 2000, the Millennium Development Goals focused on ways to reduce global poverty and hunger, illiteracy, child and maternal mortality, gender inequality, and other world development challenges by 2015. However, they were misrepresented as "one-size-fits-all" and represented only ultimate aims or ends but did not indicate the means by which the ends should be achieved (Vandemoortele 2012). To achieve the said objectives of the 2030 Agenda for Sustainable Development, 17 Sustainable Development Goals (SDGs), 169 targets, and 230 indicators that are both integrated and indivisible were devised. The SDGs were designed by deepening and extending efforts pursuant to MDGs while also addressing complex and critical elements that are required to achieve the goals of sustainable development. The SDGs were devised to keep up the momentum of the MGDs and to achieve what the MDGs failed to achieve. So, the sustainable development goals are more comprehensive in coverage and aim to achieve peace and prosperity across the world.

The SDGs are aimed at achieving sustainable development simultaneously in three aspects: economic, social, and environmental, that is, from eliminating extreme poverty, combating inequalities, promoting prosperity, strengthening global partnerships to protecting the environment, among others (Table 17.1). These goals are global in nature and are universally applicable to both developed and developing countries. The implementation of these goals and targets requires coherent policies and actions that systematically address synergies and trade-offs between the economic, social, and environmental dimensions of sustainable development. International trade is an engine for inclusive economic growth and can contribute to the promotion of sustainable development. Trade has been

B. N. Gopalakrishnan (✉)
School of Environmental and Forestry Sciences, University of Washington, Seattle, WA, USA

S. Nawani
Infinite Sum Modelling LLC, Sammamish, WA, USA

S. Hazra, A. Bhukta (eds.), *Sustainable Development Goals*, Sustainable Development Goals Series,
https://doi.org/10.1007/978-3-030-42488-6_17

Table 17.1 SDGs and their dimensions

SDGs		No. of targets	Dimension
1	No poverty	5	Social
2	No hunger, food security, sustainable agriculture	5	Social
3	Good health and well-being	9	Social
4	Quality education	7	Social
5	Gender equality	6	Social
6	Clean water and sanitation	6	Social/ environmental
7	Affordable and clean energy	3	Environmental
8	Decent work and economic growth	10	Economic
9	Industry, innovation, and infrastructure	5	Economic
10	Reduced inequalities	7	Social
11	Sustainable cities and communities	7	Social
12	Sustainable consumption and production	8	Environmental
13	Climate action	3	Social/ environmental
14	Life below water	7	Environmental
15	Life on land	9	Environmental
16	Peace, justice, and strong institutions	10	Social
17	Partnerships for the goals	19	Social

Source: Cutter et al. (2015)

identified as one of the important nonfinancial "means of implementation"[1] for the 2030 sustainable development agenda as trade-related actions can generate a positive impact upon a country's socioeconomic developmental capacity (United Nations 2015a).

Trade-related targets included in several SDGs are relevant to the achievement of other goals.

Some of the targets and goals of the SDGs are also influenced by the Addis Ababa Action Agenda (AAAA) of July 2015. Under the AAAA Agreement, the world leaders laid the foundation of a revitalized global partnership wherein international trade would act as a driver to attain inclusive economic growth and poverty reduction. The AAAA agenda also emphasized that with appropriate supporting policies, infrastructure, and educated workforce, trade could also help to promote productive employment and decent work, women's empowerment, and food security, as well as a reduction in inequality, and contribute to achieving the sustainable development goals. For this, promotion of a universal, rules-based, open, transparent, predictable, inclusive, nondiscriminatory, and equitable multilateral trading system under the WTO along with ensuring meaningful trade liberalization is one of the agendas of the AAAA (United Nations 2015b).

The SDGs included these features of the AAAA as targets and goal-specific means of implementation, thereby further enhancing and emphasizing the role that trade can play in promoting sustainable development (Narayanan et al. 2017). The SDGs have identified many priority areas that are directly related to the development-enabling power of trade. For instance, Goal 17.10 promotes a universal, rules-based, open, nondiscriminatory, and equitable multilateral trading system under the World Trade Organization (WTO), including the conclusion of negotiations within the Doha Development Agenda. Goal 17.11 seeks to significantly increase the exports of developing countries, in particular, with a view to doubling the share of least developed countries (LDCs) in global exports by 2020. Goal 8 calls for increased aid for trade support for developing countries, particularly LDCs, through the enhanced integrated framework for LDCs. Thus, the main focus of trade in the SDGs is on reducing tariff barriers, improving trade facilitation, and providing aid-for-trade to developing countries. Likewise, foreign direct investment (FDI) has been listed as the key source of financing for development in the context of the 2030 Agenda. Overall, 12 of the 17 SDGs contain targets that are closely linked to trade, while 35 of 169 targets either explicitly mention trade and investment or are closely related to it (Jacob 2016).

[1]Means for implementation include the mobilization of financial resources, capacity building, and the transfer of environmentally sound technologies to developing countries.

Trade has been identified as a main tool to stimulate sustainable development globally. Trade liberalization has long been seen as critical to economic growth and development especially for developing countries to find their way out of poverty as trade liberalization policies are known to expand employment opportunities and sustain livelihoods for all. Trade contributes to higher levels of investment, technology upgradation, and increases in productivity, which expands production and enables economic growth. As a result, there has been a steady increase in the importance of international trade for the global economy. However, there is an emerging vast literature which questions the benefits accrued from opening up trade, especially by many small developing and least developed countries (LDCs). Hence, use of trade as an enabler of sustainable development is only possible when it leaves no one behind, which requires appropriate global conditions. It is, therefore, important to fully understand how trade interacts with the various goals encompassed in the SDGs.

In this context, the objective of the study is to empirically evaluate the economic, social, and environmental impacts of alternative policy changes on SDGs in Asia and the Pacific–UNESCAP member states (including Associate Members) in particular. The region consists of 62 economies, with 57 developing ones, while only five economies of United Kingdom, the United States, Australia, Japan, and New Zealand are developed. This populous region inhabits more than four billion people, accounting for nearly 55% of the world's population in 2015. Economically, the region generated 2/5th of the global gross domestic product (GDP) in 2015, but at the same time, it houses around 1.2 billion people who live below the poverty line of $3.10 (2011 PPP) a day. The region accounts for a significant share of global trade (32.2%) and also remains the largest recipient of FDI. Thus, opening of further trade and investment is expected to play a critical role in achieving SDGs in the region (ADB 2016). Among developing economies, the Asia-Pacific region received the largest share of global FDI inflows, attracting 45% in 2018. Developing countries in the region attracted 40% of global FDI inflows or 88% of the total FDI inflows in the Asia-Pacific region (UNESCAP 2019).

With the spread of globalization, there has been an emergence of a large number of bilateral, regional trade agreements and multilateral trade agreements across the world. With this, there has been a considerable reduction of traditional trade market access barriers like tariffs, leading to a significant rise in world trade. However, existing tariff barriers nontariff measures still limit trade[2] (Narayanan et al. 2017). The small developing countries and LDCs face the brunt of these measures. Such restrictive policies may hinder the achievement of the SDGs. On the other hand, technological advances and liberal economic policies have facilitated the growth of global value chains (GVC). With it, the trading economies have become integrated and GVCs can play an important role in facilitating future trade, FDI patterns, and growth opportunities. The trading countries have become more integrated as intermediate goods and services flow across borders before being embedded in the final product. At present, around half of the world trade takes place through GVCs. As reported in, 48% of exports for developing economies, in value-added terms, involve GVCs.

The following section collates a thematic literature review on the key objective of the study. The literature review draws from Narayanan et al. (2017). The themes of the available literature relate to the importance of improved market access, reduction in trade costs, reduction in tariffs, greater trade facilitation among countries, and better access to finance by developing countries and LDCs. They not only promote greater world trade and increase technology transfers between countries but are equally pivotal to achieve the desired SDG targets.

A vast body of available literature focuses on how improved market access can act as an enabler in the implementation of sustainable develop-

[2]Other non-tariff measures (NTMs) and behind-the-border barriers in areas such as investment, competition, government procurement, and intellectual property are also important, but not the focus of this study.

ment. UNCTAD in a report published in 2016 highlighted on the presence of improved market access in both foreign markets and domestic markets as an important determinant of the effectiveness of trade as a means of implementation of sustainable development. Also, the poorer countries face higher trade costs. Any attempt toward an expansion of physical access to global markets at reasonable costs is said to benefit low-income countries from reduction in the trade barriers. While assessing the trade costs involved, the report suggests that the domestic trade costs are more restrictive than the market access conditions in home and foreign markets. Transportation costs in low-income countries are high that further limits their potential to trade more. In the process, their potential to trade more reduces and income gains from existing trade are forgone. Thus, in order to realize the potential of trade for ensuing inclusive growth and sustainable development, it becomes most relevant to balance the sustainability-enhancing effects with the trade-restricting effects of tariffs, which still remain a core challenge for trade policy for majority of the countries.

Similarly, studies by Arvis et al. (2015) and Hoekman (2016, 2017) have shown that trade costs are substantially higher in poor countries than elsewhere. Hoekman (2016, 2017) stresses on the importance of measuring trade costs as part of a comprehensive approach to reviewing the trade elements of the sustainable development. In the analysis, he predicts that the low-income countries will face a more challenging global environment for trade and investment in the coming decade than it was in the 1990s and 2000s. He further cites that a reduction in trade costs in both goods and services sectors should the most important goal for developing countries to enhance their competitiveness. A lower trade costs imply availability of cheaper inputs that positively helps to increase competitiveness and to provide households better access to products and services for improving their welfare, ranging from food security to health. Thus, sustained government efforts to reduce trade costs can help improve the availability and quality of both goods and services for developing countries.

Another set of literature points out that the lack of trade facilitation agreements is a major hurdle for the developing countries to fully exploit the available market access opportunities. The OECD-WTO Aid for Trade (2015) report notes that the lack of trade facilitation makes it tough for many firms in developing countries to fully exploit market access opportunities. Moreover, redundant and ill-adapted infrastructure and procedural delays owing to cumbersome and time-consuming border procedures dissuade many developing country firms from conducting international trade. This is particularly true for the LDCs. Lack of trade facilitation can undermine the potential gains from trade, especially in case of agricultural goods that have a short shelf-life and are perishable. Lack of effective trade measures further pushes up costs. Hence, an improvement in trade facilitation can play a significant role in increasing incomes as well as the GDP, for all countries, particularly developing countries and LDCs. This in turn will have positive impacts on export performance and on various social and environmental levels (e.g., reduction in poverty and rise in female employment).

The OECD-WTO Aid for Trade (2017) report has emphasized on the importance of both physical and digital connectivity in creating trade opportunities for developing countries. Better physical connectivity ensures a greater movement of goods and services and enables access to local, regional, and global markets, thus enhancing trade opportunities. On the other hand, digital connectivity allows businesses to develop a greater connect with the global economy and helps them to reap trade benefits from this linkage. The report advocates aid-for-trade (AfT) as a possible way to reduce trade costs. Aid-for-trade (AfT) includes both technical assistance and capacity building measures. The aid-for-trade initiative was launched in 2006, and to date, a total of USD 264.5 billion has been disbursed for financing aid-for-trade programs. Of this, more than three quarters of total aid-for-trade has been used to finance projects in four sectors that are closely related to cutting trade costs. They include transport and storage (29%), energy generation and supply (21%), agriculture (18%), and banking (11%). It may be noted that AfT has had some successes, primarily in addressing supply side and institutional constraints, such as infrastructure (e.g., roads and ports). With regard to the SDGs, the AfT initiative can be pivotal in helping to achieve SDG9 (industry, innovation, and infrastructure) and SDG8 (decent work and economic growth) (Da Silva 2017).

Besides reducing trade costs, trade facilitation is a major determinant of how developing country firms interlink themselves to the GVCs and the benefits they can draw from such participation. According to the estimates by the ESCAP-World Bank Trade Cost Database, the trade costs faced by the Asia-Pacific countries comprise of tariffs, natural trade costs (geographical and cultural factors), and nontariff policy measures (direct behind and at-the border trade costs, indirect costs of trade procedures, maritime connectivity and services, the business regulatory environment, currency fluctuations, the availability/use of ICT services, and other nontariff barriers). Of this, the share of tariffs in trade costs was in between 0% and 10%, while the share of natural trade costs is 10–30%. The share of the nontariff policy measures was the highest in between 60% and 90% (Duval 2015).

Fessehaie and Morris (2018) assess the opportunities for low-income countries in Asia-Pacific through GVCs using case studies. They sum up their research by highlighting that there is a need for economic upgrading of the economies to reap economic benefits from the GVCs. However, this alone may be insufficient as the industrial policies and technological advancements need a push. In the current global scenario, a coherent industrial and FDI policy can help bring in technological improvement benefitting domestic firms. In addition, targeted and innovative public and private policies are a must to enhance the contribution of GVC participation in the SDGs particularly in terms of income and employment. Similarly, ICTSD (2016) explores the trade and trade-related policies that are required to build and enhance the growth of GVCs in an economy. The report notes that appropriate policies that support and mutually reinforce existing policies are required to set up a productive and sound GVC network. For this, FDI inflows, gluing together fragmented production networks, lowering of trade restrictions, and strategies to enhance the competitiveness of GVCs are important. Under the appropriate conditions, the GVCs can help developing countries to progress toward several key sustainable development goals (SDGs) like the promotion of inclusive and sustainable economic growth and the reduction of inequality.

Razzaque et al. (2016) have highlighted that implementing national, regional, and multilateral trade facilitation measures will contribute to enhanced trade flows by reducing costs. They cite the example of the WTO's Trade Facilitation Agreement and note that its implementation can provide a greater and much needed thrust to global trade flows and multilateral trade cooperation. Besides this, the arrangement also represents an opportunity to streamline border procedures and reduce trade costs. This requires a sustained financial and technical support for LDCs and landlocked countries. As per the WTO, implementation of the TFA has the potential to increase global merchandise exports by up to US$1 trillion per annum, which can help bolster the role of trade as an effective means of achieving SDGs (WTO 2015). Thus, the multilateral trading system through its initiatives can help developing countries and LDCs in improving trade flows and thus contributes to achieve the SDGs.

Economic crises, breakdown of regional groupings (such as Brexit), and increased protectionism subdue global trade and economic growth. In the context of achieving the desired SDG goals, such happenings slowdown and may inhibit the pace, intent, and efforts to achieve the said objective. Evenett and Fritz (2015) note that the LDCs have incurred a loss to the tune of US$264 billion of exports as a result of protectionist measures since the global economic crisis of 2008. To address such instances, a well-designed trade policy can play a critical role toward attaining the objectives of sustainable development. Trade policy is a universal tool, which when adopted by countries in coherent national policy frameworks by identifying specific areas of trade policy reform would help to progress toward achieving various sustainable development goals. Trade policy areas that stand out in this regard include lowering tariffs, providing accessible and the affordability digital and physical connectivity, improving trade facilitation, enhancing aid-for-trade, and pursuing trade facilitation agreement, among others (Tipping 2014). The outcome depends heavily upon each country's trade policies and their interactions when the associated collective action problems are considered. The use of trade policy as a tool should be facilitated in close coordination with other sectors for achieving the SDGs. Thus, policymakers across countries can use trade policy as an instrument for achieving sustainable development by ensuring that the gains from trade are distributed widely across the economy. This can be done by ensuring that the interactions between different policy measures to achieve sustainable development in different fields (i.e., social, environmental, or economic) create synergies for each other.

Therefore, in order to determine the economic, social, and environmental impact of regional trade and investment liberalization, we consider the following aspects for our study, as potential trade policy-based facilitators of SDGs:

1. Tariff liberalization
2. Investment liberalization
3. Trade facilitation
4. Integrated liberalization

The baseline year is 2015, and all policy changes are implemented incrementally throughout the period considered, 2015–2030. Moreover, the economic impacts of the policy changes are captured through changes in GDP and trade levels; the social impact through changes in levels of inequality and undernourishment; and the environmental impact through changes in CO_2 emissions. While these aspects do not necessarily cover all dimensions of the links between trade policy and sustainability or SDGs, we delve on them illustratively to obtain a tentative picture of these links with a view to understand the trade-offs involved as well as synergies that may exist. In the next section on methodology, we explain these at length and also discuss some limitations and caveats therein.

17.2 Methodology

For the analysis, we use a global CGE model based on the Global Trade Analysis Project-Power (GTAP-POWER) model and data base, which includes comprehensive details both at a sector level and at a country level for Asia-Pacific. This particular extension of the standard GTAP model is employed as it captures rich details of different types of energy, including various forms of renewable energy. Renewable energy, energy from fossil fuels, and the CO_2 emissions associated with them are included in the model.

To suit our objective, the model is further complemented by adding equations based on models developed for other studies. Narayanan and Balie (2017) developed an econometric framework to assess the impact of food production and supply on undernourishment, which was then integrated within the GTAP model. A recursive dynamic model was developed based on Monash framework on investment dynamics and investment identified by sector and countries of origin and destination. Results from this model are used to introduce dynamic effects in our comparative static framework, so that the whole analysis is conducted in a dynamic setting. Further, the analysis focuses on inequality as an outcome of our model simulations, by looking at the divergence between the real wages of skilled and unskilled labor, to proxy intracountry inequality—this varies by country and by looking at that between the real income of rich and poor countries, to proxy global intercountry inequality—this is a single number for the whole world for a given scenario.

For the analysis, the reference year for the GTAP data base is 2011; however, the recent (2015) macrolevel data to adjust this dataset to 2015 using the GTAP Adjust tool are employed. The model assumes unemployment in labor market using a unit-elastic labor supply, which is a reasonable middle path between horizontal and vertical labor supply curves, and is also supported reasonably well by econometric literature on labor supply elasticities. These are some unique features of the modeling framework, wherein we draw from the state-of-the-art recent literature on emissions, food security, and inequality, which are also important concepts in the SDGs.

17.3 Policy Scenarios

In order to empirically analyze the impact of trade and investment along the economic, social, and environmental pillars of sustainable development from the Asia-Pacific nexus, we shock the model under the following policy changes:

Scenario 1: Enhanced tariff liberalization in Asia-Pacific ("Tariff liberalization").

In this scenario, tariffs are gradually eliminated between members of the Regional Comprehensive Economic Partnership (RCEP) and Trans-Pacific Partnership (TPP) (minus the United States). In addition, other economies within Asia-Pacific introduce a 50% tariff reduction. This is in line with the current proliferation of bilateral, plurilateral, and regional trade agreements, as well as the Regional Economic Cooperation and Integration (RECI) initiative at ESCAP.

Scenario 2: Investment liberalization in Asia-Pacific ("Investment liberalization").

This scenario closely follows the approach outlined in scenario 1. Intraregional investment liberalization is simulated by gradually eliminating investment barriers between members of RCEP and TPP (minus the United States). In addition, other Asia-Pacific countries introduce a 50% investment barrier reduction.

Scenario 3: Trade facilitation implementation in Asia-Pacific ("Trade facilitation").

In this scenario, trade costs are reduced as a result of implementation of trade facilitation and paperless trade measures across the Asia and the Pacific. Gradually, all the economies in the region reach the trade facilitation level of China. Quantifying the impact of trade facilitation using an index is a complex undertaking, requiring in this case, a concordance between trade data and the trade facilitation implementation rates available from the ESCAP-led Global Survey on Trade Facilitation and Paperless Trade Implementation.

Scenario 4: Simultaneous implementation of tariff liberalization, investment liberalization, and trade facilitation ("Integrated liberalization").

This scenario combines all above-mentioned policy changes to demonstrate the aggregate effect of tariff liberalization, investment liberalization, and trade facilitation implementation. As will be discussed, combining liberalization policies amplifies the economic benefits offered by the individual scenarios. Moreover, some adverse effects created by one policy are negated by one of the others.

Further, in order to explore how specific domestic policies can help channel trade and investment liberalization toward sustainable development, three additional policy scenarios and their economic, social, and environmental impact are analyzed, as follows.

Scenario 5: Implementation of Paris Accord.

The previous policy simulations (scenarios 1–4) do not include constraints on emissions, and as such, emissions rise in each. In this scenario, all signatories of the Paris Accord globally reduce CO_2 emissions in accordance with agreed levels in the Agreement.[3]

Scenario 6: Welfare transfers.

This scenario addresses inequality through a 3% redistribution of income from skilled to

[3]More information on the Paris Accord and nationally determined contributions (NDCs) is available from http://unfccc.int/paris_agreement/items/9485.php

unskilled labor, taking place in all Asia-Pacific economies. Specific policies to achieve this objective could include welfare payments to low-income households, government subsidies, and progressive taxation.

Scenario 7: Combined trade, environmental, and social policies.

This final scenario combines integrated liberalization (scenario 4) with the Paris Accord and welfare transfers. It shows the interlinkages between what are sometimes thought of as diametrically opposed policies and demonstrates that environmental and social goals can be achieved in parallel with trade-driven economic development.

17.4 Results

While regional and global trade prospects have been improving, the risk of a renewed wave of trade protectionism remains. Accordingly, we also model the possibility of a global trade war, where the trade protectionist measures initiated by one or a few countries ultimately lead to other countries retaliating. Specifically, the scenario simulates the effects of all countries raising import tariffs to their bound levels globally between 2015 and 2030. Table 17.2 shows the overall impact of the tariff hikes scenario.

Table 17.2 shows the overall impact of the WTO-consistent tariff hikes, and we show it separately, as it shows negative impact of a global trade war. All regions experience a reduction in GDP, varying between 0.2% annually in Latin America to 0.9% per year in Europe. This leads to an overall global reduction in GDP of $380 billion a year. The significant reduction is driven by trade, as expected, with exports and imports both plummeting globally. CO_2 emissions, on the other hand, are reduced due to overall lower economic activity. Tariff hikes do not affect inequality because the reduction in wages for skilled and unskilled labor is equally affected by the global tariff hikes, but undernourishment increases significantly across all regions.

Among Asia-Pacific subregions, North and Central Asia sees the most significant reduction in GDP of 0.7% or $16 billion annually. In absolute terms, East Asia's GDP decreases the most by over $51 billion. Overall, Asia-Pacific's GDP decreases more than $110 billion per year between 2015 and 2030. The declines in GDP are directly caused by significant reduction in trade, which vary significantly across subregions. India experiences a dramatic export decline of 38% annually. On the other hand, West Asia and the Pacific see their exports decline by only 5% a year, on average. The difference in the effect is largely due to the export product mix, that is, subregions that are most affected have exports that are more susceptible to being blocked by increases in applied rates. In terms of imports, since countries in West Asia and North and

Table 17.2 Results of a hike in tariffs globally (% deviations from the baseline)

	GDP	Exports	Imports	CO_2 emissions	Undernourishment
Asia-Pacific	−0.4	−19	−17	−0.4	1.5
East Asia	−0.3	−13	−7	−1.3	1.5
South and Southeast Asia	−0.3	−25	−14	0.1	0.9
India	−0.3	−38	−8	−0.1	0
West Asia	−0.3	−5	−38	0	3.4
North and Central Asia	−0.7	−15	−31	−0.1	1.8
The Pacific	−0.4	−5	−5	−0.4	1.2
Latin America	−0.2	−13	−22	−1.8	0.3
Sub-Saharan Africa	−0.7	−3	−27	−1.1	2.4
Other Africa	−0.5	−6	−10	−2.6	1.6
Europe	−0.9	−37	−16	−0.2	2.8
North America	−0.3	−15	−26	−1.9	1

Central Asia have the most policy space (i.e., their applied tariff levels are much lower than their bound tariff rates), they reduce imports by more than 30%. CO_2 emissions actually increase in Southeast Asia due to an increase in local production of previously imported products. Other countries experience declines in emissions as economies are in depressions. In terms of undernourishment, the most notable subregional result is in West Asia, which sees an increase of 3.4% a year, underlying the danger the tariff war poses for the poor.

In scenario 1, tariff liberalization results are comparably modest, as many of the countries in the region already have low tariff rates. Consequently, the impact of tariff liberalization on GDP is relatively small, ranging from 0.01% to 0.03% across subregions. In absolute terms, however, the effect on regional GDP is an annual increase of $6.5 billion.

The impacts of tariff reduction are more pronounced in trade. Exports increase on average 0.22% annually every year until 2030, while imports increase by 0.15%. In absolute terms, this is an annual increase of $17.8 billion and $10.7 billion for exports and imports, respectively. The almost 70% larger increase in export compared to import demonstrates the significance of the interlinkages between the economies through regional and global value chains. Decreasing import tariffs, while increasing imports, also enables countries to reduce input costs, improve variety and quality of intermediate goods, become more competitive, and increase exports even more.

Tariff liberalization's effect on the environment regionally is almost negligible. CO_2 emissions remain relatively neutral, increasing overall by less than 0.1% annually. However, driven by the redistribution of production, subregions exhibit some variations. In particular, emissions in East Asia increase 0.14% on average annually or 12.9 million tons of oil equivalent (MTOE) per year.

Turning to the social impact, the effects of intraregional tariff cuts on inequality within the region and subregions are not significant. This can be largely attributed to the tariff cuts taking place across-the-board (i.e., in all sectors and countries). Such cuts are likely to offset increased inequality in sectors displaced by trade with gains from export-oriented sectors of the economies. The impact on undernourishment is also insignificant in the region and subregions. While increased interregional trade may negatively affect some producers, the negative effects are offset by decreased prices, resulting in a net neutral effect in the region and the subregions.

The cumulative result of tariff liberalization is an increase in regional GDP of 0.3% and trade of 2.9% by 2030. This is consistent with the findings from Gilbert (2013), who found that a free trade agreement encompassing all members of ESCAP, which involved cutting all intraregional tariffs to zero, resulted in GDP increases ranging from 0.3% to 0.78%. However, tariff liberalization is only a small part of TPP, which also includes significant trade facilitation, investment liberalization, as well as services liberalization commitments.

In scenario 2, the economic impact of regional investment liberalization is significantly higher than that of tariff liberalization. This is consistent with the fact that, unlike for tariffs, FDI restrictiveness remains high in many countries of the region. Economic gains from investment liberalization largely follow the pattern of tariff liberalization outcomes at the regional and subregional level. GDP increases by 0.1%, or $19.5 billion annually, with North and Central Asia attracting much needed investment and experiencing the highest relative growth at 0.29%. Similarly, India and Southeast Asia experience significant annual GDP boosts of 0.14% and 0.11%, respectively.

However, the growth mechanism of investment liberalization differs from the one for tariff liberalization. In the case of tariff liberalization, all changes in economic performance are trade driven. In contrast, improved economic performance from investment liberalization is driven by increase in capital stock, which may or may not increase trade. As a result, the impact of investment liberalization on trade varies across subregions. For example, exports in India and Southeast Asia are below those gained through tariff liberalization. North and Central Asia imports, on the other hand, significantly increase, growing at 0.38% per year. On average, however, regional exports and imports grow at 0.26% and 0.21% annually—only slightly higher than was achieved through tariff liberalization.

Unlike tariff liberalization, investment liberalization decreases inequality in the region by 0.02% per year. Among all the subregions, inequality in Southeast Asia and West Asia declines most significantly. Similarly, undernourishment declines significantly across the region (falling 0.14% annually), and particularly in North and Central Asia and India. In these subregions, undernourishment declines 0.39% and 0.29%, respectively. Therefore, investment liberalization reduces inequality and undernourishment more than tariff liberalization. The increases in capital stock allow more utilization of unskilled labor, and the combination of lower production costs and increased incomes makes food more affordable.

In terms of environmental impact, investment liberalization increases CO_2 emissions in the region by 0.1% or 13 MTOE annually, similar to the results for tariff liberalization. At the subregional level, North and Central Asia gains the most economically, but also has the highest CO_2 emission, an increase of 0.29% annually. The subregions' CO_2 emission increase stems from increased infrastructure investment. In India and Southeast Asia, emissions also increase significantly (0.14% and 0.11%, respectively).

Turning to scenario 3, trade facilitation has large and generally positive impacts across the region. Asia and the Pacific experience an annual increase in GDP of 0.32%, which is equivalent to nearly $87 billion per year—14 times more than under enhanced tariff liberalization, and four times more than under investment liberalization. The GDP growth is driven by trade, with Asia-Pacific exports and imports annually increasing by 0.93% and 0.81%, respectively. These results are consistent with the empirical literature, which generally finds the impact of trade facilitation to be much larger than trade liberalization (ADB and ESCAP 2013; Gilbert 2013).

While trade increases are much larger than under the tariff liberalization scenario, both policy changes show regional exports expanding more than imports, resulting in an increase in net exports of $16 billion per year (exports increase by $75 billion and imports by $59 billion). These results suggest that trade facilitation and paperless trade implementation, and more generally, enhancing regional connectivity to reduce trade costs, should be a top priority among policymakers.

Although trade facilitation results in large regional trade gains, CO_2 emissions increase only marginally—and in any case less than with investment liberalization. The modest increase compared to tariff liberalization is the result of increased economic activity in North and Central Asia, India, and Southeast Asia. On the other hand, in this scenario, emissions in the Pacific decline 1% annually, helping to offset increases in other subregions.

Trade facilitation decreases inequality within the region overall. While the decrease is generally small, it is larger than those gained through tariff or investment liberalization. This result is mainly due to changes in India and, to a lesser extent, the Pacific. India has some of the most inefficient trade procedures in the region. Only those of the Small Island Developing States in the Pacific are worse. The results for the Pacific region as a whole are heavily moderated by Australia and New Zealand, who in contrast to other countries in the region have some of the highest trade facilitation implementation rates in Asia and the Pacific. Significant improvements in trade facilitation in low-income economies of India and the Pacific could reduce inequality by increasing their export of traditionally labor-intensive goods (e.g., textiles) and boosting both the employment and wages of unskilled labor. However, the rise in inequality in other subregions highlights the necessity of complementary policies.

Trade facilitation also leads to significant reduction in undernourishment in Asia and the Pacific. In this scenario, undernourishment decreases by an average of 1.4% annually. By 2030, this is a reduction of over 20% overall in the region. The results, however, are not uniform. Malnourishment increases slightly in Southeast

Asia (0.52% annually) due to the displacement of the agricultural sector by other higher value-added sectors. Nevertheless, trade facilitation appears to be a promising strategy for reducing undernourishment in the region and contributing to SDG 2 to achieve zero hunger.

In scenario 4, the policy change investigated combines tariff liberalization, investment liberalization, and trade facilitation. This "integrated liberalization" approach provides the highest overall economic benefit and increases GDP for the region by 0.34% annually or $94 billion. The increase is largely driven by gains from trade facilitation, which by itself accounts for 0.32% increase in annual GDP.

"An integrated approach to trade and investment liberalization can increase the region's exports by $101 billion and imports by $75 billion per year." In addition, integrated liberalization increases trade significantly more than any of the other stand-alone policy changes. Exports and imports are projected to increase by 1.25% and 1.03%, respectively ($101 billion and $75 billion in absolute terms, per year), and the region's increased annual net exports could reach more than $25 billion. This is about $10 billion more than with trade facilitation alone. This integrated approach facilitates the participation of countries in global value chains and significantly increases the competitiveness of regional exports. This integrated liberalization scenario provides strong evidence of the important synergies that can be achieved by liberalizing and facilitating trade and investment.

In terms of CO_2 emissions, the negative effects of the investment liberalization and tariff liberalization scenarios accumulate, resulting in an increase of 0.16% in emissions for the region or approximately 24 MTOE per year. The North and Central Asia, East Asia, and Southeast Asia regions are largely responsible for the regional increase. In the Pacific, on the other hand, there is a large relative decrease, but this does not affect the regional results significantly, given the relatively low contribution of that subregion to total regional emissions.

The integrated liberalization approach magnifies the small reductions in inequality observed under the trade facilitation and investment liberalization scenarios. Despite inequality increasing modestly in East Asia, and to a lesser extent North and Central Asia, inequality falls by an average of 0.05% per year at the regional level. However, the extent and the drivers of these reductions vary across subregions. For instance, both India and West Asia experience particularly significant inequality reductions. In India, investment liberalization reduces inequality, whereas trade facilitation leads to the reductions in West Asia.

Interestingly, the integrated liberalization approach does not show synergies between the liberalization and facilitation scenarios in terms of food security. Rather, undernourishment worsens marginally compared to the stand-alone trade facilitation scenario. Similar to the trade facilitation scenario, undernourishment still decreases significantly in most subregions, but it increases slightly in Southeast Asia. This is the result of complex interactions between trade and investment liberalization in the agriculture and food sector.

Overall, trade facilitation—which may be best understood in the context of this analysis in its broadest sense, that is, as a reduction in trade costs—promises the most economic gains compared to either tariff liberalization or investment liberalization. However, in the trade facilitation scenario, there are also substantial variations among countries and between subregions in terms of the environmental and social impacts. We find that the greatest economic, social, and environmental benefits are in the combined "integrated liberalization" scenario.

The results of the simulations for Paris Accord, income transfer, and integrated liberalization and combined scenarios are discussed herein. In scenario 5 (Paris Accord scenario), the effect on GDP across the Asia-Pacific region is, contrary to expectations, mildly positive, a modest 0.1% per annum.[4] The economies of India and

[4]This approximately equals the effects of the investment liberalization scenario in Section B in the following publication: https://www.adb.org/publications/win-win-how-international-trade-can-help-meet-sdgs

https://www.adb.org/sites/default/files/publication/327451/adbi-win-win-how-international-trade-can-help-meet-sdgs.pdf#page=28

Southeast Asia grow by 0.38% and 0.34%, respectively, driven largely by the growth in the renewable sector of these economies. However, due to the oil-centered economy of the Islamic Republic of Iran, West Asia GDP declines by 0.5% per annum.

The Paris Accord implementation includes cutting emission commitments by countries outside the region, which also affect economic growth outcomes in Asia and the Pacific. For example, relative to the status quo without emission constraints, GDP in the United States and the European Union annually contracts 0.18% and 0.62%, respectively. The biggest reductions in output growth globally are in the petrochemical industries, with an annual reduction of 2.2% of output, while the renewables sector sees a 2.4% annual output growth.

Exports in Asia and the Pacific remain unchanged under the Paris Accord scenario. The subregional breakdown shows that exports in India rise significantly by 0.5% per year and 0.1% in West Asia, whereas they decrease by 0.1% in the Pacific and Southeast Asia. At the same time, imports decline by 0.14% in the region, falling by 0.6% in India and 0.4% in South East Asia. The declines in these two subregions are partially the result of their reliance on fossil fuel imports by both subregions, which have been reduced to meet the Paris Accord obligations.

Following implementation of the Paris Accord, CO_2 emissions, as expected, fall across the region by 0.6% annually. Subregionally, North and Central Asia is projected to have the largest decline (1.8% per year). In contrast, the largest overall emitter, East Asia, which contributes nearly two thirds of the region's emissions, has the lowest relative decline of 0.13%.[5]

The Paris Accord does not have any significant effect on inequality in the region. Undernourishment also does not change significantly for the region as whole, but effects differ across subregions. While falling somewhat in North and Central Asia and India, it rises slightly in West Asia due to the dependence on oil and the subsequent loss of economic activity. Added to the decrease in GDP in the subregion, this highlights that CO_2 emissions reduction commitments and the move to renewable energies are likely to negatively impact oil-producing countries.

In scenario 6 (the welfare transfer scenario), since income is redistributed from high-skilled workers to low-skilled workers within each economy, no impact on GDP at regional and subregional levels is registered. This result is in line with economic findings that show the economic effects of tax increases on high-income earners are net neutral. Similarly, welfare transfers do not affect exports and imports at the aggregate level. However, small movements are apparent at the subregional level, with West Asia exports contracting by 0.2%, and North and Central Asia imports shrinking by 0.3%. As low-skilled workers have more disposable income, they require a different product mix and rely on fewer imports.

Emissions do not increase in the region, but there is noticeable variation among subregions. While CO_2 emissions decrease in North and Central Asia by 0.2%, the emissions in India, South East Asia, and West Asia increase by 0.2%, 0.2% and 0.4%, respectively. While not evident in the GDP effect, these increases are most likely due to the rising income of lower skilled workers and their subsequent consumption of more energy intensive products. North and Central Asia's emissions decrease indicates that the product mix by low-skilled workers is less energy intensive in that subregion. The results illustrate that even income redistribution policies have the potential to impact emissions, with the direction of the impact depending on the product mix facing unskilled/low-income consumers.

As expected, the welfare transfer policy reduces within-country inequality in all subregions as well as in the region as a whole. Inequality decreases by 0.03% per annum in Asia-Pacific, with the largest decreases in India, which has one of the highest rates of inequality in the region. Similarly, undernourishment falls across the region by 0.15%, with the strongest effects in North and Central Asia, East Asia, and

[5]At the global level, CO_2 emissions annually decline by 1.3% in emissions or 19.5% between 2015 and 2030. These results are in line with Campagnolo and Davide (2017), who similarly model a 19% decline in emissions from the status quo.

India, the three regions with the highest levels of undernourishment in the region.

The final scenario (scenario 7) estimates the impact of integrated liberalization in Asia and the Pacific (scenario 4) when countries implement the Paris Accord commitments and income transfer policies. The combined effect is a net positive result regionally on the economic, social, and environmental variables under consideration. Moreover, this combined scenario offsets most of the negative impacts seen in stand-alone policies. However, the results also show that some subregions still experience adverse effects. This underscores the need for the policy mix to take into account subregional and country-level differences, particularly in light of different policy priorities.

The overall impact on GDP is an increase in annual growth of 0.4% or $116 billion. This is larger than the effect from integrated liberalization alone, meaning that there is a high level of complementarity among the policies. Over the 15-year period considered, the policy mix results are an increase in regional GDP of $1.7 trillion or over 6% of the regional GDP.

Under the combination scenario, the subregions benefitting the most economically are North and Central Asia (annual growth of 1.4%), India (1.0%), and South East Asia (0.8%). However, GDP in oil-dependent West Asia shrinks as a result of the implementation of the Paris Accord. Moreover, none of the modeled trade policy changes (scenarios 1–4) lead to significant GDP gains for this subregion either. As such, this highlights the difficulty oil-exporting countries may face operating in a more sustainable global economy.

The effect on exports and imports in the final scenario is largely additive. Exports increase regionally by 1.3%, and imports increase by 0.8%, resulting in net annual export increases of over $60 billion. Total trade, however, is slightly lower when liberalization and complementary policies are combined than when integrated liberalization alone is implemented (scenario 4).

Carbon emissions are driven down by the Paris Accord implementation. They fall across the region and also negate some of the negative increases due to trade integration polices. Overall, there is a 0.4% reduction in the region, with only East Asia not showing significant reduction in CO_2 emissions—explained by the region's highest growth in exports in absolute terms. Campagnolo and Davide (2017) also find that emissions under the Paris Agreement simulations may increase in some countries because of weak mitigation targets in the national determined contributions.

Both inequality and undernourishment fall under the combined economic, social, and environmental policy scenario. Inequality falls all subregions as a result primarily of trade facilitation and income transfer effects. In West Asia, trade facilitation decreases inequality, whereas in the Pacific and East Asia, income transfers offset the increased inequality due to trade facilitation. Undernourishment also falls within most subregions, decreasing by 1.4% annually or a 20% reduction by 2030 from 2015 levels. The fall in undernourishment is essentially driven by trade facilitation, which facilitates the movement of agricultural and food products at lower costs.

This last scenario's superior economic, social, and environmental outcomes highlights the fact that social, environmental, and trade and investment policies can be synergetic. They can promote better development outcomes than can be achieved with stand-alone policies. However, this requires multilateral cooperation (as in the case of the Paris Accord) and deeper regional integration. Indeed, the results strongly support cooperative and coordinated implementation of environmental and social policies in regional trade and investment integration efforts, including to ensure that these policies do not create unnecessary or unintended barriers to trade. Encouragingly, new-generation regional trade agreements already do this to a large extent (e.g., RCEP and TPP in this region), encompassing an ever-wider range of economic, social, and environmental issues during negotiations.

We now turn our attention to India's sectoral results. We observe that all sectors' output increase in the first scenario of trade liberalization, the second scenario of investment liberalization, and the third one of trade facilitation. The magnitude of effects varies across sectors; manufacturing and energy sectors gain more than

Table 17.3 % changes in India's output in specific broad sectors

	Scenario 1	Scenario 2	Scenario 3	Scenario 4	Scenario 5	Scenario 6	Scenario 7
Crops	0.03	0.26	0.02	0.23	−0.06	0.21	0.38
Processed food	0.06	1.50	0.17	1.65	−0.38	0.20	1.47
Textiles and apparel	0.24	0.39	0.16	0.76	−0.39	0.32	0.68
Livestock	0.01	0.49	0.17	0.64	−0.16	0.39	0.87
Fish	0.15	0.93	0.04	1.06	1.02	0.17	2.26
Forest	0.00	0.14	0.01	0.15	0.25	0.11	0.51
Coal	0.15	0.56	0.02	0.69	−0.20	0.44	0.93
Oil	0.12	0.38	0.03	0.50	−0.26	0.30	0.55
Gas	0.00	0.45	0.10	0.52	−0.21	0.36	0.66
Petrol	0.15	1.22	0.03	1.33	−2.23	0.20	−0.69
Renewables	0.06	0.12	0.09	0.25	2.44	0.10	2.79
Light manufacturing	0.07	0.17	0.01	0.24	0.06	0.14	0.44
Heavy manufacturing	0.10	0.43	0.03	0.53	0.28	0.34	1.14
Services	0.12	1.70	0.00	1.73	1.95	0.36	4.05

crops, for example. These effects combine together to get amplified in the fourth scenario that combines all these positive trade policies of first three scenarios. Paris Accord implementation may have a negative effect on all sectors except renewables, fishing, forestry, nonfood/textile manufacturing, and services. Income transfers have a positive effect on all sectors, while the combined effect of all scenarios is positive for all sectors except petroleum products (Table 17.3).

17.5 Conclusions

This study attempts to analyze the potential effects of regional tariff liberalization, investment liberalization, and trade facilitation on economic growth, trade, CO_2 emissions, inequality, and undernourishment, both individually and when implemented in an integrated policy package using a global CGE framework based on GTAP. The effects of emission reduction commitments under the Paris Agreement, as well as of domestic income transfers in India, were also analyzed, as example of complementary policies needed to channel trade and investment into sustainable development. Our special framework combines the latest developments in the literature that are most relevant for trade and SDGs.

The analysis highlights the importance of trade and investment liberalization and facilitation as a key driver of economic growth. An integrated approach to trade and investment liberalization and facilitation is preferred, boosting trade by over $175 billion annually and partly reducing negative social and environmental impacts associated with individual liberalization policies. The results suggest that lowering trade and investment barriers regionally in an integrated manner increases the competitiveness of regional firms in the global market by enabling them to effectively participate in global value chains.

Gains from liberalization are mainly driven by trade facilitation, whose economic impact significantly outweighs that of tariff and investment liberalization. In the context of the model, trade facilitation may be interpreted broadly as reduction in trade costs, including but not limited to the implementation of trade facilitation and paperless trade measures included in the WTO Trade Facilitation Agreement and the new Framework Agreement on Facilitation of Cross-Border Paperless Trade in Asia and the Pacific.

Finally, the analysis confirms the need for complementary social and environmental policies to better channel trade and investment into sustainable development. Indeed, while the social and environmental impacts from liberalization are typically not large at the regional level, they vary substantially across subregions in both significance and direction. Importantly, the combined effect of integrated liberalization, domestic welfare transfers, and global Paris Accord implementation is a higher level of economic growth in India than what is achieved through integrated liberalization alone, while inequality and CO_2 emissions are also mitigated. Increasing multilateral and regional cooperation on trade and investment as well as social and environmental issues are the key to achieving positive outcomes across all three pillars of sustainable development.

References

ADB (2016) Key indicators for Asia and the Pacific 2016. Asian Development Bank, Manila

ADB, and ESCAP (2013) Designing and implementing trade facilitation in Asia and the Pacific. Asian Development Bank, Manila. Available from http://www.unescap.org/resources/designing-and-implementing-trade-facilitation-asia-and-pacific-2013-update

Arvis JF, Duval Y, Shepherd B, Utoktham C, Raj A (2015) Trade costs in the developing world: 1995–2010. Working paper DTC-2. The World Bank, Washington, DC

Campagnolo L, Davide M (2017) Can the Paris deal boost SDGs achievement? An assessment of climate mitigation co-benefits or side-effects on poverty and inequality. Fondazione Eni Enrico Mattei working papers. Paper 1215. Available from http://services.bepress.com/feem/paper1215

Cutter A, Osborn D, Romano J, Ullah F (2015) Sustainable development goals and integration: achieving a better balance between the economic, social and environmental dimensions, the German Council for Sustainable Development

Da Silva JM (2017) How can aid for trade help advance the sustainable development goals? Bridges Africa 6(5). ICSTD

Duval Y (2015) ESCAP-World Bank Trade Cost Database—a brief introduction. UNESCAP, Bangkok

Evenett SJ, Fritz J (2015) Throwing sand in the wheels: how protectionism slowed export-led growth for the world's poorest countries. Report prepared for the Government of Sweden (revised version)

Fessehaie J, Morris M (2018) Global value chains and sustainable development goals: what role for trade and industrial policies? ICTSD, Geneva

Gilbert J (2013) The economic impact of new regional trading developments in the ESCAP region. Asia Pac Dev J 20(1):1–32. Available from http://www.unescap.org/sites/default/files/chap-1-Gilbert.pdf

Hoekman B (2016) Trade and the SDGs: making 'means of implementation' a reality, no. 128, trade hot topics. The Commonwealth, London

Hoekman B (2017) Trade and the post-2015 development agenda. In: Helble M, Shepherd B (eds) Win–Win: how international trade can help meet the sustainable development goals. Asian Development Bank Institute, Manila

ICTSD (2016) Trade policies and sustainable development in the context of global value chains. ICTSD, Geneva

Jacob A (2016) Trade and the new global development framework. UNESCAP, Bangkok

Narayanan B, Balie J (2017) Trade policy, domestic support and food security. Paper presented at the 20th annual conference of global economic analysis, Purdue University

Narayanan B, Duval Y, Kravchenko A, Wadhwa D (2017) Sustainable development impact of trade and investment liberalization in Asia and the Pacific, working paper no. 173. UNESCAP, Bangkok

OECD/WTO (2015) Aid for trade at a glance 2015: reducing trade costs for inclusive, sustainable growth. WTO/OECD Publishing, Geneva/Paris

OECD/WTO (2017) Aid for trade at a glance 2017: promoting trade inclusiveness and connectivity for sustainable development. WTO/OECD Publishing, Geneva/Paris

Razzaque M, Vickers B, Goel P (2016) Global trade slowdown, Brexit and SDGs: issues and way forward, no. 132, trade hot topics. The Commonwealth, London

Tipping A (2014) Trade policy and the post-2015 agenda. Biores 8(1), ICSTD

Tipping A, Wolfe R (2016) Trade and sustainable development: options for follow-up and review of the trade-related elements of the Post-2015 agenda and financing for development, issue paper. ICSTD and IISD, Geneva

UNCTAD (2016) Trading into sustainable development: trade, market access, and the sustainable development goals. United Nations, New York

UNESCAP (2019) Foreign direct investment trends and outlook in Asia and the Pacific 2019/2020. https://www.unescap.org/resources/foreign-direct-investment-trends-and-outlook-asia-and-pacific-20192020

United Nations (2015a) Transforming our world: the 2030 Agenda for Development. https://sustainabledevelopment.un.org/post2015/transformingourworld

United Nations (2015b) Addis Ababa Action Agenda of the Third International Conference on Financing for Development (Addis Ababa Action Agenda). http://www.un.org/esa/ffd/wp-content/uploads/2015/08/AAAA

Vandemoortele J (2012) Advancing the global development agenda post-2015: some thoughts, ideas and practical suggestions. United Nations System Task Team on the Post-2015 United Nations Development Agenda

WTO (2015) Speeding up trade: benefits and challenges of implementing the WTO Trade Facilitation Agreement. World Trade Report 2015. WTO, Geneva

WTO (2018) Mainstreaming trade to attain the sustainable development goals. WTO, Geneva. https://doi.org/10.30875/9c96f135-en

Annex 1

S. Hazra, A. Bhukta (eds.), *Sustainable Development Goals*, Sustainable Development Goals Series,
https://doi.org/10.1007/978-3-030-42488-6

Data Snapshot

Goal 1: End poverty in all its forms everywhere		
SL	**National Indicator**	**Base Value**
Target 1.1 : By 2030, eradicate extreme poverty for all people everywhere, currently measured as people living on less than $1.25 a day		
1	1.1.1 : Proportion of population living below the national poverty line, 2011-12*(in percentage)*	21.92
2	1.1.2 : Poverty Gap Ratio, *2011-12 (in percentage)*	a) Rural – 5.05 b) Urban – 2.70
Target 1.2 : By 2030, reduce at least by half the proportion of men, women and children of all ages living in poverty in all its dimensions according to national definitions		
	National Indicator not yet evolved	
Target 1.3 : Implement nationally appropriate social protection systems and measures for all, including floors, and by 2030 achieve substantial coverage of the poor and the vulnerable		
1	1.3.1 : Percentage of households with any usual member covered by a health scheme or health insurance, 2015-16*(in percentage)*	28.70
2	1.3.2 : Number of Beneficiaries under Integrated Child Development Scheme(ICDS), 2015-16, *(in number)*	10,21,31,284
3	1.3.3 : Proportion of the population (out of total eligible population) receiving social protection benefits under Mahatma Gandhi National Rural Employment Guarantee Act(MGNREGA)	Not Available
4	1.3.4 : Number of Self Help Groups (SHGs) formed and provided bank credit linkage, 2015-16*(in lakhs)*	18.32
5	1.3.5 : Proportion of the population (out of total eligible population) receiving social protection benefits under Maternity Benefit, 2015-16*(in percentage)*	36.40
6	1.3.6 : Number of senior citizens provided institutional assistance through Old Age Homes/Day Care Centers funded by the Government, 2016-17, *(in number)*	22,050
Target 1.4 : By 2030, ensure that all men and women, in particular the poor and the vulnerable, have equal rights to economic resources, as well as access to basic services, ownership and control over land and other forms of property, inheritance, natural resources, appropriate new technology and financial services, including microfinance		
1	1.4.1 : Proportion of population (Rural) living in households with access to safe drinking water & sanitation (Toilets)	Not Available
2	1.4.2 : Proportion of population (Urban) living in households with access to safe drinking water & sanitation (Toilets)	Not Available
3	1.4.3 : Proportion of population (Urban/Rural) living in households with access to electricity Modified Indicator : Percentage of household with access to electricity, 2015-16, (in percentage)	88.20
4	1.4.4 : Proportion of homeless population to total population 2011, *(in percentage)*	0.15
5	1.4.5 : Proportion of population having bank accounts Modified Indicator : Number of accounts (include deposit and credit accounts) of scheduled commercial banks per 1,000 population, 2015-16*(in '000 population)*	1,425

6	1.4.6 : Number of mobile telephones as percentage of total population Modified Indicator : Number of telephone subscriptions as percentage of total population, 2015-16, *(in percentage)*	83.40
Target 1.5 : By 2030, build the resilience of the poor and those in vulnerable situations and reduce their exposure and vulnerability to climate-related extreme events and other economic, social and environmental shocks and disasters		
1	1.5.1 : Number of deaths, missing persons and directly affected persons attributed to disasters per 100,000 population	Not Available
2	1.5.2 : Proportion of States that adopt and implement local disaster risk reduction strategies in line with national disaster reduction strategies	Not Available
Target 1.a : Ensure significant mobilization of resources from a variety of sources, including through enhanced development cooperation, in order to provide adequate and predictable means for developing countries, in particular least developed countries, to implement programmes and policies to end poverty in all its dimensions		
1	1.a.1 : Proportion of domestically generated resources allocated by the government (Central & State) directly to poverty reduction programmes	Not Available
2	1.a.2 : Proportion of total government spending on essential services (education, health and social protection) *(in percentage)-2015-16*	29.83
Target 1.b : Create sound policy frameworks at the national, regional and international levels, based on pro-poor and gender-sensitive development strategies, to support accelerated investment in poverty eradication actions		
1	1.b.1 : Proportion of budget earmarked under pro-poor gender specific budgeting	Not Available

Goal 2:End hunger, achieve food security and improved nutrition and promote sustainable agriculture

SL	National Indicator	Base Value
Target 2.1 : By 2030, end hunger and ensure access by all people, in particular the poor and people in vulnerable situations, including infants, to safe, nutritious and sufficient food all year round		
1	2.1.1 : Percentage of children aged under 5 years who are underweight, 2015-16 *(in percentage)*	35.70
2	2.1.2 : Proportion of population (marginalized and vulnerable) with access to food grains at subsidized prices, 2013*(in percentage)*	a) Rural – 75.00 b) Urban – 50.00
Target 2.2 : By 2030, end all forms of malnutrition, including achieving, by 2025, the internationally agreed targets on stunting and wasting in children under 5 years of age, and address the nutritional needs of adolescent girls, pregnant and lactating women and older persons		
1	2.2.1 : Percentage of children under age 5 years who are stunted 2015-16*(in percentage)*	38.40
2	2.2.2 : Percentage of children under age 5 years who are wasted 2015-16*(in percentage)*	21.00
3	2.2.3 : Percentage of women whose Body Mass Index (BMI) is below normal (BMI),2015-16*(in percentage)*	22.90
4	2.2.4 : Percentage of pregnant women age 15-49 years who are anemic (<11.0g/dl) Modified Indicator : Percentage of pregnant women age 15-49 years who are anaemic (Hb<12.0 g/dl), *2015-16(in percentage)*	50.40

5	2.2.5 : Percentage of Children age 6-59 months who are anemic (<11.0g/dl) *2015-16(in percentage*	58.40
Target 2.3 : By 2030, double the agricultural productivity and incomes of smallscale food producers, in particular women, indigenous peoples, family farmers, pastoralists and fishers, including through secure and equal access to land, other productive resources and inputs, knowledge, financial services, markets and opportunities for value addition and non-farm employment		
1	2.3.1 : Agriculture productivity of wheat and rice, 2015-16 *(in kg per hectare)*	a) Wheat - 3034 b) Rice -2400
2	2.3.2 : Gross Value Added in Agriculture per worker	Not Available
3	2.3.3 : Ratio of institutional credit to agriculture to the agriculture output.	Not Available
Target 2.4 : By 2030, ensure sustainable food production systems and implement resilient agricultural practices that increase productivity and production, that help maintain ecosystems, that strengthen capacity for adaptation to climate change, extreme weather, drought, flooding and other disasters and that progressively improve land and soil quality		
1	2.4.1 : Proportion of degraded land to net sown area	Not Available
2	2.4.2 : Percentage of farmers issued Soil Health Card	Not Available
3	2.4.3 : Percentage of net area under organic farming	Not Available
Target 2.5 : By 2020, maintain the genetic diversity of seeds, cultivated plants and farmed and domesticated animals and their related wild species, including through soundly managed and diversified seed and plant banks at the national, regional and international levels, and promote access to and fair and equitable sharing of benefits arising from the utilization of genetic resources and associated traditional knowledge, as internationally agreed		
1	2.5.1 : Number of accessions conserved in the base collection (-18 degree Celsius) at National Gene Bank, *2017 (in number)*	4,34,946
2	2.5.2 : Conservation of germplasm, *2015-16 (in number)*	75,563
3	2.5.3 : Conservation of fish genetic resource (in number)	Not Available
Target 2.a : Increase investment, including through enhanced international cooperation, in rural infrastructure, agricultural research and extension services, technology development and plant and livestock gene banks in order to enhance agricultural productive capacity in developing countries, in particular least developed countries		
1	2.a.1 : Percentage share of expenditure in R&D in agriculture to GVA in agriculture Modified Indicator :Percentage share of Intellectual Property Product (Research & Development) to total GVA in Agriculture, forestry & fishing, *2015-16 (in percentage*	0.04
2	2.a.2 : Proportion of public investment in agriculture to GVA in agriculture Modified Indicator : Percentageof Government investment(Gross Fixed Capital Formation) in agriculture to GVA in agriculture, *2015-16 (in percentage*	10.55
Target 2.b : Correct and prevent trade restrictions and distortions in world agricultural markets, including through the parallel elimination of all forms of agricultural export subsidies and all export measures with equivalent effect, in accordance with the mandate of the Doha Development Round		
	National indicator yet to develop	

2.c : Adopt measures to ensure the proper functioning of food commodity markets and their derivatives and facilitate timely access to market information, including on food reserves, in order to help limit extreme food price volatility		
1	2.c.1 : Percentage of Agriculture Mandis enrolled in e-market	Not Available

	Goal 3:Ensure healthy lives and promote well-being for all at all ages	
Target 3.1 : By 2030, reduce the global maternal mortality ratio to less than 70 per 100,000 live births		
1	3.1.1 : Maternal Mortality Ratio,*2014-16 (per 1,00,000 live birth)*	130
2	3.1.2 : Percentage of births attended by skilled health personnel(Period 5 years), *2015-16 (in percentage)*	81.40
3	3.1.3 : Percentage of births attended by skilled health personnel(Period 1 year), *2015-16 (in percentage)*	84.40
4	3.1.4 : Percentage of women aged 15–49 years with a live birth, for last birth ,who received antenatal care, four times or more(Period 5 years/1 year), *2015-16 (in percentage)*	51.20
Target 3.2 : By 2030, end preventable deaths of newborns and children under 5 years of age, with all countries aiming to reduce neonatal mortality to at least as low as 12 per 1,000 live births and under-5 mortality to at least as low as 25 per 1,000 live births		
1	3.2.1 : Under-five mortality rate,*2016 (per 1000 live births)*	39
2	3.2.2 : Neonatal mortality rate, *2016(per 1000 live births)*	24
3	3.2.3 : Percentage of children aged 12-23 months fully immunized (BCG, Measles and three doses of Pentavalent vaccine), *2015-16(in percentage)*	62.00
Target 3.3 : By 2030, end the epidemics of AIDS, tuberculosis, malaria and neglected tropical diseases and combat hepatitis, water-borne diseases and other communicable diseases		
1	3.3.1 : Number of new HIV infections per 1,000 uninfected population, *2015 (per 1000 population)*	0.06
2	3.3.2 : Tuberculosis incidence per 100,000 population, *2015 (per 1,00,000 population)*	217
3	3.3.3 : Malaria incidence per 1,000 population, *2015(per 1000 population)*	0.92
4	3.3.4 : Viral Hepatitis (including A & B) incidence per 100,000 population	Not Available
5	3.3.5 : Dengue: Case Fatality Ratio (CFR), *2015 (in percentage)*	0.20
6	3.3.6 : Number of Chikungunya cases,*2015 (in number)*	27,553
7	3.3.7 : Number of new cases of Kalaazar/ V Leishmaniasis, *2015(in number)*	8,500
8	3.3.8 : Number of new cases of Lymphatic Filariasis(LF), 2015 *(in number)*	12,21,191
9	3.3.9 : The proportion of grade-2 cases amongst new cases of Leprosy, *2015-16 (in percentage)*	4.60
10	3.3.10 : HIV Prevalence Rate, *2015(in percentage)*	0.26
Target 3.4 : By 2030, reduce by one third premature mortality from noncommunicable diseases through prevention and treatment and promote mental health and wellbeing		
1	3.4.1 : Number of deaths due to cancer, *2012-14 (in maximum and minimum numbers as per PBCRs Report as national value is not available)*	a) Male(max.) : 4,591 b) Male(Min.): 39 c) Female(max.) : 3,915 d) Female(min.) : 28

2	3.4.2 : Suicide mortality rate, *2015(per 1,00,000 population)*	10.60
3	3.4.3 : Percentage distribution of leading cause groups of deaths, *2016(in percentage)*	19.00
Target 3.5 : Strengthen the prevention and treatment of substance abuse, including narcotic drug abuse and harmful use of alcohol		
1	3.5.1 : Percentage of adults (15+ years) who have had at least 60 milliliter or more of pure alcohol on at least one occasion weekly (approximately equivalent to standard alcoholic drinks)	Not Available
2	3.5.2 : Number of persons treated in de-addiction centres, *2015-16 (in number)*	1,46,124
3	3.5.3 : Percentage of population (men (15-54 years) and women (15-49 years)) who consume alcohol,*2015-16 (in percentage)*	a) Male – 29.50 b) Female – 1.20
Target 3.6 : By 2020, halve the number of global deaths and injuries from road traffic accidents		
1	3.6.1 : Death rate due to road traffic accidents, *2015 (per 1,00,000 population)*	11.81
Target 3.7 : By 2030, ensure universal access to sexual and reproductive health-care services, including for family planning, information and education, and the integration of reproductive health into national strategies and programmes		
1	3.7.1 : Percentage of currently married women (15-49 years) who use any modern family planning methods, *2015-16 (in percentage)*	47.70
2	3.7.2 : Percentage of women aged 15-19 years who were already mothers or pregnant,*2015-16 (in percentage)*	a) Women already mothers : 5.20 b) Pregnant women with first child : 2.70
3	3.7.3 : Percentage of Institutional Births(5 years/1 years),*2015-16 (in percentage)*	a) During last 5 years : 78.90 b) During last 1 years : 82.60
Target 3.8 : Achieve universal health coverage, including financial risk protection, access to quality essential health-care services and access to safe, effective, quality and affordable essential medicines and vaccines for all		
1	3.8.1 : Percentage of currently married women (15-49 years) who use any modern family planning methods, *2015-16 (in percentage)*	47.70
2	3.8.2 : Percentage of TB cases successfully treated (cured plus treatment completed) among TB cases notified to the national health authorities during a specified period, *2015-16(in percentage)*	84.00
3	3.8.3 : Percentage of people living with HIV currently receiving ART among the detected number of adults and children living with HIV*(in percentage)*	Not Available
4	3.8.4 : Proportion of population in age group 15-49 years who are currently taking antihypertensive medication among age group 15- 49 with systolic blood pressure $\geq$ 140 mmHg, or with diastolic blood pressure $\geq$ 90mmHg, *2015-16 (in percentage)*	a) Male : 14.80 b) Female : 11.00
5	3.8.5 : Proportion of population in age group 15-49 years who are currently taking medication for diabetes (insulin or glycaemic control pills) among number of adults 15-49 years who are having random blood sugar level – high (>140 mg/dl), *2015-16 (in percentage)*	Among Male a) 141-160 mg/dl (high) : 3.00 b) >160 mg/dl (very high) : 2.80 Among Female a) 141-160 mg/dl (high) : 4.10 >160 mg/dl (very high) : 3.90

6	3.8.6 : Proportion of women aged 30-49 years who report they were ever screened for cervical cancer and the proportion of women aged 30-49 years who report they were screened for cervical cancer during the last 5 years	Not Available
7	3.8.7 : Prevalence of current tobacco uses among men and women aged 15 - 49 years, *2015-16 (in percentage)*	a) Male : 44.50 b) Female : 6.80
8	3.8.8 : Total physicians, nurses and midwives per 10,000 population, *2015 (per 10,000 population)*	35.80
Target 3.9 : By 2030, substantially reduce the number of deaths and illnesses from hazardous chemicals and air, water and soil pollution and contamination		
1	3.9.1 : Mortality rate attributed to unintentional poisoning, *2015 (per 100,000 population)*	2.10
2	3.9.2 : Proportion of men and women reporting Asthma 15-49 years *(in percentage)-2015-16*	a) Male : 1.20 b) Female : 1.90
Target 3.a : Strengthen the implementation of the World Health Organization Framework Convention on Tobacco Control in all countries, as appropriate		
1	3.a.1 : Prevalence of current tobacco uses among men and women aged 15 - 49 years, *2015-16(in percentage)*	a) Male : 44.50 b) Female : 6.80
Target 3.b : Support the research and development of vaccines and medicines for the communicable and non-communicable diseases that primarily affect developing countries, provide access to affordable essential medicines and vaccines, in accordance with the Doha Declaration on the TRIPS Agreement and Public Health, which affirms the right of developing countries to use to the full the provisions in the Agreement on Trade-Related Aspects of Intellectual Property Rights regarding flexibilities to protect public health, and, in particular, provide access to medicines for all		
1	3.b.1 : Total net official development assistance to medical research and basic health sectors	Not Available
Target 3.c : Substantially increase health financing and the recruitment, development, training and retention of the health workforce in developing countries, especially in least developed countries and small island developing States		
1	3.c.1 : Total physicians, nurses and midwives per 10,000 population, *2015 (per 10,000 population)*	35.80
2	3.c.2 : Percentage of public investment in health as proportion to GDP Modified Indicator : Percentage of government spending on gross capital formation(GCF) in health sector to GDP,*2015-16 (in percentage)*	1.17
Target 3.d : Strengthen the capacity of all countries, in particular developing countries, for early warning, risk reduction and management of national and global health risks		
	National Indicator not yet evolved	

Goal 4:Ensure inclusive and equitable quality education and promote lifelong learning opportunities for all		
SL	**National Indicator**	**Base Value**
Target 4.1 : By 2030, ensure that all girls and boys complete free, equitable and quality primary and secondary education leading to relevant and effective learning outcomes		
1	4.1.1 : Net Enrolment Ratio in primary and upper primary education,*2015-16(in percentage)*	a) Primary : 87.30 b) Upper Primary : 74.70
2	4.1.2 : Adjusted Net Enrolment Ratio in primary, upper primary and secondary education,*2015-16 (in percentage)*	a) Primary : 91.64 b) Upper Primary : 84.36 c) Secondary : 63.37

3	4.1.3 : Gross Enrolment Ratio in higher secondary education, *2015-16 (in percentage)*	56.16
4	4.1.4 : Percentage of students in grade 3, 5, 8 and 10 achieving at least a minimum proficiency level in terms of nationally defined learning outcomes to be attained by pupils at the end of each of above grades, *2017-18 (in percentage)*	a) Language (Class-3/5/8) : 90.29/85.55/82.82 b) Mathematics(Class-3/5/8) : 88.69/79.81/60.93
5	4.1.5 : Gross intake ratio to the last grade (primary, upper primary and secondary)	Not Available
6	4.1.6 : Proportion of students enrolled in Grade 1 who reaches last grade or primary/upper primary/secondary levels	Not Available
7	4.1.7 : Out of school ratio (primary, upper primary, elementary, secondary and higher secondary)	Not Available
8	4.1.8 : Number of years (i) free and (ii) compulsory education guaranteed in legal frameworks, *2009 (in number)*	a) Free : 8 Years b) Compulsory : 8 Years
Target 4.2 : By 2030, ensure that all girls and boys have access to quality early childhood development, care and pre-primary education so that they are ready for primary education		
1	4.2.1 : Participation rate in organized learning one year before official primary entry, *2015-16(in percentage)*	34.08
2	4.2.2 : Gross early childhood education enrolment ratio	Not Available
Target 4.3 : By 2030, ensure equal access for all women and men to affordable and quality technical, vocational and tertiary education, including university		
1	4.3.1 : Participation rate of youth and adults in formal and non-formal education and training in the previous 12 months	Not Available
2	4.3.2 : Proportion of male-female enrolled in higher education, technical and vocational education,*2015-16 (in Ratio)*	a) Higher- 0.92 b) Technical : Not Available c) Vocational : Not Available
3	4.3.3 : Gross enrolment ratio for tertiary education, *2015-16 (in percentage)*	24.50
Target 4.4 : By 2030, substantially increase the number of youth and adults who have relevant skills, including technical and vocational skills, for employment, decent jobs and entrepreneurship		
1	4.4.1 : Proportion of computer literate adults	Not Available
Target 4.5 : By 2030, eliminate gender disparities in education and ensure equal access to all levels of education and vocational training for the vulnerable, including persons with disabilities, indigenous peoples and children in vulnerable situations		
1	4.5.1 : Enrolment ratio of children with disabilities, *2015-16 (in percentage)*	a) Primary : 1.18 b) Upper Primary : 1.13 c) Elementary : 1.16 d) Secondary : 0.56 e) Higher Secondary : 0.25
2	4.5.2 : Gender Parity indices for Primary/Secondary/Higher Secondary/Tertiary education, *2015-16 (in percentage)*	a) Primary : 1.03 b) Secondary : 1.02 c) Higher Secondary : 1.01 d) Tertiary education : 0.93
Target 4.6 : By 2030, ensure that all youth and a substantial proportion of adults, both men and women, achieve literacy and numeracy		
1	4.6.1 : Literacy rate of youth in the age group of 15-24 years,*2011-12 (in percentage)*	86.10

Target 4.7 : By 2030, ensure that all learners acquire the knowledge and skills needed to promote sustainable development, including, among others, through education for sustainable development and sustainable lifestyles, human rights, gender equality, promotion of a culture of peace and non-violence, global citizenship and appreciation of cultural diversity and of culture's contribution to sustainable development		
	National Indicator not yet evolved	
Target 4.a : Build and upgrade education facilities that are child, disability and gender sensitive and provide safe, non-violent, inclusive and effective learning environments for all		
1	4.a.1 : Proportion of schools with access to: (a) electricity; (b) computers for pedagogical purposes; (c) adapted infrastructure and materials for students with disabilities/ disabled friendly ramp and toilets; (d) basic drinking water; (e) single-sex basic sanitation facilities; and (f) basic hand washing facilities (as per the WASH indicator definitions), *2015-16(in percentage)*	(a) Electricity : 62.81 (b) Computers for pedagogical purposes : 27.31 (c) Adapted infrastructure and materials for students with disabilities/ disabled friendly ramp and toilets : 81.99 (d) Basic drinking water : 96.81 (e) Single-sex basic sanitation facilities (Girl's only) : 97.52 (f) Basic hand washing facilities : 48.94
Target 4.b : By 2020, substantially expand globally the number of scholarships available to developing countries, in particular least developed countries, small island developing States and African countries, for enrolment in higher education, including vocational training and information and communications technology, technical, engineering and scientific programmes, in developed countries and other developing countries		
	National Indicator not yet evolved	
Target 4.c : By 2030, substantially increase the supply of qualified teachers, including through international cooperation for teacher training in developing countries, especially least developed countries and small island developing States		
1	4.c.1 : Proportion of trained teachers, by education level (pre-primary, primary, upper primary, elementary, secondary and higher secondary education), *2015-16 (in percentage)*	a) Pre-Primary : Not Available b) Primary : 75.49 c) Upper Primary : 78.18 d) Elementary : 79.15 e) Secondary : 78.52 f) Higher Secondary : 64.45
2	4.c.2 : Pupil/trained teacher ratio by education level, *2015-16 (in percentage)*	(a) Pre-Primary : Not Available (b) Primary : 23 (c) Upper Primary - 27 (d) Elementary : N/A (e) Secondary : 27 (f) Higher Secondary : 37

Goal 5:Achieve gender equality and empower all women and girls		
SL	**National Indicator**	**Base Value**
Target 5.1 : End all forms of discrimination against all women and girls everywhere		
1	5.1.1 : Rate of crimes against women per every 1,00,000 female population, 2015 *(per 100000 female population)*	53.90
2	5.1.2 : Proportion of women subjected to dowry related offences to total crime against women,*2015(in percentage)*	5.30
3	5.1.3 : Sex Ratio at Birth,*2014-16 (per 1000 male births)*	898
4	5.1.4 : Whether or not legal framework are in place to promote, enforce and monitor equality and non-discrimination on the basis of sex, *2017 (in percentage)*	Area 1(Overarching legal frameworks and public life) : 76.90 Area 2(Violence against women) : 92.30 Area 3(Employment and economic benefits) : 87.50 Area 4(Marriage & family): 100
Target 5.2 : Eliminate all forms of violence against all women and girls in the public and private spheres, including trafficking and sexual and other types of exploitation		
1	5.2.1 : Proportion of crime against women to total crime reported in the country during the calendar year, *2015 (in percentage)*	7.00
2	5.2.2 : Proportion of sexual crimes against women to total crime against women during the calendar year, *2015 (in percentage)-*	51.00
3	5.2.3 : Proportion of cruelty/ physical violence on women by husband or his relative to total crime against women during the calendar year, *2015 (in percentage)*	34.60
4	5.2.4 : Proportion of sexual crime against girl children to total crime against children during the calendar year*(in percentage)-2015*	34.80
5	5.2.5 : Proportion of trafficking of girl children to total children trafficked during the calendar year, *2015(in percentage)*	90.30
6	5.2.6 : Percentage of currently partnered girls and women aged 15-49 years who have experienced physical and / or sexual violence by their current intimate partner in the last 12 months Modified Indicator : Percentage of currently partnered girls and women aged 15-49 years who have experienced physical and / or sexual violence by their current or previous husband in the last 12 months, 2015-16 *(in percentage)*	a) Physical or Sexual violence : 28.60 b) Emotional, physical or sexual violence : 31.00
7	5.2.7 : Child Sex Ratio (0-6 Years)*2011, (per 1000 male births)*	919
Target 5.3 : Eliminate all harmful practices, such as child, early and forced marriage and female genital mutilation		
1	5.3.1 : Proportion of cases reported under the Prohibition of Child Marriage Act (early marriage of children below 18 years of age) to total crime against children,*2015 (in percentage)*	0.30
2	5.3.2 : Proportion of women aged 20-24 years who were married or in a union before age 18, *2015-16 (in percentage)*	26.80

Target 5.4 : Recognize and value unpaid care and domestic work through the provision of public services, infrastructure and social protection policies and the promotion of shared responsibility within the household and the family as nationally appropriate		
1	5.4.1 : Proportion of time spent on unpaid domestic and care work.	Not Available
Target 5.5 : Ensure women's full and effective participation and equal opportunities for leadership at all levels of decision-making in political, economic and public life		
1	5.5.1 : Proportion of seats held by women in national Parliament, State Legislation and Local Self Government *2014-16, (in percentage)*	a) Lok Sabha(2014):12.80 b) Rajya Sabha(2014):11.40 c) PRI(2016) : 46.14
2	5.5.2 : Number of women in Board of listed companies Modified Indicator : Proportion of women in managerial positions including women in Board of Director, in listed companies,*2015-16 (per 1000 persons)*	173
Target 5.6 : Ensure universal access to sexual and reproductive health and reproductive rights as agreed in accordance with the Programme of Action of the International Conference on Population and Development and the Beijing Platform for Action and the outcome documents of their review conferences		
1	5.6.1 : Percentage of currently married women (15-49 years) who use modern methods of family planning, *2015-16 (in percentage)*	47.70
2	5.6.2 : Unmet need for family planning for currently married women aged 15-49 years, *2015-16 (in percentage)*	12.90
3	5.6.3 : Proportion of population aged 15-24 years with comprehensive correct knowledge of HIV / AIDS, *2015-16 (in percentage)*	a) Male – 87.40 b) Female – 76.30
Target 5.a : Undertake reforms to give women equal rights to economic resources, as well as access to ownership and control over land and other forms of property, financial services, inheritance and natural resources, in accordance with national laws		
1	5.a.1 : Operational land holdings - gender wise, *2015-16 (* Area in '000' Hectare and Number in '000)	a) Male (Number/ Area) – 1,25,245/ 1,37,430 b) Female (Number/ Area) 20,218/ 18,187
2	5.a.2 : Proportion of female agricultural labourers Modified indicator: Percentage of female agricultural workers among female workers, *2011-12 (in percentage)*	62.8

3	5.a.3 : Wages of casual labourers (gender wise), *2011-12 (in Rupees)*	1) Casual labour in public works other than MGNREGA public works in rural area – 121.46 a) Rural (M/F) – 127.39/110.62 2) Casual labour in MGNREGA public works in rural area – 106.71 a) Rural (M/F) - 112.46/101.97 3) Casual labour in other type of works in rural area – 138.62 a) Rural(M/F) – 149.32/103.28 4) Casual labour in other type of works Urban area – 170.1 a) Urban(M/F) – 182.04/110.62
4	5.a.4 : Agricultural wages (gender wise)	Not Available
5	5.a.5 : Exclusive women SHGs in Bank linked SHGs, *2015-16 (in percentage)*	88.90
6	5.a.6 : Percentage of adult having an account at a formal financial institution	Not Available
7	5.a.7 : Percentage of women having an account at a formal financial institution	Not Available
8	5.a.8 : No. of borrowers per 1,00,000 adults(Male & Female – wise)	Not Available
Target 5.b : Enhance the use of enabling technology, in particular information and communications technology, to promote the empowerment of women		
1	5.b.1 : Percentage of women employed in IT and ITeS industry, *2017-18 (in percentage)*	34.00
Target 5.c : Adopt and strengthen sound policies and enforceable legislation for the promotion of gender equality and the empowerment of all women and girls at all levels		
1	5.c.1 : Number of Central Ministries and States having Gender Budget Cells (GBCs),*2015-16 (in number)*	g) Ministries/Departments : 57 h) States :14

Goal 6: Ensure availability and sustainable management of water and sanitation for all		
SL	**National Indicator**	**Base Value**
Target 6.1 : By 2030, achieve universal and equitable access to safe and affordable drinking water for all		

1	6.1.1 : Percentage of population having safe and adequate drinking water within their premises.	Not Available
2	6.1.2 : Percentage of population using an improved drinking water source(Rural)	Not Available
Target 6.2 : By 2030, achieve access to adequate and equitable sanitation and hygiene for all and end open defecation, paying special attention to the needs of women and girls and those in vulnerable situations		
1	6.2.1 : Proportion of households having access to toilet facility (Urban & Rural),*2015-16 (in percentage)*	61.12 a) Rural – 45.90 b) Urban – 89.50
2	6.2.2 : Percentage of Districts achieving Open Defecation Free (ODF) target	Not Available
3	6.2.3 : Proportion of schools with separate toilet facility for girls,*2015-16 (in percentage)*	97.52
6.3 : By 2030, improve water quality by reducing pollution, eliminating dumping and minimizing release of hazardous chemicals and materials, halving the proportion of untreated wastewater and substantially increasing recycling and safe reuse globally		
1	6.3.1 : Percentage of sewage treated before discharge into surface water bodies	Not Available
2	6.3.2 : Percentage of industries(17 category of highly polluting industries/grossly polluting industry/red category of industries) complying with waste water treatment as per CPCB norms.	Not Available
3	6.3.3: Proportion of waste water treatment capacity created vis-à-vis total generation	Not Available
6.4 : By 2030, substantially increase water-use efficiency across all sectors and ensure sustainable withdrawals and supply of freshwater to address water scarcity and substantially reduce the number of people suffering from water scarcity		
1	6.4.1 : Percentage ground water withdrawal against availability,*2015-16 (in percentage)*	61.50
2	6.4.2 : Per capita storage of water, 2015-16 *(in m3/person)*	204.75765
3	6.4.3 : Per capita availability of water,2011 *(in m3/person)*	1545
6.5 : By 2030, implement integrated water resources management at all levels, including through trans- boundary cooperation as appropriate		
1	6.5.1 : Percentage area of river basins brought under integrated water resources management	Not Available
6.6 : By 2020, protect and restore water-related ecosystems, including mountains, forests, wetlands, rivers, aquifers and lakes		
1	6.6.1 : Area under over-exploited blocks Modified Indicator: Number of over-exploited blocks, 2010-11 (in number)	1071

2	6.6.2 : Percentage sewage load treated in major rivers	Not Available
3	6.6.3 : Biological assessment information of surface water bodies.	Not Available
6.a : By 2030, expand international cooperation and capacity-building support to developing countries in water- and sanitation-related activities and programmes, including water harvesting, desalination, water efficiency, wastewater treatment, recycling and reuse technologies		
1	6.a.1 : Amount of water- and sanitation-related official development assistance that is part of a government-coordinated spending plan	Not Available
2	6.a.2 : Number of MoU/Co-operation agreements for capacity building and technology transfer	Not Available
6.b : Support and strengthen the participation of local communities in improving water and sanitation management		
1	6.b.1 : Percentage of developed Irrigated Command Area brought under Water Users Association(WUAs)	Not Available
2	6.b.2 : Proportion of villages with Village Water & Sanitation Committee [VWSC]	Not Available

Goal 7:Ensure access to affordable, reliable, sustainable and modern energy for all		
SL	**National Indicator**	**Base Value**
7.1 : By 2030, ensure universal access to affordable, reliable and modern energy services		
1	7.1.1 : Percentage of Households electrified Modified Indicator: Percentage of household with access to electricity, *2015-16 (in percentage)*	88.20
2	7.1.2 : Percentage of household using clean cooking fuel, *2015-16 (in percentage)*	63.08
7.2 : By 2030, increase substantially the share of renewable energy in the global energy mix		
1	7.2.1 : Renewable energy share in the total final energy mix	Not Available
7.3 : By 2030, double the global rate of improvement in energy efficiency		
1	7.3.1 : Energy intensity measured in terms of primary energy and GDP, *2015-16 (in mega joules per rupee)*	0.2489
7.a : By 2030, enhance international cooperation to facilitate access to clean energy research and technology, including renewable energy, energy efficiency and advanced and cleaner fossil-fuel technology, and promote investment in energy infrastructure and clean energy technology		
1	7.a.1 : Official Development Assistance on clean energy	Not Available

7.b : By 2030, expand infrastructure and upgrade technology for supplying modern and sustainable energy services for all in developing countries, in particular least developed countries, small island developing States and landlocked developing countries, in accordance with their respective programmes of support		
	National Indicator not yet evolved	

Goal 8:Promote sustained, inclusive and sustainable economic growth, full and productive employment and decent work for all

SL	National Indicator	Base Value
8.1 : Sustain per capita economic growth in accordance with national circumstances and, in particular, at least 7 per cent gross domestic product growth per annum in the least developed countries		
1	8.1.1 : Annual growth rate of GDP (adjusted to price changes) per capita,*2015-16 (in percentage)*	6.80
8.2 : Achieve higher levels of economic productivity through diversification, technological upgrading and innovation, including through a focus on high-value added and labour-intensive sectors		
1	8.2.1 : Annual growth rate of GDP per employed person	Not Available
2	8.2.2 : Total number of patents issued (granted), 2015-16 *(in number)*	6,326
3	8.2.3 : Annual growth in manufacturing sector,*2015-16 (in percentage)*	12.08
4	8.2.4 : Annual growth in agriculture sector,*2015-16 (in percentage)*	0.60
8.3 : Promote development-oriented policies that support productive activities, decent job creation, entrepreneurship, creativity and innovation, and encourage the formalization and growth of micro-, small- and medium-sized enterprises, including through access to financial services		
1	8.3.1 : Proportion of unorganized employment in non-agricultural sectors	Not Available
2	8.3.2 : Coverage under ESI and EPS	Not Available
3	8.3.3 : Coverage of NPS	Not Available
4	8.3.4 : No. of MSME units registered under the online Udyog Aadhaar registration,*2016-17 (in number)*	23,73,564
5	8.3.5 : Number of start-ups recognized under Start-up India*(in number)-2016*	516
6	8.3.6 : Total number of patents issued (granted), 2015-16 *(in number)*	6,326
7	8.3.7 : Growth of Registered Micro, Small and Medium Size Enterprises, *2017-18 over 2016-17 (in percentage)*	-36.01

8	8.3.8 : Total loans outstanding to micro, small and medium enterprises	Not Available
9	8.3.9 : Outstanding Credit to Micro, Small and Medium Enterprises	Not Available
8.4 : Improve progressively, through 2030, global resource efficiency in consumption and production and Endeavour to decouple economic growth from environmental degradation, in accordance with the 10-Year Framework of Programmes on Sustainable Consumption and Production, with developed countries taking the lead		
1	8.4.1 : Renewable energy share in the total final energy mix	Not Available
2	8.4.2 : Per capita fossil fuel consumption	Not Available
3	8.4.3 : Proportion of waste recycled vs. waste generated	Not Available
4	8.4.4 : Proportion of sewage recycled vs. sewage generated	Not Available
8.5 : By 2030, achieve full and productive employment and decent work for all women and men, including for young people and persons with disabilities, and equal pay for work of equal value		
1	8.5.1 : Unemployment rate, *2011-12 (in percentage)*	2.30
2	8.5.2 : Workforce Participation Ratio (WPR), 2011-12 *(in percentage)*	38.60
3	8.5.3 : Wages earned by male-female in regular / casual employment*(in rupees)*	Regular wage/ salaried employees a) Rural: 298.96 b) Urban: 449.65 Casual labour engaged in public works other than MGNREG public works a) Rural: 121.46 b) Urban: Not Available Casual labour engaged in MGNREG public works a) Rural: 106.71 b) Urban: Not Available Casual labour engaged in works other than public works a) Rural: 138.62 b) Urban: 170.10
4	8.5.4 : Number of employed persons with disabilities in public services.	Not Available
5	8.5.5 : Total population with disabilities covered under social protection schemes	Not Available
6	8.5.6 : Share of unemployed persons in population aged 15-24,*2011-12 (in percentage)*	2.90

8.6 : By 2020, substantially reduce the proportion of youth not in employment, education or training		
1	8.6.1 : Unemployment Rate (15-24 years),2011-12 *(in percentage)*	8.10
2	8.6.2 : Proportion of youth (15-24 years) not in education, employment or training (NEET)	Not Available
8.7 : Take immediate and effective measures to eradicate forced labour, end modern slavery and human trafficking and secure the prohibition and elimination of the worst forms of child labour, including recruitment and use of child soldiers, and by 2025 end child labour in all its forms		
1	8.7.1 : Total crimes relating to human trafficking, *2015 (in number)*	6,877
2	8.7.2: Number of missing children, *2015 (in number)*	60,443
8.8 : Protect labour rights and promote safe and secure working environments for all workers, including migrant workers, in particular women migrants, and those in precarious employment		
1	8.8.1 : Number of workers covered under Employees State Insurance (ESI) Act	Not Available
2	8.8.2 : Number of migrant workers	Not Available
3	8.8.3 : Number of accidents in factories, *2013(in number)*	2,445
4	8.8.4 : Employment generated under Mahatma Gandhi National Rural Employment Guarantee Act(MGNREGA), *2015-16 (in number)*	7,22,59,479
8.9 : By 2030, devise and implement policies to promote sustainable tourism that creates jobs and promotes local culture and products		
1	8.9.1 : Percentage change in number of tourists (domestic and foreign), 2016 *(in percentage)*	12.70
2	8.9.2 :Direct contribution of Tourism to total GDP and in growth rate, 2015-16 *(in percentage)*	-3.00
8.10 : Strengthen the capacity of domestic financial institutions to encourage and expand access to banking, insurance and financial services for all		
1	8.10.1 : Indicator on Financial Inclusion	Not Available
2	8.10.2 : Proportion of population having Bank accounts Modified Indicator: Number of accounts (include deposit and credit accounts) of scheduled commercial banks per 1,000 population,*2015-16 (per 1,000 population)*	1,425
3	8.10.3 : Number of banking outlets per 1,00,000 population, *2015-16 (per 1,00,000 population)*	10.73
4	8.10.4 : Automated Teller Machines (ATMs) per 1,00,000 population, *2015-16 (per 1,00,000 population)*	15.69
5	8.10.5 : No. of accounts with Nil/1-5/more than 5 transactions	Not Available

8.a : Increase Aid for Trade support for developing countries, in particular least developed countries, including through the Enhanced Integrated Framework for Trade-related Technical Assistance to Least Developed Countries		
	National Indicator not yet evolved	
8.b : By 2020, develop and operationalize a global strategy for youth employment and implement the Global Jobs Pact of the International Labour Organization		
1	8.b.1 : Number of man days created under Mahatma Gandhi National Rural Employment Guarantee Act(MGNREGA), *2015-16 (in number)*	2,35,14,19,465

Goal 9: Build resilient infrastructure, promote inclusive and sustainable industrialization and foster innovation		
SL	**National Indicator**	**Base Value**
Target 9.1 : Develop quality, reliable, sustainable and resilient infrastructure, including regional and trans-border infrastructure, to support economic development and human well-being, with a focus on affordable and equitable access for all		
1	9.1.1 : Proportion of the rural population who live within 2 km of an all-season road	Not Available
2	9.1.2 : Passenger and freight volumes, by mode of transport, *2015-16 (passengers in number billion / freight in billion Tonne)*	a) Road Transport(Passenger/ freight)- 15.415/2,026.10 b) Scheduled Indian Airlines (Passenger/ Cargos) – 0.10380/0.0009 c) Indian Railways(Passenger/ freight) – 8107/655605
3	9.1.3 : Gross Capital Formation by industry of use, *2015-16 (in Rupee crore)*	a) At Current Price- 41,73,023 b) At Constant Price-6,80,642
Target 9.2 : Promote inclusive and sustainable industrialization and, by 2030, significantly raise industry's share of employment and gross domestic product, in line with national circumstances, and **double its share in least developed countries**		
1	9.2.1 : Annual mean levels of fine particulate matter (e.g. PM2.5 and PM10) in cities (population weighted)	Not Available
2	9.2.2 : Manufacturing employment as a proportion of total employment, *2011-12 (in percentage)*	12.60
Target 9.3 : Increase the access of small-scale industrial and other enterprises, in particular in developing countries, to financial services, including affordable credit, and their integration into value chains and markets		
1	9.3.1 : Share of household sector in total industry value added,*2015-16 (in percentage)*	43.50
2	9.3.2 : Percentage/ Proportion of Credit flow to MSMEs (as a Percentage of Total Adjusted Net Bank Credit)	Not Available
Target 9.4 : By 2030, upgrade infrastructure and retrofit industries to make them sustainable, with increased resource-use efficiency and greater adoption of clean and environmentally sound technologies and industrial processes, with all countries taking action in accordance with their respective capabilities		
1	9.4.1 : CO2 equivalent emission per unit of value added	Not Available

2	9.4.2 : Energy use intensity of manufacturing value added, *2015-16* (Toe/000' rupees)	0.00940
Target 9.5 : Enhance scientific research, upgrade the technological capabilities of industrial sectors in all countries, in particular developing countries, including, by 2030, encouraging innovation and substantially increasing the number of research and development workers per 1 million people and public and private research and development spending		
1	9.5.1 : Percentage share of expenditure in Research & Development to total GDP Modified Indicator: Percentage share of government spending on intellectual properties product (Research and Development) to total GDP,*2015-16 (in percentage)*	3.52
2	9.5.2 : Researchers (in full time equivalent) per million inhabitants	Not Available
3	9.5.3 : Total number of patents issued (granted) *2015-16 (in number)*	6,326
Target 9.a :Facilitate sustainable and resilient infrastructure development in developing countries through enhanced financial, technological and technical support to African countries, least developed countries, landlocked developing countries and small island developing States		
1	9.a.1 : Total official international support (official development assistance plus other official flows) to infrastructure	Not Available
Target 9.b : Support domestic technology development, research and innovation in developing countries, including by ensuring a conducive policy environment for, inter alia, industrial diversification and value addition to commodities		
1	9.b.1 : Share of Intellectual Property Products in total Gross Fixed Capital Formation, *2015-16 (in percentage)*	12.38
2	9.b.2 : Share of GVA of companies with research & development as main activity in total GVA from Private Corporate Sector	Not Available
3	9.b.3 : Share of GVA of Information and Computer related activities in total GVA, *2015-16 (in percentage)*	a) At Current Price- 4.18 b) At Constant Price-4.55
Target 9.c : Significantly increase access to information and communications technology and strive to provide universal and affordable access to the Internet in least developed countries by 2020		
1	9.c.1 : Proportion of population covered by a mobile network, by technology Modified Indicator : Number of Internet Subscriptions as percentage of total population, *2015-16(in percentage)*	26.98
2	9.c.2 : No. of broadband subscribers per 10000 persons,*2015-16(number per 10,000 population)*	1,178.81

Goal 10:Reduce inequality within and among countries		
SL	**National Indicator**	**Base Value**
Target 10.1 : By 2030, progressively achieve and sustain income growth of the bottom 40 per cent of the population at a rate higher than the national average		
1	10.1.1 : Growth rates of household expenditure per capita among the bottom 40 per cent of the population and the total population	Not Available
2	10.1.2 :Gini Coefficient of Household Expenditure.	Not Available

Target 10.2 : By 2030, empower and promote the social, economic and political inclusion of all, irrespective of age, sex, disability, race, ethnicity, origin, religion or economic or other status		
1	10.2.1 : Proportion of people living below 50 per cent of median household expenditure	Not Available
2	10.2.2 : Proportion of seats held by women in national Parliament, State Legislation and Local Self Government, *2014-16 (in percentage)*	d) Lok Sabha(2014):12.80 e) Rajya Sabha(2014):11.40 f) PRI(2016) : 46.14
3	10.2.3 : Proportion of persons from vulnerable groups in elected bodies. Modified Indicator : Proportion of SC/ST persons in LokSabha, *2014(in percentage)*	a) SCs – 15.50 b) STs – 8.70
Target 10.3 : Ensure equal opportunity and reduce inequalities of outcome, including by eliminating discriminatory laws, policies and practices and promoting appropriate legislation, policies and action in this regard		
	National Indicator not yet evolved	
Target 10.4 : Adopt policies, especially fiscal, wage and social protection policies, and progressively achieve greater equality		
1	10.4.1 : Proportion of budget allocated to North Eastern States	Not Available
Target 10.5 : Improve the regulation and monitoring of global financial markets and institutions and strengthen the implementation of such regulations		
	National Indicator not yet evolved	
Target 10.6 : Ensure enhanced representation and voice for developing countries in decision-making in global international economic and financial institutions in order to deliver more effective, credible, accountable and legitimate institutions		
	National Indicator not yet evolved	
Target 10.7 : Facilitate orderly, safe, regular and responsible migration and mobility of people, including through the implementation of planned and well-managed migration policies		
	National Indicator not yet evolved for this target.	
Target 10.a : Implement the principle of special and differential treatment for developing countries, in particular least developed countries, in accordance with World Trade Organization agreements		
	National Indicator not yet evolved	
Target 10.b : Encourage official development assistance and financial flows, including foreign direct investment, to States where the need is greatest, in particular least developed countries, African countries, small island developing States and landlocked developing countries, in accordance with their national plans and programmes		
	National Indicator not yet evolved	
Target 10.c : By 2030, reduce to less than 3 per cent the transaction costs of migrant remittances and eliminate remittance corridors with costs higher than 5 per cent		
1	10.c.1 : Remittance costs as a proportion of the amount remitted	Not Available

Goal 11:Make cities and human settlements inclusive, safe, resilient and sustainable		
SL	**National Indicator**	**Base Value**

11.1 : By 2030, ensure access for all to adequate, safe and affordable housing and basic services and upgrade slums		
1	11.1.1 : Percentage of Slums/Economically Weaker Sections (EWS) households covered through formal/affordable housing.	Not Available
2	11.1.2 : Percentage of Slum Area Covered with basic Services	Not Available
3	11.1.3 : Proportion of Urban Population Living in Slums, informal Settlements or Inadequate Housing	Not Available
11.2 : By 2030, provide access to safe, affordable, accessible and sustainable transport systems for all, improving road safety, notably by expanding public transport, with special attention to the needs of those in vulnerable situations, women, children, persons with disabilities and older persons		
1	11.2.1 : Proportion of cities with efficient urban mobility and public transport	Not Available
11.3 : By 2030, enhance inclusive and sustainable urbanization and capacity for participatory, integrated and sustainable human settlement planning and management in all countries		
1	11.3.1 : Proportion of cities with integrated development plans.	Not Available
2	11.3.2 : Share of Mixed Land Use Area in overall city land use	Not Available
3	11.3.3 : Net Density	Not Available
11.4 : Strengthen efforts to protect and safeguard the world's cultural and natural heritage		
1	11.4.1 : Restoration and Reuse of Historic Buildings	Not Available
11.5 : By 2030, significantly reduce the number of deaths and the number of people affected and substantially decrease the direct economic losses relative to global gross domestic product caused by disasters, including water-related disasters, with a focus on protecting the poor and people in vulnerable situations		
1	11.5.1 : Number of deaths, missing persons and directly affected persons attributed to disasters per 100,000 population	Not Available
11.6 : By 2030, reduce the adverse per capita environmental impact of cities, including by paying special attention to air quality and municipal and other waste management		
1	11.6.1 : Proportion of urban solid waste regularly collected and with adequate final discharge out of total urban solid waste generated, by cities	Not Available
2	11.6.2 : Annual mean levels of fine particulate matter (e.g. PM2.5 and PM10) in cities (population weighted)	Not Available
3	11.6.3 : Number of days the levels of fine particulate matter (PM 2.5 and PM 10) above mean level	Not Available
11.7 : By 2030, provide universal access to safe, inclusive and accessible, green and public spaces, in particular for women and children, older persons and persons with disabilities		
1	11.7.1 : Per Capita Availability of Green Spaces	Not Available
11.a : Support positive economic, social and environmental links between urban, peri-urban and rural areas by strengthening national and regional development planning		
1	11.a.1 : Proportion of population living in cities that implement urban and regional development plans integrating population projections and resource needs, by size of city	Not Available
11.b : By 2020, substantially increase the number of cities and human settlements adopting and implementing integrated policies and plans towards inclusion, resource efficiency, mitigation and adaptation to climate change, resilience to disasters, and develop and implement, in line with the Sendai Framework for Disaster Risk Reduction 2015-2030, holistic disaster risk management at all levels		

1	11.b.1 : Whether the country has adopted and implemented national disaster risk reduction strategies in line with the Sendai Framework for Disaster Risk Reduction 2015-2030	YES
2	11.b.2 : Proportion of State and local governments that adopt and implement local disaster risk reduction strategies in line with national disaster risk reduction strategies	Not Available
11.c : Support least developed countries, including through financial and technical assistance, in building sustainable and resilient buildings utilizing local materials		
	National Indicator not yet evolved	

Goal 12:Ensure sustainable consumption and production patterns		
SL	**National Indicator**	**Base Value**
12.1 : Implement the 10-Year Framework of Programmes on Sustainable Consumption and Production Patterns, all countries taking action, with developed countries taking the lead, taking into account the development and capabilities of developing countries		
1	12.1.1 : Formulation of national SCP framework and integration of SCP with national/State planning process	Not Available
12.2 : By 2030, achieve the sustainable management and efficient use of natural resources		
1	12.2.1 : Percentage variation in per capita use of natural resources	Not Available
12.3 : By 2030, halve per capita global food waste at the retail and consumer levels and reduce food losses along production and supply chains, including post-harvest losses		
1	12.3.1 : Per capita food availability, *2015 (Kg per year per person)*	169.80
2	12.3.2 : Post harvest storage and distribution losses of central/states pool stocks of wheat and rice, *2015-16 (in percentage)*	1) Storage loss : (-0.08) 2) Transit loss : 0.30
12.4 : By 2020, achieve the environmentally sound management of chemicals and all wastes throughout their life cycle, in accordance with agreed international frameworks, and significantly reduce their release to air, water and soil in order to minimize their adverse impacts on human health and the environment		
1	12.4.1 : Developing national secondary resource policy framework	Not Available
2	12.4.2 : Development of national policy for environmentally sound management of hazardous chemical and waste	Not Available
3	12.4.3 : Implementation of National Action Plan for fulfilling obligations of various Multilateral Environmental Agreements (MEA) ratified.	Not Available
12.5 : By 2030, substantially reduce waste generation through prevention, reduction, recycling and		

reuse		
1	12.5.1 : Number of waste recycling plants installed	Not Available
2	12.5.2 : Number of municipal corporations using waste segregation techniques	Not Available
3	12.5.3 : Number of municipal corporations banning use of plastic.	Not Available
12.6 : Encourage companies, especially large and transnational companies, to adopt sustainable practices and to integrate sustainability information into their reporting cycle		
1	12.6.1 : Proportion of companies publishing sustainability reports.	Not Available
12.7 : Promote public procurement practices that are sustainable, in accordance with national policies and priorities		
1	12.7.1 : Green public procurement policy developed and adopted by the Central Ministries/States/UTs (Numbers)	Not Available
12.8 : By 2030, ensure that people everywhere have the relevant information and awareness for sustainable development and lifestyles in harmony with nature		
1	12.8.1 : Develop icon on sustainable development	Not Available
2	12.8.2 : Government to Celebrate year on Sustainable development	Not Available
12.a : Support developing countries to strengthen their scientific and technological capacity to move towards more sustainable patterns of consumption and production		
	National Indicator not yet evolved	
12.b : Develop and implement tools to monitor sustainable development impacts for sustainable tourism that creates jobs and promotes local culture and products		
1	12.b.1 : Number of sustainable tourism strategies or policies and action plans implemented with agreed monitoring and evaluation tools.	Not Available
12.c : Rationalize inefficient fossil-fuel subsidies that encourage wasteful consumption by removing market distortions, in accordance with national circumstances, including by restructuring taxation and phasing out those harmful subsidies, where they exist, to reflect their environmental impacts, taking fully into account the specific needs and conditions of developing countries and minimizing the possible adverse impacts on their development in a manner that protects the poor and the affected communities		
1	12.c.1 : Subsidy per unit of fossil fuel consumption.	Not Available
2	12.c.2:Tax per unit of fossil fuel consumption.	Not Available

Goal 13:Take urgent action to combat climate change and its impacts		
SL	**National Indicator**	**Base Value**
13.1 : Strengthen resilience and adaptive capacity to climate-related hazards and natural disasters in all countries		
1	13.1.1 : Number of States with strategies for enhancing adaptive capacity and dealing with climate extreme weather events.	Not Available
13.2 : Integrate climate change measures into national policies, strategies and planning.		
1	13.2.1 : Pre 2020 action achievements of pre 2020 Goals as per country priority	Not Available
2	13.2.2 : Achievement of Nationally Determined Contribution(NDC) Goals in post 2020 period.	Not Available
13.3 : Improve education, awareness-raising and human and institutional capacity on climate change mitigation, adaptation, impact reduction and early warning		
1	13.3.1 : Number of States that have integrated climate mitigation and adaptation in education curricula and outreach programs	Not Available
13.a : Implement the commitment undertaken by developed-country parties to the United Nations Framework Convention on Climate Change to a goal of mobilizing jointly $100 billion annually by 2020 from all sources to address the needs of developing countries in the context of meaningful mitigation actions and transparency on implementation and fully operationalize the Green Climate Fund through its capitalization as soon as possible		
	National Indicator not yet evolved	
13.b : Promote mechanisms for raising capacity for effective climate change-related planning and management in least developed countries and small island developing States, including focusing on women, youth and local and marginalized communities		
	National Indicator not yet evolved	

Goal 14:Conserve and sustainably use the oceans, seas and marine resources for sustainable development		
SL	**National Indicator**	**Base Value**
14.1 : By 2025, prevent and significantly reduce marine pollution of all kinds, in particular from land-based activities, including marine debris and nutrient pollution		
1	14.1.1 : Health index of area of coastal water (percentage change)	Not Available
2	14.1.2 : Number of sewage treatment plants installed along the coast and construction of toilets under Swachh Bharat Mission	Not Available

3	14.1.3 : Percentage change in use of nitrogen fertilizers in the coastal States,*2015-16 over 2014-15 (percentage change)*	0.35
14.2 : By 2020, sustainably manage and protect marine and coastal ecosystems to avoid significant adverse impacts, including by strengthening their resilience, and take action for their restoration in order to achieve healthy and productive oceans		
1	14.2.1 : Percentage change in area under mangroves, *2015 over 2013 (percentage change)*	2.43
2	14.2.2 : Implementation of Coastal Zone Regulation Notification of 2011.	Not Available
3	14.2.3 : Percentage change in Marine Protected Areas (MPA)	Not Available
14.3 : Minimize and address the impacts of ocean acidification, including through enhanced scientific cooperation at all levels		
1	14.3.1 : Coral health index of Exclusive Economic Zone(EEZ)	Not Available
14.4 : By 2020, effectively regulate harvesting and end overfishing, illegal, unreported and unregulated fishing and destructive fishing practices and implement science-based management plans, in order to restore fish stocks in the shortest time feasible, at least to levels that can produce maximum sustainable yield as determined by their biological characteristics		
1	14.4.1 : Maximum Sustainable Yield (MSY) in fishing.	Not Available
14.5 : By 2020, conserve at least 10 per cent of coastal and marine areas, consistent with national and international law and based on the best available scientific information		
1	14.5.1 : Coverage of protected areas in relation to marine areas.	Not Available
2	14.5.2 : Percentage change in area under mangroves, *2015 over 2013.(in percentage)*	2.43
14.6 : By 2020, prohibit certain forms of fisheries subsidies which contribute to overcapacity and overfishing, eliminate subsidies that contribute to illegal, unreported and unregulated fishing and refrain from introducing new such subsidies, recognizing that appropriate and effective special and differential treatment for developing and least developed countries should be an integral part of the World Trade Organization fisheries subsidies negotiation		
	National Indicator not yet evolved	
14.7 : By 2030, increase the economic benefits to small island developing States and least developed countries from the sustainable use of marine resources, including through sustainable management of fisheries, aquaculture and tourism		
	National Indicator not yet evolved	
14.a : Increase scientific knowledge, develop research capacity and transfer marine technology, taking into account the Intergovernmental Oceanographic Commission Criteria and Guidelines on the Transfer of Marine Technology, in order to improve ocean health and to enhance the contribution of marine biodiversity to the development of developing countries, in particular small island developing States and least developed countries		

1	14.a.1 : Allocation of budget resources for research as per the EEZ or coastal line.	Not Available
14.b : Provide access for small-scale artisanal fishers to marine resources and markets		
1	14.b.1 : Assistance to the traditional / artisanal fishers for procurement of FRP boats and other associated fishing implements.	Not Available
14.c : Enhance the conservation and sustainable use of oceans and their resources by implementing international law as reflected in the United Nations Convention on the Law of the Sea, which provides the legal framework for the conservation and sustainable use of oceans and their resources, as recalled in paragraph 158 of "The future we want"		
1	14.c.1 : Percentage compliance of international laws.	Not Available

Goal 15:Protect, restore and promote sustainable use of terrestrial ecosystems, sustainably manage forests, combat desertification, and halt and reverse land degradation and halt biodiversity loss		
SL	**National Indicator**	**Base Value**
15.1 : By 2020, ensure the conservation, restoration and sustainable use of terrestrial and inland freshwater ecosystems and their services, in particular forests, wetlands, mountains and dry lands, in line with obligations under international agreement		
1	15.1.1 : Forest area as a proportion of total land area Modified Indicator: Forest Area as a proportion of total geographical area, *2015-17 (in proportion)*	0.23
2	15.1.2 : Percentage of Tree Outside Forest (TOF) in total forest cover	Not Available
15.2 : By 2020, promote the implementation of sustainable management of all types of forests, halt deforestation, restore degraded forests and substantially increase afforestation and reforestation globally		
1	15.2.1 : Percentage change in Forest Area coverage, *2015 over 2013 (percentage change)*	0.37
2	15.2.2 : Total area covered under different afforestation schemes	Not Available
3	15.2.3 : Total tree cover achieved outside forest area	Not Available
4	15.2.4 : Number of Nagar-vans and School Nurseries created	Not Available
15.3 : By 2030, combat desertification, restore degraded land and soil, including land affected by desertification, drought and floods, and strive to achieve a land degradation-neutral world		
1	15.3.1 : Percentage of degraded area restored.	Not Available
2	5.3.2 : Increasing Tree / forest cover in degraded area	Not Available

3	15.3.3 : Percentage increase in net sown area, *2013-14 over 2012-13(percentage change)*	-0.92
15.4 : By 2030, ensure the conservation of mountain ecosystems, including their biodiversity, in order to enhance their capacity to provide benefits that are essential for sustainable development		
1	15.4.1 : Increase in forest / vegetative cover in mountain areas	Not Available
2	15.4.2 : Restoration of water bodies / stream in mountain areas	Not Available
3	15.4.3 : Conservation of local wildlife species	Not Available
4	15.4.4 : Increase in per capita income of mountain dwellers	Not Available
15.5 : Take urgent and significant action to reduce the degradation of natural habitats, halt the loss of biodiversity and, by 2020, protect and prevent the extinction of threatened species		
1	15.5.1 : Red List Index	Not Available
15.6 : Promote fair and equitable sharing of the benefits arising from the utilization of genetic resources and promote appropriate access to such resources, as internationally agreed		
1	15.6.1 : Number of Access and Benefit Sharing (ABS) agreements signed	Not Available
15.7 : Take urgent action to end poaching and trafficking of protected species of flora and fauna and address both demand and supply of illegal wildlife products		
1	15.7.1 :Percentage reduction in traded wildlife that was poached or illicitly trafficked.	Not Available
15.8 : By 2020, introduce measures to prevent the introduction and significantly reduce the impact of invasive alien species on land and water ecosystems and control or eradicate the priority species		
1	15.8.1 : Percentage change in prevention and control of invasive alien species	Not Available
15.9 : By 2020, integrate ecosystem and biodiversity values into national and local planning, development processes, poverty reduction strategies and accounts		
1	15.9.1 : Progress towards national targets established in accordance with Aichi Biodiversity Target 2 of the Strategies Plan for Biodiversity 2011-2020	Not Available
15.a : Mobilize and significantly increase financial resources from all sources to conserve and sustainably use biodiversity and ecosystems		
1	15.a.1 : Official development assistance and public expenditure on conservation and sustainable use of biodiversity and eco system.	Not Available
15.b : Mobilize significant resources from all sources and at all levels to finance sustainable forest management and provide adequate incentives to developing countries to advance such management, including for conservation and reforestation		

1	15.b.1 : Percentage of fund utilized for environmental conservation Modified Indicator: Percentage of government spending on environmental protection, *2015-16 (in percentage)*	0.05
15.c : Enhance global support for efforts to combat poaching and trafficking of protected species, including by increasing the capacity of local communities to pursue sustainable livelihood opportunities		
1	15.c.1 : Number of detection and prevention of traded wildlife that was poached or illicitly trafficked.	Not Available

Goal 16: Promote peaceful and inclusive societies for sustainable development, provide access to justice for all and build effective, accountable and inclusive institutions at all levels		
SL	**National Indicator**	**Base Value**
Target 16.1 : Significantly reduce all forms of violence and related death rates everywhere		
1	16.1.1 : Number of victims of intentional homicide per 100,000 population, *2015 (per 1,00,000 population)*	2.60
2	16.1.2 : Proportion of population subjected to physical, psychological or sexual violence in the previous 12 months, *2015 (per 1,00,000 population)*	43.70
Target 16.2 : End abuse, exploitation, trafficking and all forms of violence against and torture of children		
1	16.2.1 : Number of victims of human trafficking per 100,000 population, by sex, age and form of exploitation Modified Indicator: Number of victims rescued from human trafficking per 100,000 population, by sex, age and form of exploitation, *2015 (per 1,00,000 population)*	1.99
2	16.2.2 : Proportion of Crime Committed against Children during the year, *2015(Rate i.e. per 1,00,000 children)*	22.10
3	16.2.3 : Number of Missing Children, *2015 (in number)*	60,443
Target 16.3 : Promote the rule of law at the national and international levels and ensure equal access to justice for all		
1	16.3.1 : Number of courts per lakh population	Not Available
2	16.3.2 : Number of Judges (all levels) per lakh population	Not Available
Target 16.4 : By 2030, significantly reduce illicit financial and arms flows, strengthen the recovery and return of stolen assets and combat all forms of organized crime		
1	16.4.1 : Number of cases under the Arms Act, *2015(in number)*	51,158
2	16.4.2 : Value of Property Stolen & Recovered and Percentage Recovery, *2015 (in percentage)*	1) Property Stolen(in INR Crore) – 8210.40 2) Property Recovered(in INR Crore) – 1350.20

		3) Recovery(in percentage) – 16.40
Target 16.5 : Substantially reduce corruption and bribery in all their forms		
1	16.5.1 : Persons Arrested In Total Cognizable Crime Cases under Offences under Prevention of Corruption Act and Related Sections of Indian Penal Code(IPC), *2015(in number)*	6,223
Target 16.6 : Develop effective, accountable and transparent institutions at all levels		
1	16.6.1 : Number of Government services provided online to citizens, *2015-16 (in number)*	3577
2	16.6.2 : Percentage of RTI queries responded, *2015-16(in percentage)*	81.76
3	16.6.3 : Number of applications filed with institutions coming under Right To Information Act, *2015-16(in number)*	11,65,217
Target 16.7 : Ensure responsive, inclusive, participatory and representative decision-making at all levels		
1	16.7.1 : Proportion of seats held by women in national Parliament, State Legislation and Local Self Government *2014-16, (in percentage)*	g) Lok Sabha(2014):12.80 h) Rajya Sabha(2014):11.40 i) PRI(2016) : 46.14
2	16.7.2 : Proportion of SC/ST persons in the elected bodies Modified Indicator : Proportion of SC/ST persons in LokSabha, *2014(in percentage)*	a) SCs - 15.50 b) STs – 8.70
Target 16.8 : Broaden and strengthen the participation of developing countries in the institutions of global governance		
	Indicator is yet to develop for this target	
Target 16.9 : By 2030, provide legal identity for all, including birth registration		
1	16.9.1: Percentage of births registered, *2015(in percentage)*	88.30
2	16.9.2 : Proportion of population covered under Aadhaar, *2015-16 (in percentage)*	82.60
Target 16.10 : Ensure public access to information and protect fundamental freedoms, in accordance with national legislation and international agreements		
1	16.10.1 : Percentage of RTI queries responded, *2015-16(in percentage)*	81.76
Target 16.a : Strengthen relevant national institutions, including through international cooperation, for building capacity at all levels, in particular in developing countries, to prevent violence and combat terrorism and crime		
	Indicator is yet to develop for this target	
Target 16.b : Promote and enforce non-discriminatory laws and policies for sustainable development		
	Indicator is yet to develop for this target	

Goal 17: Strengthen the means of implementation and revitalize the Global Partnership for Sustainable Development
National indicator is yet to be developed for this Goal.

Annex 2

S. Hazra, A. Bhukta (eds.), *Sustainable Development Goals*, Sustainable Development Goals Series,
https://doi.org/10.1007/978-3-030-42488-6

Table: Performance of state/UT on each SDG

State / UT	SDG1	SDG2	SDG3	SDG4	SDG5	SDG6	SDG7	SDG8	SDG9	SDG10	SDG11	SDG15	SDG16	Composite SDG
	Index Score													
Andhra Pradesh	67	[illegible]	68	77	44	[illegible]	76	81	31	78	26	87	90	[illegible]
Arunachal Pradesh	[illegible]	[illegible]	38	44	32	[illegible]	44	72	16	47	44	73	77	[illegible]
Assam	[illegible]	[illegible]	30	[illegible]	36	42	18	[illegible]	35	75	32	100	[illegible]	49
Bihar	45	39	40	36	24	31	67	[illegible]	38	82	43	[illegible]	[illegible]	40
Chhattisgarh	[illegible]	46	42	[illegible]	49	96	36	[illegible]	30	73	[illegible]	100	85	[illegible]
Goa	[illegible]	80	65	71	35	65	[illegible]	90	0	[illegible]	71	100	87	[illegible]
Gujarat	48	49	[illegible]	87	31	100	67	80	65	79	[illegible]	71	73	[illegible]
Haryana	[illegible]	[illegible]	[illegible]	65	31	80	[illegible]	72	[illegible]	[illegible]	30	43	78	[illegible]
Himachal Pradesh	[illegible]	[illegible]	[illegible]	82	42	96	[illegible]	71	43	98	41	93	91	69
Jammu and Kashmir	[illegible]	[illegible]	[illegible]	[illegible]	39	[illegible]	[illegible]	43	35	71	23	74	89	[illegible]
Jharkhand	37	35	40	[illegible]	32	[illegible]	20	[illegible]	47	72	[illegible]	96	[illegible]	[illegible]
Karnataka	[illegible]	[illegible]	69	76	41	[illegible]	77	72	[illegible]	68	36	88	74	[illegible]
Kerala	66	72	92	87	[illegible]	[illegible]	[illegible]	[illegible]	88	72	46	75	82	69
Madhya Pradesh	44	41	38	49	33	[illegible]	[illegible]	[illegible]	27	75	39	91	[illegible]	[illegible]
Maharashtra	47	47	[illegible]	74	43	81	89	74	[illegible]	78	34	86	82	[illegible]
Manipur	44	74	67	65	25	44	39	33	72	98	31	100	70	[illegible]
Meghalaya	66	43	[illegible]	38	36	40	11	[illegible]	42	100	39	94	[illegible]	[illegible]
Mizoram	71	69	[illegible]	[illegible]	43	67	78	65	0	100	32	69	71	[illegible]
Nagaland	[illegible]	69	34	45	42	[illegible]	45	40	0	80	32	75	87	[illegible]
Odisha	[illegible]	40	[illegible]	46	43	46	23	[illegible]	32	78	34	100	[illegible]	[illegible]
Punjab	[illegible]	71	71	[illegible]	43	[illegible]	[illegible]	[illegible]	48	[illegible]	36	67	84	[illegible]
Rajasthan	[illegible]	45	49	73	37	43	[illegible]	[illegible]	[illegible]	79	45	66	81	[illegible]
Sikkim	[illegible]	67	[illegible]	47	[illegible]	78	47	[illegible]	1	67	[illegible]	98	86	[illegible]
Tamil Nadu	76	[illegible]	77	75	38	66	89	71	46	85	33	74	[illegible]	66
Telangana	[illegible]	[illegible]	73	66	43	[illegible]	[illegible]	75	16	100	44	83	66	[illegible]
Tripura	71	[illegible]	[illegible]	[illegible]	38	38	32	[illegible]	36	89	36	86	71	[illegible]
Uttar Pradesh	48	43	25	[illegible]	27	[illegible]	23	[illegible]	29	36	37	[illegible]	[illegible]	42
Uttarakhand	65	[illegible]	36	68	41	78	[illegible]	67	33	[illegible]	41	100	86	[illegible]
West Bengal	[illegible]	[illegible]	66	[illegible]	40	[illegible]	40	[illegible]	45	76	25	88	72	[illegible]
A & N Islands	[illegible]	36	[illegible]	69	[illegible]	71	[illegible]	[illegible]	0	69	[illegible]	84	72	[illegible]
Chandigarh	39	70	23	85	[illegible]	100	96	82	76	[illegible]	40	83	90	68
D & N Haveli	21	40	32	77	41	100	73	84	0	100	6	100	[illegible]	[illegible]
Daman and Diu	[illegible]	42	47	46	38	99	84	91	0	100	49	84	79	[illegible]
Delhi	30	72	47	[illegible]	37	[illegible]	[illegible]	86	100	60	39	77	88	[illegible]
Lakshadweep	43	47	[illegible]	[illegible]	35	100	[illegible]	[illegible]	0	100	Null	100	74	[illegible]
Puducherry	[illegible]	71	66	69	27	45	[illegible]	86	100	94	27	[illegible]	92	65
India	[illegible]	48	[illegible]	[illegible]	36	[illegible]	[illegible]	65	44	71	39	90	71	[illegible]
Target	100	100	100	100	100	100	100	100	100	100	100	100	100	100

Note: A break-up of SDG India Index score by the component goals has been shown in the table for each State and UT. The table can be read both horizontally and vertically. Horizontal view helps to gauge a State/UT's performance across the 13 Goals. The vertical view enables a reader to compare the distance to target achieved by a State/UT relative to other States/UTs

Source: SDG India Index: Baseline Report 2018; NITI Aayog

Index

S. Hazra, A. Bhukta (eds.), *Sustainable Development Goals*, Sustainable Development Goals Series,
https://doi.org/10.1007/978-3-030-42488-6

The manufacturer's authorised representative in the EU is Springer Nature Customer Service Centre GmbH, Europaplatz 3, 69115 Heidelberg, Germany. If you have any concerns regarding our products, please contact ProductSafety@springernature.com

Printed and bound by CPI Group (UK) Ltd, Croydon, CR0 4YY
30/09/2026
02225491-0020